PHENOMENOLOGY OF THE CULTURAL DISCIPLINES

CONTRIBUTIONS TO PHENOMENOLOGY

IN COOPERATION WITH
THE CENTER FOR ADVANCED RESEARCH IN PHENOMENOLOGY

Volume 16

Editor:

William R. McKenna, Miami University

Editorial Board:

Scope

The purpose of this series is to foster the development of phenomenological philosophy through creative research. Contemporary issues in philosophy, other disciplines and in culture generally, offer opportunities for the application of phenomenological methods that call for creative responses. Although the work of several generations of thinkers has provided phenomenology with many results with which to approach these challenges, a truly successful response to them will require building on this work with new analyses and methodological innovations.

PHENOMENOLOGY
OF THE CULTURAL DISCIPLINES

edited by

MANO DANIEL

and

LESTER EMBREE

Department of Philosophy, Florida Atlantic University

KLUWER ACADEMIC PUBLISHERS

DORDRECHT / BOSTON / LONDON

A C.I.P. Catalogue record for this book is available from the Library of Congress.

142.7
P54126P

ISBN 0-7923-2792-6

Published by Kluwer Academic Publishers,
P.O. Box 17, 3300 AA Dordrecht, The Netherlands.

Kluwer Academic Publishers incorporates
the publishing programmes of
D. Reidel, Martinus Nijhoff, Dr W. Junk and MTP Press.

Sold and distributed in the U.S.A. and Canada
by Kluwer Academic Publishers,
101 Philip Drive, Norwell, MA 02061, U.S.A.

In all other countries, sold and distributed
by Kluwer Academic Publishers Group,
P.O. Box 322, 3300 AH Dordrecht, The Netherlands.

Printed on acid-free paper

Printed in the Netherlands

Table of Contents

Preface

The chapters in the volume arose from a research symposium held at Delray Beach in May 1992 under the auspices of the Center for Advanced Research in Phenomenology and Florida Atlantic University. Don Ihde's article "Technology and Cultural Revenge," was previously published as "La technologia e la vendetta della culturea," *Paradigmi: Rivista di critica filosofia* 10.30 (1992) and is published here in English for the first time. The editors thank the contributors for their original essays, their spirited participation and discussion during the symposium, and their prompt cooperation in the preparation of the manuscript for publication.

Mano Daniel
Lester Embree

August 1993
Delray Beach

Introduction

Reflection on the Cultural Disciplines

Lester Embree
Florida Atlantic University

Abstract: *The generic concept of cultural discipline and the specific concepts of theoretical, practical, and axiotic disciplines are clarified and the attempt made to show how such disciplines can be and are already related to in philosophy that is phenomenological.*

There has been philosophical examination of combinations of human practices ever since, while defending his life, Socrates engaged in the critical comparison of politics, poetry, craftsmanship, and philosophy. The proliferation of combinations of practices in the millennia since has been great, although the philosophical coverage may not always have been as comprehensive. The present reflection as well as this volume (and also the research symposium from which this volume has been developed) originated when the present writer came to doubt the adequacy of the widely accepted expression "human science(s)" and came to recognize that there is a number of phenomenological efforts currently underway, his own included, that reflect on one or another "cultural discipline" and might benefit from an attempt at some eidetic clarification.

The following exposition is arranged in relation to five questions: (1) What is a "discipline"? (2) When can a discipline best be deemed "cultural"? (3) What sorts of cultural disciplines are there? (4) When is philosophical reflection on an individual, a species, or the genus of cultural discipline "phenomenological"? and (5) Does the phenomenology of the cultural disciplines exist?

M. Daniel and L. Embree (eds.), Phenomenology of the Cultural Disciplines, 1–37.

I. What is a "Discipline"?

There are matters that appear best called "cultural *practices*." Mowing the lawn is an example even if it may not obviously be a component of a form of high culture such as religion, art, myth, or science. Cultural practices can be classified in many ways. If there is innate or instinctual behavior, mowing the lawn would hardly be a case of it, for lawn mowing does involve at least a modicum of learning and skill. Not only does lawn mowing not occur in all societies, but also not throughout those societies in which it does occur. In some places the lawn is mowed to keep snakes from coming out of the bush into the house. In American suburbia, however, the practice seems less practical.[1] Variation according to society and by group and area within societies can help the difference between the learned and the instinctual, e.g., breathing, be recognized. Lawn mowing is also not a matter that occurs but once; rather, it is something that is repeatable, it can recognizably be done again and again. This is what "practice" chiefly connotes. Being learned and being repeatable suffice preliminarily to define a behavior as a cultural practice.

Cultural practices thus defined can be divided into the *personal* or idiosyncratic (the present writer habitually changes to an old pair of sandals when he comes home, but knows not and cares less whether others also do so) and the *social*, which at least two subjects perform with definite awareness that others engage in practices of the same sort. That social practices attract the most attention in public discourse does not preclude recognition of idiosyncratic practices, some of which are considered matters of creative style. Finally, some cultural practices are *reflective* and some are not; the former presuppose the latter. It is difficult to contend that lawn mowing is an essentially reflective practice. One may consider how to do it, perhaps through watching others, before first trying to perform the task, use recollection in order to describe the technique to others, and even consider alternative techniques and equipments, e.g., electric mowers, when presented with them, and such glimpses might be considered at the root of systematic reflective investigation of the practice, but they are not the tree.

---

[1] Regarding the cultural character of lawns, cf. Michael Pollan, "Why Mow? The Case against Lawns," in Scott H. Slovic and Terrel F. Dixon, eds., *Being in the World* (New York: Macmillan, 1993), 433ff.

Reflection upon cultural practices is of course possible. This can be done cognitively. (Non-cognitive reflectiveness will be returned to at the end of Part IV.) For example, one may ask whether lawn mowing is performed by an individual acting alone, by a group, or somehow and sometimes both; one may ask whether it is a cognitive effort, ultimately a practical effort, or something else; one may ask about the mowing effort as an activity directed at objects and, correlatively, about the lawn as it presents itself to the mower as to be, being, or having been mowed; one may ask how equipment is used in lawn mowing; one may ask how a practice such as lawn mowing combines with others, such as tree pruning, into combinations, e.g., gardening; and so on. Such questions are external to lawn mowing itself because lawn mowing can be performed without worrying about them, while they imply that the unreflective practices such questions are about can at least be feigned and are independently possible, and because they are raised out of concerns different from those essential to the straightforward or unreflective practice reflected upon.

Enough may now have been said for cultural *practices* to be recognized. This recognition is important if only because the cultural *disciplines*, the theme of this essay, can be characterized in terms of the types of cultural practices that pertain to them, lawn mowing again being, for example, a type of landscape maintenance pertinent to the discipline of landscape architecture. If cultural practices can be combined in distinct ways, then the questions can be raised of whether all combinations of cultural practices are *disciplines* and then whether all disciplined combinations of cultural practices are *cultural* disciplines.

A weekday morning's "amateur" practices may consist of getting up, using practice-specific equipments in the bathroom, dressing, using other practical-specific equipments to prepare food, eating it, and traveling to work. Amateur lawn mowing is probably done on the weekend. Besides being a part of the gardening or yard maintenance that a house holder might engage in as an amateur, lawn mowing can also be done by a *professional*. As in erotics, athletics, and academics, the professionals are the ones who are, as a rule, rewarded in money. There are many types of professionals thus defined, but two broad kinds, each of which contains many sorts, can be distinguished according to degree of competency. Some professionals might be termed *craftspeople*. They have preparation, experience, equipment perhaps, motivation, and skill that is beyond what is involved in an amateur cultural practice. In this signification, technicians would be among the more skilled and thus well trained of the crafts-people. The aspect of preparation deserves to be dwelt upon further.

Basic professional preparation can occur through relatively passive on-the-job learning in which one simply sees how others perform tasks and imitates them. The difference between this and the preparation of an amateur consists in the quantity and quality of this learning. Probably there is less trial and error and probably some coaching comes from already skilled fellow workers or supervisors, but essentially the skill is acquired through participation. Since skill acquisition of this type occurs on all levels, it ought not to be overlooked or underestimated when practices also involving other types of preparation than on-the-job learning through participation are focally examined.

In the type of profession that might emphatically be called a craft, preparation would seem often to involve something that might be called "apprenticeship," which does not need to be formal or official, as it is in some circumstances. It can merely consist in working with a selected individual or individuals for a period of time, perhaps called the "training period." What is essential is that the learner or trainee and those charged with her preparation be recognized as such within the professional context. This can happen during the preparation of musicians, for example, who can afterwards claim to have studied with somebody of stature. The result of preparation of this sort might ultimately be called "mastery" and the process might best be called "training." Preparation of this sort will also be found on all levels. Even philosophers are sometimes prepared to some extent in this way.

Once the vast quantities of the amateur and "crafty" combinations of cultural practices are appreciated, it is easy to see that, regardless of elitist conceits and self-deceptions, proportionally quite few combinations are *disciplinary*. As sciences sometimes pretend to be technologies and technologies sometimes pretend to be sciences, crafts can pretend to be disciplines and vice versa. Preparation again appears the best aspect through which to approach the matter and perhaps "advanced training" is the best generic label for specifically disciplinary preparation. Depending on the discipline, the advanced training may occur in the first four years of college, at least the Western systems of so-called "higher education." Often it requires so-called graduate training at a university, but sometimes this occurs in such institutions as art, business, divinity, law, medical, public administration, or other professional schools not affiliated with a university. Nevertheless, there is a marked increase in specialized teachers and courses, concentrated programs, classroom instruction, textbooks, journals, disciplinary jargon, and finished students who are certifiable whether or not certificates are officially granted.

During professional preparation in some disciplines there are often courses in "theory," which are often combined with the history of the discipline, something that is apt because history can be effective in making what had been taken for granted in a disciplinary perspective problematical. Theory courses can include the purposes of the discipline, its methods, its relations with other disciplines, its socio-historical conditions, and the practices combined in it. Usually such courses reinforce established orientations, but they can also be used to oppose old ways and advocate new ones. Such reflections can be conducted from a standpoint *within* the discipline or they can be conducted from outside it. Thus the sociology of natural science is typically done by sociologists, who are outsiders of the discipline reflected upon, as are philosophers with respect to the special cultural disciplines. Since the affinity of reflection from within and reflection from outside is great, they may both be named with an expression formed by adding "meta-" as a prefix, e.g., metanursing, provided this exposes rather than hides the inside/outside distinction within the affinity.[2]

Rather than the roles of trainees or apprentices and coaches or masters, one can speak of "students" and "teachers," some of the latter of whom are even professors. If there were need for a concise generic title for the product of the disciplinary preparation process, there might be difficulty because "disciple" appears too narrow easily to broaden. "Professional" is too broad since there are also professionals on the craft level. Although it might usefully foster reflection on the cultural disciplines in general, a generic title does not seem needed by the advanced-prepared or disciplined practitioners, who typically participate in the particular disciplinary self-consciousnesses that not only produce titles like "engineer" but even "hydraulic engineer" and "computer engineer." Some leading film makers have earned doctorates in what it is difficult in present terms not to call a discipline and indeed a cultural discipline. To return to lawn mowing, it can be part of the combination of practices called landscape *gardening* as a craft, and also be somehow of concern to landscape *architecture* as a cultural discipline.

---

[2] Cf. Lester Embree, "The Future and Past of Metaarchaeology," in Lester Embree, ed. *Metaarchaeology* (Dordrecht: Kluwer Academic Publishers, 1992), 3-50.

A chart can summarize the taxonomy of levels of competency to some extent clarified above (See Figure 1). It assumes the yet to be completed definition of cultural practices and is open to further specification. While the disciplines are in various respects the most structured and the crafts more structured than amateur practices, it should be borne in mind that there are far more crafts than disciplines and far more amateur practices than the crafty and disciplinary combinations of practices put together. (If "crafty" does not currently signify "of or pertaining to a craft or crafts," it can be made to do so.)

(Figure 1)

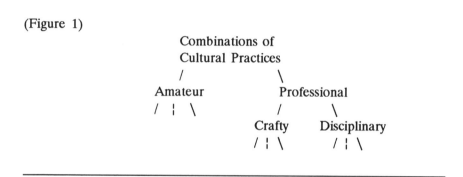

II. When is a Discipline "Cultural"?

Theoretical disciplines, which are "sciences" in the strict signification, have been played down in the foregoing sketch in order to counter-balance tendencies, at least among academics, to think of them first. Whether landscape archicture is a theoretical, practical, or some other sort of cultural discipline still remains to be seen. A brief and surely incomplete list can quickly show, however, that there are numerous *theoretical* disciplines concerned with different aspects of cultural worlds: archaeology, art history, biography, communications, economics, ethnology, geography, history, linguistics, political science, psychology, and sociology. Many of these, e.g., history, have many divisions within them that at least call for the recognition of cultural scientific *subdisciplines*. Why the combinations of cultural practices called physics, chemistry, astronomy, mathematics, logic, etc., are not on this list will be answered presently.

As part of the reception of Phenomenology and, more broadly, Continental Philosophy in North America during the last several decades, there has been a struggle to render into English the concept Wilhelm Dilthey expressed with *"die Geisteswissenschaften"* and the French with *"les sciences humaines."* The reference can be said preliminarily to be to the psychological, social, and historical sciences as they thematize aspects of human life, but the need is for a concise and apt expression. Suggestions that the so-called "humanities and social sciences" are combined in the category in question may omit the role of psychology, as does the expression "socio-historical sciences."³ It is ironic that Dilthey was explicitly accepting Schiel's 1849 German translation of Mill's "moral sciences," which only the book series edited by John O'Neill seems to have attempted recently to revive in English.

In reaction to the positivistic tendency whereby all science is considered to be or at least to be like the natural sciences, the first widely accepted rendering was "human studies," which is most conspicuously reflected in the title of the journal edited by George Psathas and in the subtitle of a major recent book.⁴ Amadeo Giorgi, however, had prominently used "human science" in 1970,⁵ the present writer used it as a subtitle in 1988,⁶ and it is now standard in the series of volumes of Dilthey in English translation coming from Princeton University Press: "As given its classic formulation by Dilthey, this theory has been entitled in English as that of the 'human studies' in order to differentiate it from the positivist ideal of a 'unified science'. Currently, the more forthright title, 'human sciences,' has been adopted . . . ."⁷

Without challenging the suitability of "human science" as a translation of Dilthey, it is now clear that there is a need for an even broader concept and expression. There are two reasons for this. Firstly, much that

---

³ Dorion Cairns, *Guide to Translating Husserl*, (The Hague: Martinus Nijhoff, 1973), 60.

⁴ Rudolf A. Makkreel, *Dilthey: Philosopher of the Human Studies* (Princeton: Princeton University Press, 1975).

⁵*Psychology as a Human Science* (New York: Harper & Row, 1970). Cf. Donald E. Polkinghorne, *Methodology for the Human Sciences* (Albany: State University of New York Press, 1983).

⁶ Lester Embree, ed., *Worldly Phenomenology: The Continuing Influence of Alfred Schutz on North American Human Science* (Washington, D.C.: Center for Advanced Research in Phenomenology & University Press of America, 1988).

⁷ Wilhelm Dilthey, *Introduction to the Human Sciences*, edited by Rudolf A. Makkreel and Frithjof Rodi (New Jersey: Princeton University Press, 1989), Preface to All Volumes, xiii.

is currently being called human science is not science *sensu strictu*, i.e., is not theoretical, although it may be "applied science," a notion not without difficulties. Secondly, there is reason to doubt that the extension should be confined to the human, prominent as the cultural practices of *Homo sapiens sapien* may be within the class designated. These reasons need elaboration.

For anyone who views science strictly speaking as theoretical, "nursing as a human science" or even "psychology as a human science," where "psychology" chiefly designates counseling and psychotherapy, are misleading expressions. They may be defended by recourse to the distinction between pure and applied science and the suggestion that expressions of the form "X as a human science" might better be expressed as "human-scientific X," the latter formulation not claiming that X is a science but only that it is science-based. Both of these moves raise significant points about how theory relates to practice. For example, it could be a mistake to believe that there is a theoretical discipline that is simply applied, because most if not all science-based practices are based not upon one but upon a variety of theoretical disciplines. It would be another mistake to believe that, temporally speaking, the theoretical sciences came first and then practical efforts were developed, which does sometimes happen currently. In most cases, however, there are ancient practices, such as agriculture, to which science has only been applied relatively recently. Then again, it needs to be recognized that no practice is entirely science-based because, on the one hand, it also includes skill acquired through experience, coaching, apprenticeship, etc., i.e., amateur and crafty skill and, on the other hand, there is the crucial question of the values and valuing that move and perhaps even justify the characteristic actions of the profession in question.[8]

As for other expressions, "humane disciplines" may have even more of a moralistic connotation than "moral sciences." "Humanistic disciplines" is better in this respect, but the first word in that expression seems too narrow insofar as sociology, for example, is seldom considered one of the so-called humanities. (Some deep thinkers play with "human science" not

---

[8] When practice-specific equipment is involved, it is appropriate to speak of "technology," but when psychotherapy, for example, can be performed while jogging in the park, as allegedly happens in California, or the equipment relied upon by nurses is not different from that used by doctors, so that the specific difference lies in the practice rather than the equipment, then that expression seems inappropriate.

as "science of the human" but as "science as done by humans," without, however, going very far into the question of just which *non-humans* might engage in the practice of science.) Those of competence and good will can learn to express and comprehend any of the mentioned expressions with the proper signification, but if the referent of the expression is considered, then a broader qualifier is called for.

It has proven difficult to define the common subject matter of the class of disciplines in question if they must be construed as addressing aspects of specifically human life and the objects, the situations, and indeed the worlds that life of that sort relates to. Efforts to define what is specifically human have not been encouraging. The most popular current view has humans as *the language using animals*. There are problems with this. Studies of the great apes show a capacity for language at least comparable with that of young humans. If language is the difference, one must then deny that the latter are human or include the former in the class thus defined. Going further, would human infants less than a year old be not human if they had no genuine linguistic comprehension? As prominent as language is, particularly in the lives of intellectuals and academics, it appears not to differentiate the human.[9]

Other attempts concerning technology or equipment-using have also not been successful in marking off the human from the non-human. Many animals of non-human species use tools. Chimpanzees plainly make tools. Actually, if the focus is removed from relatively small and movable equipment, such as hand tools, then the fact is that birds and gophers not only live and rear young in nests and holes but also construct them. These can then be considered practice-specific equipments. (Whether objects need to be artifactual in order to be equipment is an interesting question, but regularly built objects are readily discerned.) Concerning the building and using of nests, etc., one may then distinguish between behavior that is learned and behavior that is instinctual. The latter emerges in the same form regardless of the parents and group in which an individual is raised and the former can vary enormously under such conditions. A useful initial definition of "culture" is "learned behavior"

---

[9] In "Social Theory and the Second Biological Revolution" (*Social Research*, Vol. 57 (1990)), Alan Wolfe appreciates a wealth of previously thought exclusively human traits now recognized in subhumans but fails to clarify what "meaning" is or how there is a difference in kind rather than degree regarding the meaning producing, the meaning attributing, and the meaning interpretation in humans.

and of course the suggestion urged here is that the word "culture" replace the word "human" just as the word "discipline" replace the word "science," so that "cultural discipline(s)" replace "human science(s)." It may be best now to urge the proposed expression directly.

To begin with, "cultural discipline" readily designates "disciplines concerned with cultural matters," whereas "cultural discipline" as "discipline that is cultural" is unhelpful because it is difficult to see how there might be a non-cultural combination of cultural practices. In other words, all disciplines as such are cultural by virtue of being made up of learned practices even if not all of them are cultural by virtue of that which is dealt with in them! How the matters or objects addressed in cultural disciplines are cultural will be returned to presently. On the linguistic level, it would furthermore seem that the expression "cultural discipline(s)" is readily translatable into at least the Western European languages. But clearly the principal advantage of "cultural discipline" is the inclusion of the worlds of non-humans among the cultural matters investigated and otherwise related to and thus the inclusion of at least primate ethology (and conservation) among the cultural disciplines.[10]

Research on chimpanzees, bonobos, and gorillas has now shown that they are the most similar of known non-humans to humans in anatomy and behavior. The behavioral affinities include "friendship, love, aggression, language, and tool use."[11] Chimpanzees specifically can comprehend human language better than human two-year-olds, use plants as medicines, use stones and sticks to crack nuts and collect insects and teach these

---

[10] Cf. W. C. McGrew, *Chimpanzee Material Culture* (Cambridge: Cambridge University Press, 1992). Adding two to those Alfred Krober proposed in 1928 (in reflection on Köhler's work), McGrew proposes the following eight criteria, the whole set of which would need to be met for "recognizing cultural acts in other species" (except for the last two, which do not apply to humans beyond the foraging type of life-way, these are met by lawnmowing!):

Innovation	New pattern is invented or modified
Dissemination	Pattern acquired by another from innovator
Standardisation	Form of pattern is consistent and stylised
Durability	Pattern performed without presence of demonstrator
Diffusion	Pattern spreads from one group to another
Tradition	Pattern persists from innovator's generation to next one
Non-Subsistence	Pattern transcends subsistance
Naturalness	Pattern shown in absence of direct human influence

[11] Eugene Linden, "A Curious Kinship: Apes and Humans," *National Geographic*, 181 (1992), 10. Cf. Eugene Linden, "Bonobos, Chimpanzees with a Difference," *idem*.

technologies to their young, hunt individually and in groups and bring back meat to share with the larger group, engage in warfare as well as murder and infanticide, vary by group culturally inasmuch as different groups exploit different resources in similar ecosystems, and, as the television programs of Jane Goodall show, have definitely individual personalities. There are now even questions concerning how to distinguish archaeologically between what are very early human sites and what may actually be early chimpanzee sites! Plainly, these apes engage in cultural practices at least like those classified as amateur above.

Phenomenological interest in the anthropoid apes goes back to the early work of Gurwitsch, who was responding to research by the Gestalt psychologists, especially Köhler, into the practical problem-solving intelligence that chimpanzees and humans share.[12] This would seem part of the structure common not only to the anthropoid apes and human infants, but also the brain-injured adults investigated by Gelb and Goldstein, the so-called primitive humans investigated by Lévy-Bruhl, and "the practical attitude of normal [human] subjects."[13] Altogether, Gurwitsch presents the world of the natural attitude (of humans, although probably subhumans have only this attitude) as originally practical and habitual and made up of situations filled with interrelated functional or practical objects (A chimpanzee world is cultural in these terms):

> The life-world is essentially a socio-historical, that is a cultural, world. In the life-world we do not encounter—at least not in the first place—mere corporeal objects, pure perceptual things, which can be exhaustively described in terms of what traditionally are called primary and secondary qualities. What we encounter are cultural objects, objects of value, e.g., works of art, buildings which serve specific purposes, like abodes, places for work, schools, libraries, churches, and so on. Objects pertaining to the life-world present themselves as tools, instruments, and utensils related to human needs and desires; they have to be handled and used in appropriate ways to satisfy those needs and to yield desired results . . . . It is the specific sense of their instrumentality which essentially defines these objects and makes them be what they are, that

---

[12] "Some Aspects and Developments of Gestalt Psychology" (1936), translated by Richard Zaner in Aron Gurwitsch, *Studies in Phenomenology and Psychology* (Evanston, Ill.: Northwestern University Press, 1966), 50-52.

[13] Aron Gurwitsch, *Phenomenology and the Theory of Science*, edited by Lester Embree (Evanston, Ill.: Northwestern University Press, 1974), 171 n. 23. In *The Structure of Behavior*, translated by Alden Fischer, (Boston: Beacon Press, 1963), 113-128, Maurice Merleau-Ponty also reflected at length on anthropoid life.

is, what they mean to the members of the socio-historical group to whose life-world they belong.[14]

This notion of culture that can include the practices of apes is certainly not the same as "high culture," which can be said to include the cultural disciplines of art, literature, organized religion, science, and high technology. It may best be called "basic culture." *Basic culture* can—but does not need to—include high culture within it, and, while high culture requires basic culture, the contrary is not, again, the case. This is crucial for any approach to culture from below rather than from above, an approach that thus does not leave "lower" combinations of cultural practices under-appreciated if not ignored. As members of the élite, intellectuals can be tempted to look down on the practical disciplines, e.g., engineering, as well as the sub-disciplinary practices of the crafty and amateur sorts, such as plumbing and housework. If one prefers to approach culture from above, the reasons for doing so should be explicit and well-argued. Also, one should recall in which type of profession Socrates was able to discern some non-philosophical wisdom (and the folly it was prone to) and think about where his scheme might be expanded to include the disciplined combinations of cultural practices.

Contributions to the phenomenology of cultural life or cultural worlds thus viewed can be found before Gurwitsch in the early Heidegger,[15] Scheler,[16] and even Husserl, who in 1913 writes,

---

[14] *Phenomenology and the Theory of Science*, 143, cf. 92 ff. Cf. also Lester Embree, "Some Noetico-Noematic Analyses of Action and Practical Life," in John Drummond and Lester Embree, (editors) *The Phenomenology of the Noema*, (Dordrecht: Kluwer Academic Publishers, 1992), Part II. and "A Gurwitschean Model for the Explanation of Culture or How Spear Throwers are Used," forthcoming in the volume of essays edited by J. Claude Evans from the symposium in memory of Aron Gurwitsch held at the Graduate Faculty of the New School in December 1991.

[15] Martin Heidegger, *Being and Time*, translated by John Macquarrie & Edward Robinson (London: SCM Press, 1962). Some later works are also relevant.

[16] Works such as Max Scheler, *The Nature of Sympathy*, translated by Peter Heath, (London: Routledge & Kegan Paul, 1954), *Formalism in Ethics and Non-Formal Ethics of Values*, translated by Manfred S. Frings and Roger L. Funk, (Evanston, Ill.: Northwestern University Press, 1973), and *Selected Philosophical Essays*, translated by David L. Lachterman, (Evanston, Ill.: Northwestern University Press, 1973) contain many insights relevant for the cultural disciplines.

I am conscious of a world endlessly spread out in space, endlessly becoming and having become in time. I am conscious of it: that signifies, above all, that intuitively I find it immediately, that I experience it. By my seeing, touching, hearing, and so forth, and in the different modes of sensuous perception, corporeal physical things with some spatial distribution or other are *simply there for me, "on hand"* in the literal or figurative sense, whether or not I am particularly heedful of them and busied with them in my considering, thinking, feeling, or willing. Animate beings too—human beings, let us say—are immediately there for me: I look up; I see them; I hear their approach; I grasp their hands; talking with them I understand immediately what they objectivate and think, what feelings stir within them, what they wish or will.

. . . .

In my waking consciousness I find myself in this manner at all times, and without ever being able to alter the fact, in relation to the world which remains one and the same, though changing with respect to the composion of its contents. It is continually "on hand" for me and I myself am a member of it. Moreover, this world is there for me not only as a world of mere things, but also with the same immediacy as a *world of objects with values, a world of goods, a practical world.* I simply find the physical things in front of me furnished not only with merely material determinations but also with value-characteristics, as beautiful and ugly, pleasant and unpleasant, agreeable and disagreeable, and the like. Immediately, physical things stand there as Objects of use, the "table" with its "books," the "drinking glass," the "vase," the "piano," etc. These value-characteristics and practical characteristics also belong *constitutively to the Objects "on hand" as Objects,* regardless of whether or not I turn to such characteristics and Objects. Naturally, this applies not only in the case of the "mere physical things," but also in the case of humans and brute animals belonging to my surroundings. They are my "friends" or "enemies," my "servants" or "superiors," "strangers" or "relatives," etc.[17]

---

[17] Edmund Husserl, *Ideas pertaining to a Pure Phenomenology and to a Phenomenological Philosophy*, First Book, translated by Fred Kersten (Dordrecht: Kluwer Academic Publishers, 1983), 51-53. Cf. Second Book the recently translated, by Richard Rojcewicz and Andre Schuwer (Dordrecht: Kluwer Academic Publishers, 1989), Part Three, "The Constitution of the Spiritual [*geistes* can also be rendered as "cultural"] World," Edmund Husserl, "Philosophy as a Rigorous Science," translated by Quentin Lauer, in Edmund Husserl, *Phenomenology and the Crisis of Philosophy* (New York: Harper & Row, 1965), Edmund Husserl, *The Crisis of the European Sciences and Transcendental Phenomenology* translated by David Carr (Evanston, Ill: Northwestern University Press, 1970), etc.

What some have called "human existence," but which may be better called "cultural life," can be sketched in general terms. The matter in question involves collective as well as individual subjectivity or intentive life and is in society and in history as well as in nature, but the distinctive thing is that it is cultural. Negatively speaking, culture is, as intimated, not confined to humans. It is also not inherently linguistic, which is not to say that cultural practices and cultural life may not be extensively speech-accompanied and profoundly speech-affected, but only that they are not in essence always infected by speech.[18] If cultural life as such were always speech-infected, speaking would have to be going on at all times, rather as perceiving and valuing and even striving are to be found always to go on in life. Recourse to unconscious language, to the potential capacity for language, or to the broadening of speech to include, curiously, non-linguistic processes, such as perception, is to hypothesize epicycles. Seeing is not comprehending, interpreting, or reading.

Positively speaking, culture is a matter of what is learned especially with respect to valuing and willing. "Willing" is used here in a broad signification and is thus not confined to cases in which an I engages in making or executing a decision. Rather, it includes all broadly volitional or practical processes in which objects making up situations and worlds are constituted as ends and means or, in general, as useful. Indeed, the emphasis is also on the routine and habitual rather than the spontaneous. Valuing also can be positive, negative, or neutral and the positive, negative, and neutral values of objects are correlatively constituted in this stratum. Believing and the belief characteristics that objects have correlative to it can also be learned, as are, it can further be contended, awarenesses of various types, the perceptual included, but *recognition of the strata of* learned *or* habitual *valuing and willing and their correlates is sufficient to distinguish cultural practices as cultural.*

Since the formal and natural sciences are disciplines that are cultural in that they essentially include learned valuing and willing but are not cultural disciplines, it needs to be recalled that the cultural disciplines are cultural by virtue of the cultural objects, situations, and worlds dealt with in them. Where the objects as they are intended to or as they present themselves in these practices are concerned, cultural characteristics can

---

[18] Lester Embree, "zoon logon ekhon," in Lester Embree, (editor) *Essays in Memory of Aron Gurwitsch* (Washington, D.C.: Center for Advanced Research in Phenomenology & University Press of America, 1983).

be recognized as belonging to the species of positional characteristics, above all values and uses, that are constituted in learned, perhaps skilled (but there are bad habits), and thus habitual and routine life. The automatically or instinctually valued and willed is not cultural.

Not all combinations of cultural practices are cultural disciplines. Besides those combinations that are amateur and crafty, there are disciplines made up of cultural practices that are not cultural disciplines. In these the cultural characteristics that objects, situations, and worlds are always already equipped with are set aside. In the cultural disciplines, by contrast, these characteristics are not only not set aside but are central to what the objects dealt with are. When reflected upon, cultural practices can be recognized to be intentive to focal objects that are in situations that themselves are in worlds, all three of which, by virtue of the values and uses they have always already acquired in habitual life, i.e., their cultural characteristics, are cultural. Thus generically characterized, cultural life can be thematized in different respects and in different manners by different individual cultural disciplines and species. This general statement may become clearer when the question of the three species of cultural disciplines is discussed presently.

To close these remarks about the subject matter in general of the cultural disciplines, it may be pointed out that in his *Guide to Translating Husserl* Cairns also alternatively proposes "cultural sciences" as a translation of *Geisteswissenschaften* and that the title of what is arguably the central essay of the philosopher who has written the most in the phenomenological philosophy of such sciences, i.e., Alfred Schutz, which title is translated as "Phenomenology and the Social Sciences," was originally entitled "Phaenomenologie und Kulturwissenschaft (Larchmont 6.8/1939)"[19] Schutz uses *Geisteswissenschaft* and *Sozialwissenschaft* as well as *Kulturwissenschaft* repeatedly in his German writings from the beginning and, although his emphasis was on Weberian or *verstehende* sociology and so-called Austrian economics, he additionally recognized archaeology, art history, ethnology, history, linguistics, philology, psychology, and political science as belonging to the same class of sciences.

---

[19] Schutz Nachlass, 13,758-13,790. The original of this Nachlass is held at Yale's Beineke Library and copies are held in the Sozialwissenschafts Archiv, Konstanz, and The Center for Advanced Research in Phenomenology, Inc., Florida Atlantic University. The English translation of this essay is reprinted in Alfred Schutz, *Collected Papers*, Vol. I, edited by Maurice Natanson (The Hague: Martinus Nijhoff, 1962).

Schutz's notion is thus broader than that of "social sciences" usual in North American usage, which typically excludes the psychological as well as the historical sciences. Nevertheless, while Schutz's authority can be used to defend the concept as well as expression "cultural sciences," there are more than the cultural *sciences* in the cultural *disciplines*.

III. Three Species of Cultural Discipline

There seem to be three species of cultural disciplines. This is best recognized by considering the combinations of cultural practices that make them up teleologically or, in other words, by asking what the overall effort in the discipline culminates in. This needs to be done carefully. Thus, one might hold that lawn mowing culminates in having a mowed lawn, but a modicum of reflection can show that having a mowed lawn is more than merely a matter of perception and indeed that it is the positive valuing colloquially called "liking" or enjoying that predominates in "having a mowed lawn" and that the lawn thus "had" can then be said to be "good," "beautiful," or at least "nice." One might go further and ask "for whom" mowed lawns are valued and recognize that not only the householder but also her neighbors value a mowed lawn positively, which does not change it that the cultural practice of lawn mowing is best classified as evaluational or *axiotic*. (Lawn mowing is thus not unlike hair cutting.)

Lawn mowing is not predominantly an effort merely to know about lawns, although some cognition at least in a broad signification, e.g., some justifiable believing about the effects of lawn mowers on blades of grass, is necessarily involved. Moreover, while the effort does have a real effect in that many leaves and stems are cut, making that effect into a purpose and even fulfilling it is also not the culmination. Perceiving the lawn and how lawn mowers function may immediately justify believing in ways in which to do it and mediately justify the valuing of a nice lawn, but it is also not what lawn mowing overall culminates in.

Those operating the equipment or managing the company in the trade or craft of lawn mowing or, more generally, "yard care" or gardening, may be seeking as efficiently as possible to make money, which is useful for paying bills, but that too is not what the practice overall aims at, which is the pleasure or enjoyment, the valuing, that takes place in suitably prepared people who encounter the mowed lawn and, along with that, the pruned trees, weeded flower beds, etc. "Suitably prepared" refers to how we learn to value neatly trimmed lawns positively. A society can be feigned in which people are enculturated in such a way that a lawn, neatly trimmed or not, is a strange, unnatural, even frightening and thus negatively valued object. After all, plants do not normally grow in homogeneous, flat, and rectangular forms like outdoor imitations of carpets and thus lawns are as unnatural as trees lined up in orchards or in tree farms like soldiers on parade. For the suitably prepared, however, these arrangements are not unnatural.

To take another example and to look at it differently, the owners, coaches, players, etc., in professional sports may all be out themselves to make money, but the combination of practices that is the sport is a failure rather than a success if the fans do not enjoy the game. Baseball as much culminates in evaluation as fine art does. Whether art and sport are crafts or disciplines is a different question. The cultural is, of course, often taken for granted, but reflective examination, particularly when stimulated by comparisons or history, can challenge whether mowed lawns, like artfully slaughtered cattle (i.e., bull fights), are universally and necessarily beautiful things. And the highly trained landscape architect may also be seeking money and reputation and pursuing a career, but if that discipline is about the design, creation, and maintenance of landscapes (within the standards of the given cultural world), then lawn mowing is sometimes part of something that culminates in the appreciation of beauty in gardens, parks, boulevards, etc.

The trifurcation of cultural practices offered above in Part I and the trifurcation of the cultural disciplines in the present section can be combined into a cross-classification (See Figure 2). As has been intimated, the distinguishing of the species of cultural disciplines is best done with respect to the type of positionality that predominates in the culmination of the practice that pertains to the discipline. Cultural life and cultural practices always concretely include types of awareness, such as perception, recollection, and representational awareness, but they also concretely include positionality of all three already mentioned types, which can alternatively be referred to as the doxic, praxic, and pathic strata and positional characteristics.

When the doxic component, i.e., the stratum of believing that always occurs in the concrete intentive process and the correlative belief characteristic in the concrete object as it presents itself, predominates, then the cultural practice can be called, in a broad signification, *cognitive*.

(Figure 2)

	Theoretical	Practical	Axiotic
Disciplinary			
Crafty			
Amateur			

When the pathic component, i.e., the stratum of valuing and the correlative values that also always occur, predominates, then the practice can be said to be *axiotic*, an expression broader than "aesthetic." And when the praxic component, i.e., the ever-present stratum of willing (in the broad signification) and the correlative means- and end-uses in objects as intended to, predominates, the practice can be said to be *practical.* That the word "practice," which can also signify a phase of careful preparation, and "practical" are cognate might help to counteract the intellectualist tendency to focus on the cognitive foundations that are indeed recognizable in every practice.[20] The points here, however, are (a) that there are pathic and praxic as well as doxic strata in all concrete cultural practices and cultural characteristics correlatively in all of the concrete cultural objects, situations, and worlds constituted in them, with sometimes one and sometimes another predominating, and (b) that cultural disciplines can be classified according to the positionality that predominates in the practices in which they culminate.

a. Theoretical Cultural Disciplines.

Disciplines of the cognitive or theoretical sort can be called "cultural sciences." The cultural sciences are not the only species of science, for the formal and naturalistic sciences can of course also be recognized.[21] If

---

[20] Phenomenological foundation, which is a relation between founding and founded strata within a concrete intentive process, has, plainly, no relation to foundationalism, which seems a matter of propositions and logic.

[21] Cf. Thomas Seebohm, Dagfinn Føllesdal, and J. N. Mohanty, edd., *Phenomenology and the Formal Sciences* (Dordrecht: Kluwer Academic Publishers, 1991) and Lee Hardy and Lester Embree, (editors) *Phenomenology of the Natural*

science in general seeks theoretical knowledge, one species can be concerned with form and include logic, mathematics, and grammar as subspecies. These rely on a procedure called formalization by which the contents of assertions are abstracted from leaving the form. Most if not all cultural sciences employ or apply formal techniques today, so they are no longer purely "qualitative," but this methodological addition is also not essential to what they are. That would be analogous to claiming that road building was not road building until naturalistic science arose and began to be applied in it. And if formal techniques were essential, then disciplinary efforts historically or structurally prior to the addition of the formal techniques would not be scientific. Was Aristotle's physics not physics?

In contrast, one may speak of material or contentual sciences, which require no such abstraction from all content. Then it can be recognized that the *cultural* sciences are the most concrete among the contentual sciences in that the cultural characteristics that objects have for the cultural practice investigated in the cultural sciences are not abstracted from, while in the *natural* sciences the cultural characteristics that all objects always already have in non-scientific life are abstracted from. (The qualifier "non-scientific" protects the possibility of objects having discipline-specific scientific value and use, e.g., a good experiment as good from the standpoint of the community of naturalistic scientists.)

The cultural sciences can be regarded philosophically in many ways, only some which are mentioned in this essay. One perspective concerns the difference of the cultural and other kinds of positive science, which has just been addressed, but that the allegation whereby Dilthey contrasted the natural sciences as explaining with the *Geisteswissenschaften* as understanding is false deserves mention; sciences of both kinds both explain and understand.[22] Other positions here would contrast the particularizing and empirical with the generalizing and nomothetic disciplines, something that seems more to pertain historically to an alleged difference between the social and the historical *subspecies* of the cultural sciences and, in any case, it can now be seen as more a difference of emphasis than of essence because all contentual sciences are concerned with both the empirical and particular and the general and nomic.

Sciences (Dordrecht: Kluwer Academic Publishers, 1992).
    [22] Makkreel, *loc. cit.*

There is, however, a clear distinction between the cultural sciences that investigate individual life and those that investigate collective life, the former being called "psychological sciences," while the latter can be called the "communal sciences," at least while the differences between the social sciences and the historical sciences within the *communal* type of science are explored. Some believe that the historical sciences are confined to textual data and thus different from the social sciences as focused on behavior. But this is to fail to appreciate that prehistoric archaeology is an historical science (in the broad signification) that relies on non-verbal remains and that the participant observation relied on in various social sciences is typically accompanied with, if not dominated by, the interpretation of written as well as spoken expressions. There can be no doubt that many cultural disciplines, both social-scientific and historical-scientific, rely to vast extents on the comprehension of language and thus that the methods of interpretation of linguistic data are of great importance for all the cultural sciences. Phenomenologists have not ignored this.[23] The overlooking the non-verbal is more likely the problem.

The cultural social sciences, such as economics, ethnology, geography, linguistics, sociology, and political science, are focused on the contemporary world and are thus *synchronic*, while the historical cultural sciences are *diachronic*, relating events of different times, the future even, and having, also, no preference necessarily for the contemporary. It is not uninteresting that evolutionary biology, geology, paleontology, etc., are also diachronic and thus historical *naturalistic* sciences and to be compared as well as contrasted with the historical *cultural* sciences. Whether there are synchronic natural sciences is another question.

Social, historical, and psychological scientists who happen to read the present essay might at this point be thinking that, like the character in Molière, they are only hearing a fancy new name for what they are already doing. Nevertheless, it can still be asked what makes their disciplines cultural. This question needs to be asked because a certain naturalism is often observable in these disciplines. This naturalism is

---

[23] See, e.g., Thomas M. Seebohm, "The New Hermeneutics, Other Trends, and The Human Sciences from the Standpoint of Transcendental Phenomenology," in Hugh Silverman, John Sallis, and Thomas M. Seebohm, *Continental Philosophy in America* (Pittsburgh: Duquesne University Press, 1983), "Boeckh and Dilthey: The Development of Methodical Hermeneutics," *Man and World* 17 (1984), "Falsehood as the Prime Mover of Hermeneutics," *The Journal of Speculative Philosophy*, 4 (1992), etc.

possible because psychic life and social and even historical relations can also be viewed through an habitual abstraction from the values and purposes they have for the individuals and groups in the relevant cultural worlds. This would seem characteristic, for example, of psychology as a naturalistic science in contemporary academe.[24]

(Figure 3)

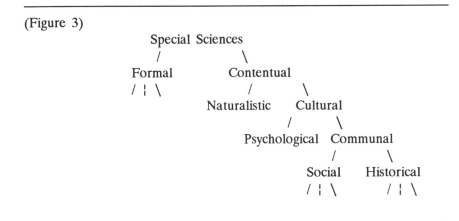

This sketchy survey of the cultural sciences would not be complete without it being emphasized again that there are various particular cultural sciences within the subspecies and species discussed. Something in the way of a list of particular *social* sciences has been offered above. It and the following can be related to Figure 3, where the opening for further specification is shown at the bottoms of the branching root system. The *historical* cultural sciences include cultural history in a perhaps narrower signification, diplomatic history, economic history, military history, history of science, social history, history of technology, gender history, ethnic history, etc. But philosophical reflection that focuses on one or another individual discipline or subdiscipline would do well to be cognizant of the species and genus that the combination of cultural practices focused on belongs to and what can be learned by doing so.

---

[24] Advocacy for psychology as a cultural science can be traced back to Dilthey and appears clearly within phenomenology in Edmund Husserl, *Phenomenological Psychology*, translated by John Scanlon (The Hague: Martinus Nijhoff, 1977).

Finally, it may bear brief repetition that a scientific discipline aims at and culminates not in valuing or doing but in knowing. All else is subordinate, extra-disciplinary consequences and conditions as well as other component cultural practices included.

b. Cultural Disciplines that are Practical.

A practical discipline is one in which people of advanced preparation and in the relevant disciplinary attitude strive to affect courses of events in some pertinent part of the world or other. Striving or willing characterizes a concrete practice in which the praxic component predominates (but there are pathic and doxic strata and positional characteristics as well as awareness and objects as awared also essentially involved). Their objects are the ends and means for their actions. Willing can be positive or negative and, depending on whether the objects that are willed already exist or not, willing can be destructive or preservative, creative or preventative, or it can seek to foster, impede, or merely alter them. On some occasions, the action consists in an abstaining from intervening, so that one simply lets happen what will. Besides striving to affect, in the broad signification, the course of events, which is different from both the believing in and the valuing of objects, practical practices differ from the cognitive and axiotic in focusing only on the future. In other words, one can know and also value the present and the past as well as the future (and even ideal or atemporal matters), but one cannot will objects other than in the future.

Without the effort being made adequately to subclassify the disciplines of the practical sort, it can be said that education, law, medicine, nursing, politics, psychiatry, psychotherapy, and social work are practical disciplines since they culminate in less ignorant students, enforced laws, preserved or improved political systems, and physically and/or mentally healthier people. Engineering, furthermore, is above all a practical *cultural* discipline with many subspecies. One can even argue that business is practical in seeking profits as goals and then consider whether it is sometimes disciplinary. To be sure, action is founded upon evaluation as well as cognition in all these cases, but what predominates in the culmination of the cultural practice is success or failure (which have degrees) with respect to affecting a course of events.

Also, that the use to which the results are ultimately put does not pertain especially to the designers, makers, performers, or sellers, but rather to the users of the product does not affect it that the culmination is in willing and the object being willed or used. The craft of house building as such is thus practical, although it can pertain to architecture as an axiotic discipline. Nevertheless, to enjoy a building in a more or less aesthetic way is different from using it to keep warm and dry in a cold and wet winter (or cool and humidified in a hot and dry summer), which are practical purposes of building and maintaining efforts and of the building as a means.

The cognition that action is founded upon within the combination of concrete practices at the disciplinary level may be scientific, and has no doubt ever increasingly become so in this age of scientific technology. In that case, as mentioned, the scientific basis may be made up of results and methods from more than one theoretical discipline, e.g., physical, chemical, and biological knowledge are used in physical medicine and history, sociology, ethnology, etc., are, hopefully, used along with cultural psychology in psychiatry and psychotherapy. That natural-scientific knowledge is developed in an attitude in which the cultural characteristics with which objects are always already constituted are somehow set aside does not limit the use of such knowledge in what might be called "scientific action" and in which a scientifically ascertained causal connection is then constituted volitionally as a means-end relationship.

But there is additionally, however, always some non-scientific cognition involved in action. This includes the common-sense of the amateur level. It also includes the know-how and lore of the craft level, which is the level where warfare was for the millenia before it began also to be science-based or scientific. "Military science" is now arguably a *discipline* of the practical sort, conquest and defense being practical purposes and advanced preparation occurring in military schools and war colleges. The same might be the case with law enforcement, although the present writer is not sure that he has yet heard of "scientific policing" and "police colleges." Finally, it is not impossible that philosophical cognition sometimes function in the foundations of action and its justification, philosophy, even if merely cognitive in project, being no special science with a restricted subject matter.

Theoretical science and scientific action are often confused. In part this is because science-based practical endeavors, technologies included, can find it advantageous to distinguish themselves from the non-scientific and even to go so far as to call themselves sciences and because theoretical endeavors can analogously find it on occassion advantageous to claim practical consequences. It might be argued, however, that both the scientific action and the theoretical endeavor need to be distinguished if only because what is involved in making an action scientific as well as how theoretical results are made useful can then be better understood, evaluated, and accomplished. Contrariwise, if one chose to impede the responsible application of theoretical science, obscuring this difference might be useful.

Evaluation or the valuing of believed-in possibilities and relations, especially the pathic constitution of effects with intrinsic values and of causes with extrinsic values, is central to the motivation of action. Since it is positive valuing that motivates action that is positive, i.e., creative and preservative, and negative valuing that motivates action that is negative, i.e., destructive and preventative, it is possible to confuse these strata and the objects constituted in them. There is a parallelism between the end- and the means-characteristics of objects and the intrinsic and extrinsic values that the same objects also typically have and this can similarly lead to confusion. Nevertheless, if one looks to whether the practice is directed at affecting a course of events rather than merely liking or disliking this or that possible future sequence, the distinction between the habitual positional strata in the intentive process and the parallel distinction between the cultural characteristics, the values and uses, in the objects as they are intended to can be made.

Intentive life can seem more conspicuously involved with *somatic or bodily states and processes* in predominantly practical than in predominately evaluational or predominately cognitive life, for bodily movements are necessary to affecting a course of events, but it takes only a little reflection to recognize that bodily movement is also involved in the sensuous awareness underlying the rest of life. The eyes, for example, need to be open and pointed in a direction for something to be seen. Even the expressing of a proposition requires movement of lips or fingers. The danger in investigating the role of movement (and staying still is, in this signification, a movement, just as non-intervention is a willing to affect a course of events) may not be to omit the body but rather to concentrate on somatic movement to the behavioristic disregard of the willing that makes it action and in relation to which objects are ends and means.

c. The Axiotic Disciplines.

That there are theoretical and practical disciplines that can be considered cultural qua combinations of skilled practices bearing on objects that are neither naturalized nor formalized is probably not difficult to accept, but as the sketch of landscape architecture in relation to amateur and crafty engagements in lawn-mowing above intimates, there seem to be cultural disciplines of a third sort that culminate neither in cognition nor in action but rather in valuing and values. These may be called "value disciplines" on occasions when the more technical-sounding "axiotic disciplines" would not communicate as well. Analogously, there may be occasions when it is better to speak of "knowledge disciplines" and "action disciplines." Disciplines of all three sorts are combinations of cultural practices themselves as well as related to cultural matters. Thus, while the mowing of lawns is not a practice that the landscape architect performs, it is a practice, specifically a maintenance practice, necessary, for some, to a beautiful landscape, so that a landscape including lawns in a lawn-appreciating society is less beautiful if maintenance of this sort is not performed and hence the architect needs to know about, value, and plan for it in her landscaping projects. Cultural disciplines combine many cultural practices.

Enough may have been conveyed above positively about how there are positive and negative values that can also be intrinsic or extrinsic, how these are constituted in the pathic or valuing stratum of intentive life, how this stratum is a non-self-sufficient positional part of the pertinent intentive process or concrete cultural practice, and how values can be reflectively differentiated and described when objects as intended to are reflectively observed. This analysis may help any who hesitate about an approach that takes axiotic crafts and disciplines seriously and notes that they relate not only to refined appreciation by esthetes but also to enjoyment, pleasure, even fun, etc., and additionally the negative and neutral opposites of them. "Positive valuing" is a more technical term than "liking," negative valuing is disliking, and then there is apathy. And an enormous amount of valuing, whatever the modality, is deeply habitual or, in other words, cultural. (It also deserves mention that the value characteristics of objects as they present themselves are not the same as the monetary prices that some objects have in commerce.)

For many, "value" is what talk about "culture" chiefly connotes. From within a social group values often seem all utterly unlearned or "natural" and "objective" or non-relative and the different values of other groups, which groups are sometimes "other" chiefly by virtue of those different values, then seem thoroughly pathological or somehow worse, i.e., demonic. The acquired and group-relative status of values can be recognized, especially from without, and upon doing so some people sink into sceptical relativism, while others pursue the possibility that, while many values are relative, some important ones may be non-relative. Those still resisting both naïve absolutism and sceptical relativism are concerned because of the role of values and valuing especially if one strives for more than knowledge. The volitional positing of purposes and also the means that serve them is regularly motivated immediately by evaluation and inasmuch as valuing fairly obviously has foundations in cognition (provided "cognition" is comprehended to cover not only unjustified but even unjustifiable believing as well as justifiable and justified believing). It is only a matter of emphasis. The characteristics of objects as believed in and also as willed are cultural characteristics equally.

While no serious attempt at a classification will be made here, it is clear that there are many combinations of cultural practices that are axiotic. Some pertain to so-called fine art, including not only architecture, but also dance, film, photography, literature of various sorts, painting, sculpture, and theater, and others are practices culminating in less allegedly refined enjoyments such as football, basketball, and other sports, not to speak of lawns and haircuts and *haute couture*. In such cases, the first question concerns whether the discipline culminates in something axiotic, e.g., the pleasure of the audiences at a sports car race or a rock-and-roll concert, and the second question concerns whether the leading practitioners in the practice have the degree of preparation and competency that distinguishes a professional from an amateur and, beyond that, a discipline from a craft. Elitist prejudices may also be countered by recognizing that much fine art is essentially crafty and that sometimes sport, e.g., when pursued at the Olympic level, seems rather disciplinary.

"Criticism" can take a variety of forms in relation to axiotic practices, e.g., literary criticism, painting and sculpture criticism, music criticism, and architectural criticism, but also much of what one reads on the sports pages in newspapers (or is that parallel to reviewing? And what is the status of, e.g., film reviewing?). Again there is the question of whether the preparation of the practitioner is crafty or disciplinary and whether the whole combination in which such criticism is one of the component practices culminates in cognition, evaluation, or action. Does literary criticism properly culminate in greater enjoyment of literature among readers, in getting more books sold, or in knowledge of what a text is ontologically?

A place for valuing and values within phenomenology was shown at least as early as the "Prolegomena" of the *Logical Investigations*, where Husserl discusses logic as a normative as well as a theoretical and as a practical discipline.[25] This is a discussion of three types of proposition the forms of which are of interest to formal logic. Logic as a *theoretical* discipline would culminate in cognitive claims. It is not concerned with what ought to be or with affecting events. Logic as a *practical* discipline has to do with how argument forms are involved in attempts to persuade. Thus there is practical as well as theoretical logic.

As for logic as a *normative* discipline, "A warrior ought to be brave" is the central example and Husserl interprets it as the assertion that "A warrior who is brave is good," i.e., as a judgment in which a value is grasped through the predicate term, a value judgment. This example can of course be formalized with letters substituted for "warrior" and "brave." While Husserl's analysis is about three types of propositional form, the referents of propositions of these types of form can be seen to be cognitive, practical, and evaluational *matters* distinguished in the way of the present essay. Furthermore, while Husserl's analysis is itself cognitive, culminating in cognitive expressions about three types of propositional form, practical and axiotic expressions can also be distinguished according to whether they participate in actions or evaluations, something that will be returned to presently.

Of the three sorts of cultural disciplines distinguished in this section, the axiotic is the least well understood by the present writer. Nevertheless, that there are these three species seems clear. Furthermore, it can be wondered whether all actual and possible cultural practices can be subsumed under the three species distinguished here, which might be considered a philosophical concern because outside the scope of any particular or species of cultural discipline but nevertheless relevant to their ultimate intelligibility.

---

[25] Edmund Husserl, *Logical Investigations*, translated by J. N. Findlay (London: Routledge & Kegan Paul, 1970), "Prolegomena to Pure Logic," Chapters 1 and 2.

IV. When is Reflection on the Cultural Disciplines Phenomenological?

Although the names, preparation processes, etc., have varied, there have been what are here called cultural disciplines for millennia. In ancient and medieval times there were the seven liberal arts, at least some of which are cultural disciplines, and an examination of them in relation to the analysis proposed here might be interesting.[26] It can furthermore be argued that there has been philosophical reflection, at least of the cognitive and evaluational sorts, on such disciplines at least since Socrates, Plato, and Aristotle, but comprehensive efforts in modern times only became emphatic with the efforts of Comte, Mill, Dilthey, and others during the 19th Century and seem no older than Hume's "attempt to introduce the experimental method of reasoning into moral subjects."[27]

How does what can be called *"philosophy* of the cultural disciplines" approach the world? A philosopher interested in nature can approach the natural world directly but she can also and, as history has shown, more effectively approach nature *obliquely* through reflection on the natural sciences. In parallel fashion, a philosopher of culture can approach culture, cultural life, or cultural worlds directly in a philosophy of culture, but she can also approach it obliquely through a reflection on the cultural disciplines, and that may also prove more effective.

---

[26] Actually, there seem once to have originally been nine such arts, but architecture and medicine were later omitted. Cf. Stephen H. Mason, *A History of the Sciences* (New York: Collier Books, 1962), 62. Such a comparative analysis ought to include Edward G. Ballard, *Philosophy and the Liberal Arts* (Dordrecht: Kluwer Academic Publishers, 1989).

[27] "Hume's work as a social scientist has been largely ignored by philosophers, both because of its glaring incompatibility with his positivist image and because it develops a line of thought which, until recently, was more likely to be appreciated in Continental intellectual circles than in Anglo-American philosophy. For Hume, man is not just a natural object but a cultural product., and this means that the science of man must be conceived of as a normative moral science of action, not a descriptive natural science of behavior. In social science [sic], Hume moved beyond the narrowly mechanical, and even the organic levels of explanation, to the historical and personal. What emerged was the image of man as a role-playing or rule following agent, whose comprehension and self-comprehension requires the use of *Verstehen*. Not only does this make Humean social science *Geisteswissenschaft*, but it requires that we view philosophy as a form of social science, and it requires that we reformulate our notions of what consitutes science and explanation. Thus Hume's philosophy reemerges in contemporary debate but at a most unexpected place." Nicholas Capaldi, "Hume as Social Scientist," *Review of Metaphysics* 32 (1978), 99.

When the oblique strategy is adopted, it is becomes clear that the cultural presents itself in specifically and particularly different ways in relation to the different specific and particular perspectives of the various disciplines and this raises questions about the common subject matter the different aspects of which are thus thematized. Some efforts toward answering these questions are offered above. There is additional variation with respect to the school of thought or orientation out of which any reflective efforts proceed. There have been Hermeneutical, Logical Empiricist, and NeoKantian, as well as Phenomenological orientations employed with regard to individuals and groups of the cultural sciences and other disciplines during the 20th Century. Since polemics against other schools are always possible but seldom do more than reinforce attitudes, a broad notion of phenomenological philosophy and how and why it ought to relate to the cultural disciplines will instead be sketched here. At least phenomenologists of good will interested in the cultural disciplines will centainly consider it seriously.

Phenomenology began at the turn of the 20th Century, it exists in non-philosophical as well as philosophical forms, and several tendencies have developed within it as a broad movement. Without dwelling on differences between the Constitutive, Existential, Hermeneutical, and Realistic tendencies, which can be seen as chiefly different in emphasis, an attempt at a generic statement must still solve the problem of the diversity of terminologies that stems from the variety of major figures and texts. Three solutions are possible: One can begin from one of these terminologies and make adjustments as needed, which would seem to favor the tendency begun from, one can attempt a balanced mixture of the different terminologies and spread the irritation equally, which would have benefits not worth the difficulty, or one can venture a new and hopefully neutral generic terminology, which is the tactic used here.

The present thesis is that an endeavor is phenomenological when it is (a) descriptive and (b) reflective. An effort is *descriptive* in a broad signification when it avoids positing unobservable but presumably still thinkable entities, something which, as will be seen presently, can be taken too narrowly and thus phenomenalistically exclude *representationally* observable objects in, say, archaeology.[28] In a different but not unrelated

---

[28] Lester Embree, "Phenomenology of a Change in Archaeological Observation," in Lester Embree, ed., *Metaarchaeology: Reflections by Archaeologists and Philosophers* (Dordrecht: Kluwer Academic Publishers, 1992).

signification, thought is descriptive when it recognizes describing as prior to explaining, which it can do when it recognizes that knowledge of what matters are is presupposed by attempts to account for matters in terms of other matters. There are at least four types of explanation, namely, teleological, aitiological, justificatory, and grounding, but, again, descriptions of "what is what" logically precedes attempts to answer questions of "why."

"Descriptive" furthermore connotes an epistemology in which cognition is believing that is justifiable by evidencing, "evidencing (*Evidenz*)" being the awareness in which objects are given in the most originary way possible for objects of the type. If the objects are sensuously perceivable, for one example, sensuously intuitive awareness is the most originary awareness and recollection, expectation, feigning, and, of course, empty or blind awareness and thinking, are less originary. For another example, the most originary awareness of another conscious life is had on the basis of a perceptual awareness of features of the animal's body. Like observation in archaeology, there is then a non-linguistic representational evidencing founded on that which is represented. Phenomenological method is thus concerned positively with the evidencing of objects and how evident objects are given, which evidencing can be representational as well as presentational, so that some objects are evidenced that are not directly, but indirectly given. Descriptive thought is not phenomenalism.

An effort is *reflective* (the opposite is "unreflective" or "straight-forward") when the focus is not simply on objects but rather on objects-as-they-present-themselves (or objects-as-they-are-intended-to) and, correlatively, intentive processes (*Erlebnisse*) as directed at or as intentive to them. Once this so-called "noetico-noematic correlation" became considerably clearer with Husserl's reflections, the work of anticipators, beginning with Brentano and James, could be appreciated. Nevertheless, since everyday life and most crafts and disciplines are straightforward in their primary intention, lapses into unreflectiveness are remarkably easy, as easy as speculation that is not undergirded by evidencing of the matters thought about. Metadisciplinary efforts as practiced within a discipline, i.e., intramural metadisciplinary efforts, are, like the philosophical ones, reflective with respect to the straightforward basic intention of what they reflect upon.

An entire conscious life, even a collective one, can be a reflectively considered an intentive process, but analysis will disclose particular processes or process strata within life streams. One genus of stratum within the intentive process seems best called *awareness* and, as intimated, includes species that are presentational (perceiving, remembering, expecting) and representational (e.g., awareness of objects as depicted on the basis of the perceiving of depictions). If the awkwardness be tolerated, there are thus modes of "awaring" correlative to objects that are, in general, "awared." Specifically, the awared needs to be described, according to the case, as perceived, as remembered, as depicted, etc., just as awareness needs to be specified as perceptual, memorial, etc. In contrast with awareness in the concrete practice or process there are also thetic or *positional* strata, which include the doxic believing, the pathic or axiotic valuing, and the praxic willing and, correlatively, objects as believed in, valued, and willed, all of which have also been referred to in previous parts of the present essay.

If being thus descriptive and reflective defines what it is to be *phenomenological*, something still needs to be said about how phenomenology can be *philosophical*. This is necessary because there are phenomenological tendencies, intra-disciplinary criticism included, within many of the cultural sciences.[29] It is here proposed that phenomenology is philosophical when it seeks not merely knowledge in a particular region but general and ultimate justification. *Justification* begins the move toward ultimacy when it goes beyond the limits of all individual and even all the species of discipline, i.e., when it is quite general in scope. Beyond this, a broadly *realistic* ultimate justification would involve the grounding of all positions in the widest objective context. Alternatively, an ultimately *transcendental* justification would involve the grounding of all objects in an intersubjectivity whose insertion in the world has provisionally been suspended to avoid absurdities. These strategies evolved in Ancient and Modern philosophy respectively and both have been pursued in phenomenology. Whichever is pursued, the aim is clearly philosophical.

If the foregoing is sufficient to distinguish the phenomenology that is philosophical from scientific phenomenology, then it can now be asked how, in general, the phenomenological philosophy of the cultural disciplines approaches its subject matter. As intimated, cultural matters can be approached in a direct fashion and with the highest purposes, but, because the matters approached are so complex and confusing, there is great danger of arbitrariness and floundering in that case. The oblique

---

[29] E.g., Richard L. Lanigan, *The Human Science of Communicology* (Pittsburgh: Duquesne University Press, 1992).

approach may be preferable, provided one is not distracted or one's results distorted by the specialization of the disciplines reflected upon. As with the natural and formal sciences, the considerable accomplishments of the cultural disciplines of all three sorts appear to have been possible through specialization. "Specialization" signifies that a perspective has been adopted by which but one part and not others of the world is approached. It is thus an abstraction. One can remember that some disciplinary perspectives are synchronic and others diachronic, some focus on individualized life and some on collectivities, some focus on social life and others on economic life or linguistic life, etc., and this omits consideration of which species of animals is under consideration.

When cultural life is approached obliquely for philosophical purposes, the perspectives of the individual, species, and genus of the cultural disciplines, when justified, can be used for guidance. But then it is important to look beyond the philosophy merely of the social sciences or of the historical sciences, or of architectural or medical disciplines specifically or in particular, if only because a well developed albeit circumscribed and straightforward perspective can be seductive. Phenomenologists are not the only philosophers who appreciate that the same matter can present itself in a number of perspectives none of which is privileged and all of which are disclosive, but they do seem the ones most keenly involved with this approach, which is sometimes characterized as an effort to go through subjectivity to reach objectivity. If they are, then they will be interested in discussions concerning, for example, what the historical sciences have in common, how they resemble as well as differ from the social sciences, how the axiotic disciplines compare and contrast with the cognitive and practical, what cultural life or the cultural world in general is, etc.

The affinities between the cultural sciences, the axiotic disciplines, and the practical disciplines and the traditional parts of philosophy called epistemology, axiology, and praxiology, respectively, can also be used to protect the philosopher from naïveté. The general philosophical concern with cognition transcends concern with what, e.g., particularly sociological knowledge is, the philosophical concern with evaluation goes beyond the particular aesthetics of architecture, and the philosophical concern with the ultimate goals of action are beyond those of business or medicine. It is philosophical now to ask whether there is something beyond these three parts of the philosophical task, which amount to specialization of at least a high level of generality.

The trifurcation based on the three types of positing and objects as posited that has been central to the present essay plainly extends to philosophy itself. While philosophy before Kant tended to bifurcate into theoretical and practical philosophy, there has been growing acceptance of the three parts to philosophy just related to the three species of cultural disciplines. Sir William Hamilton, for example, introduced it into English speaking philosophy and it conspicuously structured the thought of the great Alexander Bain. Husserl clearly recognizes the theory of value and the theory of practice beside the theory of knowledge, but these "theories" are cognitive disciplines and thus parts of philosophy as a strict science and Gurwitsch and Schutz, for example, follow Husserl in this intellectualizing respect.

Cairns, however, challenges the adequacy of the view whereby first philosophy is merely a science, albeit the primal science in which the positive sciences and the world are grounded.[30] That task of grounding is not itself rejected, but is considered a part of something larger. Beyond epistemology, there is first of all axiology, but this can be seen as culminating not in knowledge about correct valuing but in habits of correct valuing itself. Then one might be better off to speak of the philosophy of evaluation. Thus the philosopher seeks not merely to have ultimately justified believing but also ultimately justified valuing, to be a justified valuer as well as a justified knower. This is still not sufficient, however, for the philosopher can also go beyond the *theory* of justified action to justify her action itself and thus be a justified doer or actor. Beyond praxiology as *theory* of practice there would be the philosophy of action. Then the wisdom ultimately sought is not cognitive, nor even axiotic, but practical. Ultimately, one must act wisely. Philosophy is then not only a cultural discipline but also the ultimate specifically practical cultural discipline, for it culminates in action.

If Cairns's reform of the project of phenomenological philosophy is accepted, then the phenomenological philosophy of the cultural disciplines can serve the purposes of wisdom. Reflection on them includes not only using them to know about the world but also using them to evaluate how the world may be related to in various ways, reflections that can justify

---

[30] Dorion Cairns, "Philosophy as a Striving toward Universal *sophia* in the Integral Sense," in Lester Embree, ed., *Essays in Memory of Aron Gurwitsch* (Washington, D.C.: Center for Advanced Research in Phenomenology & University Press of America, 1983).

reflective actions, which is to say actions that bear not only upon ways in which to relate to the world but also upon how the world is as related to; either way, the course of events is affected. Can an attempt to show what the species of cultural disciplines have in common as well as how they are differentiated serve to foster those disciplines? Can a clarification of the background for their convergencies enhance the efforts of phenomenologists of the cultural disciplines? It will remain to be seen (or evidenced) whether a literary action of editing and introducing a volume of essays chiefly from a conference can fulfill practical purposes with either or both of these effects.

V. The Existence of the Phenomenology of the Cultural Disciplines

The chapters in this volume as well as other efforts by the same and other authors show that philosophical reflection of a phenomenological sort with respect to the cultural world or cultural life exists. Some examination of the forms of that reflection is in order. While the above characterizations of what it is to be phenomenological and to be philosophical can be left for others to examine, one can still ask about the background concern, the obliqueness of the approach, the relevance of the proposed tri-specification of the cultural disciplines, and whether or not the requisite mutual awareness and sophisticated preparation has been attained such that a philosophical *discipline* must be recognized.

There is a general concern with the cultural among upwards of a score of creative phenomenologists. Besides the chapters below, some of the many "cultural issues," as they may be called, that arose in the round-table session at the research symposium may be mentioned to show the richness of the region that the cultural disciplines and their philosophy address. (The symposium was held in May 1992, when ethnically charged conflicts had recently occurred in Los Angeles and in the former Soviet Union, the former Yugoslavia, South Africa, in Southwest, South, and Southeast Asia, and elsewhere.) Firstly, there are questions about the ethnic identities of individuals and whether there are non-ethnic individuals, as some in the United States curiously believe the so-called W.A.S.P.s, i.e., the White Anglo-Saxon Protestants, to be. Are racial differences not more adequately and thus properly considered ethnic differences? The same question can be asked regarding the gender and also the class identities of individuals. In such respects, the need to empower individuals ("subjective spirit" in Hegelian terms) can be maintained in the face of so-called Post-Modern thinking.

Then again, ethnic or cultural groups exist within and overlapping political structures such as nation states. Cultural groups can be approached in terms of how individuals relate to their own groups, how they are rooted in or alienated from them, or in terms of the relations, positive and negative, between cultural groups. Within societies there are ideals such as ethnic purity and multi-culturalism and policies (and not only in education) that support and undermine cultural groups and there are also processes of integration and disintegration. And, finally, there is the question of how current approaches—particular, specific, and generic—to the cultural disciplines as well as to cultural practices and phenomena can be related to phenomenological philosophy. It seems difficult to doubt that a phenomenological philosophy of the cultural disciplines, were it to exist, would be relevant in the contemporary situation.

The essays in this volume are explicitly or implicitly oblique in their approach, i.e., informed by work in cultural disciplines of one or another sort and sensitive to how the cultural world can present itself in disciplinary perspectives. Thus David Carr's discussion of Schutz implies Schutz's reflections on the interpretive sociology of Max Weber, Mano Daniel discusses Biography explicitly as a cultural discipline in which the stories of lives lived within cultures are told, Jim Hart uses "religious studies" as a point of view in considering part of Husserl's thought, Don Ihde's chapter relies on the phenomenological analyses of how engineers and others use equipment in cultural worlds that are more explicit in his various books, Stanford Lymnan and Lester Embree explore how the academic multidiscipline called Ethnic Studies might be approached phenomenologically, the field of Woman's Studies as well as that of Environmental Studies can be seen as presupposed in Don Marietta's chapter on ecofeminism, Ullrich Melle's philosophical view of the environment is also informed by ecological research, Algis Mickunas is quite explicit in discussing the work of a number of social scientists, Tom Nenon explores Connectionism as a new line of scientific inquiry that bears on cultural phenomena, Maxine Sheets-Johnson relies on paleoanthropology in her attempt to establish the invariants across vast ranges of time of the human body and behavior, Osborne Wiggins, like Nenon, reflects on the research frontier of a type of special science concerned

centrally with cultural phenomena, namely, Cognitive Science, and Richard
Zaner continues his long-term project of reflection on medicine.

It seems more likely that the mentioned figures would accept that
their philosophizing about culture generally proceeds through a consideration of cultural worlds in relation to work of specialized disciplinary
approaches than it seems that they would readily accept the taxonomy
offered above for these disciplines. Nevertheless, while there might be
disagreements about how to distinguish them, it would seem difficult to
deny that medicine and technology are *practical disciplines*[31] or that most
of the other chapters chiefly rely on *cultural sciences.*[32] Husserl's views on
religion presented by Hart presents an interesting case: Are religious
practices cognitive, practical, or axiotic? Certainly, all three components
are to be found in them, but which one predominates? And is the
combination of practices of advanced preparation provided in Religion
Departments and divinity schools a theoretical, an axiotic, or a practical
discipline? The same questions can be asked about environmental,
ecofeminist, ethnic studies and other practices and disciplines, and if these
are decided to be either practical or theoretical in their culminations,
what of architecture, alluded to above, and other fine arts phenomenological work on which is, unfortunately, not represented in this volume?

Whether or not attempting to relate their efforts to the trifurcation of
the cultural disciplines would be helpful for the phenomenology of the
cultural disciplines is for the mentioned authors and other readers to
decide in word and deed. If some phenomenologists concerned with some
cultural disciplines consider the efforts of others, which has happened,
then the further question can be asked of whether it is at the amateur,
crafty, or disciplinary level, which reduces to a question of the degree of
preparation in the form of close study of the efforts of others, i.e., the
keeping up with, e.g., Carr's work on history, Ihde's work on technology,

---

[31] Cf. also S. Kay Toombs, *The Meaning of Illness; A Phenomenological Account
of the Different Perspectives of Physician and Patient* (Dordrecht: Kluwer Academic
Publishers, 1992). Furthermore, see Anne H. Bishop and John R. Scudder, *Nursing:
The Practice of Caring* (New York: National League for Nursing Press, 1991)
pertains to another practical discipline and shows that at least bi-disciplinary efforts
can work.

[32] Recent work by Gary Madison relative to another cultural science, namely
economics, can also be mentioned, e.g., "Beyond Theory and Practice: Hayek on the
Logic of Cultural Dynamics," *Cultural Dynamics* 3 (1990).

Seebohm's work on interpretive methods, or Zaner's work on medicine, which all plainly pertain to the phenomenological philosophy of the cultural disciplines. This last question can be answered through cognitive reflection on the past, but it can also be answered through practical practices in future.[33]

[33] I thank Mano Daniel for the discussions that improved this essay in many substantial respects. The persisting blunders are on my head.

Chapter 1

Phenomenology and the Clinical Event

Richard M. Zaner
Vanderbilt University Medical Center

Abstract: *Attention to the clinical event for its own sake and as an exemplification are distinguished. Several of its essential features are delineated: situational and contextual determinants, complex asymmetry, inherently moral character, and reflexive nature of any clinical event. Dialogue is suggested as illuminating each of these features.*

Over the past ten or more years, I've often been asked to consult on individual cases. Which means that I am seen, and have served, as an "ethicist." At the same time, these and other facets of health care have attracted my philosophical interests. Accordingly, a distinction is helpful. To consult as an ethicist on any case is to be focused on the individual situation (people, setting, circumstances, issues, etc.) itself, *for its own sake.* The consultant's concerns are strictly therapeutic: attempting, for instance, to help a couple understand what they face when they are confronted with a highly problematic pregnancy, or to assist a family when continued treatment for one of their members is thought to be futile. The ethics consultant seeks to help such people become more aware of their own moral views so that they can more likely reach decisions commensurate with those views.

On the other hand, such clinical encounters may also be submitted to philosophical reflection—which will concern me here—whether to gain better purchase on ethics or some other matter. This and other examples of clinical encounters are considered strictly *as examples* in order to make certain common themes prominent. In turn, these themes may be systematically considered in further reflective work, leading ultimately to

M. Daniel and L. Embree (eds.), Phenomenology of the Cultural Disciplines, 39–66.
© 1994 *Kluwer Academic Publishers. Printed in the Netherlands.*

a more embracing philosophical understanding. Several of these themes may be delineated here [43].

I. Situation and Context

(1) Whatever the clinically presented problems may be, they are strictly problems facing the people whose situation it is—for instance, that specific couple and their physicians.[1] By the same token, the problems, alternatives, decisions, and outcomes, are strictly theirs. Any encounter presents its own set of issues, moral and other, and these are *context-specific* in the sense that working with and on behalf of such persons, helping them appreciate and advising them regarding specific issues needing resolution, and the like, requires a strict focus on the *situational definitions* of each involved person ([29]; [43]). To understand the clinical situation, there is nothing for it but to try one's best to get at the concrete ways in which the participants themselves experience and understand their situation, and endow its various components (objects, people, things, relationships) with meaning.

Thus in one case, although the attending physician had told me that "abortion" was the "problem" needing attention, this was not in fact an issue—neither for the couple nor for the attending physician, as each of them were prepared to accept the possibility of early induction and fetal demise. However, when the dismal prognosis was mentioned and the couple seemed to become "angry," the attending physician thought they were angry at him for using the word "abortion." One issue was thus obvious: to straighten out the different understandings in order to identify precisely what was at issue for each of those involved, so as to work toward a common understanding of problems, needed decisions, and, hopefully, acceptable solutions.

(2) This suggests that in such clinical situations, *moral issues are presented for deliberation, decision, and resolution solely within the contexts of their actual occurrence*. To find out and understand what's going on in any clinical encounter—what's troubling the people, what's on their minds, and thus to know what has to be addressed and how—requires cautious, attentive probing of their ongoing discourse, conducts, the setting, and other matters presented as constituting this specific context. For instance,

---

[1] As I note later, others—persons, professions, and institutions (with their departments, units, etc.)—are also invariably involved, albeit in different ways.

the couple's puzzlement about the meaning of "statistically significant" was central to their "agitation" and "anger," and thus indicated (at least in part) what theme to be addressed. But these matters could not be considered in abstraction from the actual circumstances: what each person understood, what this led them to think about, etc.

(3) The ethics consultant enters into an already constituted, ongoing clinical encounter between the couple and their physician (and others: nurses, medical and radiological consultants, etc.). The couple was responding to what they understood the physicians were telling them about the fetus' condition, and the physicians to what they thought the couple was saying. Thus, *every situational constituent, including any moral issue, is presented solely within an ongoing relationship between patient and physician*—at least, in its most minimal form. That relationship is not itself the focal or primary theme for either patient or physician.[2] The physician is instead concerned to help the patient (or at least to do no harm), and is accountable for whatever is and is not done and said. The patient's concern, on the other hand, is to have distress relieved, injuries healed, disease resolved, or at the very least to be comforted and cared for.[3] The clinical ethics consultant, however, has to address that *relationship itself*, attending to each of its integral constituents within that temporally ongoing contexture [39].

II. The Physician-Patient Relationship

Only a little reflection is needed, however, to realize that this relationship is quite special. Consider, for instance, Eric Cassell's observation:

> I remember a patient, lying undressed on the examining table, who said quizzically, "Why am I letting you touch me?" It is a very reasonable question. She was a patient new to me, a stranger, and fifteen minutes after our meeting. I was poking at her breasts! Similarly, I have access to the homes and darkest secrets of people who are virtual strangers.

---

[2] Although it may become so at some point—for instance, if the couple were later asked how they got along with their physician, or if, after changing physicians, their first physician were asked by the new one for pertinent information about his relationship.

[3] The issues posed by such phenomena as Munchausen's syndrome, factitious illness, and hypochondriasis, must be considered on their own.

In other words, the usual boundaries of a person, both physical and emotional, are crossed with impunity by physicians ([3, I], p. 119).

Or consider Edmund Pellegrino and David Thomasma's emphasis that the ground for an ethics of trust—the heart of the "clinical event"—is the patient's "existential vulnerability." Although primarily focused on the patient's threatened organic systems (disease, injury, handicap), that very vulnerability obligates the physician to be cognizant of and to respect the patient's moral agency—an obligation that increases as the invasiveness of medical procedures increases. Thus, when they contend that "it is the body of the patient that grounds this obligation, not merely social and legal structures," it cannot be ignored that this "body" is the specific embodiment of this or that specific person. While a patient's "body is probed and violated in closer proximity and more intimately than is usually permitted even to those the patient loves" ([24], p. 185), it is that patient herself who is "probed and violated."

In any event, one prominent characteristic of the physician-patient relationship is its *asymmetry*. Although the relation is in a sense reciprocal, it is unbalanced with power on the side of the physician: the physician, not the patient, has knowledge (about pregnancy, fetal development), technical skills (delivery, cephalocentesis, c-section), access to resources (ultrasound, operating room), social legitimation and legal authorization (license to practice medicine) [18]. Patients do not have this range of knowledge and skills, access and legitimation. Moreover, very often they do not even know how to assess whether the professed healer is capable of healing or how to assess the abilities of those who profess to have them. Patients are thus *existentially vulnerable* not only by virtue of their bodily conditions, but also by the relationship's asymmetry.

On the other hand, the relationship invariably involves profound *intimacies* regarding the body, the self, the family, their particular lives, beliefs, and circumstances. As the relationship is most often among strangers, furthermore, the asymmetry and intimacy of contact can be especially tense, trust often problematic, and treatment open to compromise.

III. Morality and Power

These reflections suggest that the physician-patient relationship is distinctively *valorized*: *medicine is an inherently moral enterprise* [2].

(1) To be a patient is to be disadvantaged by the very condition that brought one to the physician in the first place. Impairment compromises in multiple ways ([22]; [26]; [41]): not only by its special ways of capturing and focusing the person's attention, but also by the fact that the patient cannot "do for herself" but must rely on others. To be impaired is to experience oneself as uniquely vulnerable, exposed to the actions of others. The patient is thus disadvantaged by the very asymmetry of the relationship, as well as by the fact that those with power on their side are commonly strangers, because of which the social conditions for trust are commonly not at hand even while trust is essential to the relationship.

From the patient's perspective, the relationship is marked by the experience of having to rely on a host of affairs: instruments, medications, procedures, arrangements, and, most importantly, people. To experience impairment is to find oneself in situations marked by multiple forms of *unavoidable trust*—especially regarding people with respect to whom, being strangers who possess the knowledge and skills to engage in highly intimate contacts, trust is itself a serious and ongoing issue [46].

(2) On the other side of the asymmetrical relationship is the physician—who has the power, skills, knowledge, resources, and socio-legal authority to judge what can and should be done, and to act. Here, several things are evident.

(i) Many physicians and traditions in medicine's long history seem to have taken this asymmetry as a rationale for construing the relation to patients as *unilateral* and have thus called for solitary decision-making—a view strongly enhanced by the 20th century marriage of medicine to the biomedical sciences and the many discoveries consequent to that marriage. The realities of clinical work, however, typically force the recognition that patient encounters are *reciprocal*, in that patient trust and compliance is necessary; indeed, patients often do not agree with physicians, refuse to comply with "doctor's orders," and insist on making their own decisions[4]—including the decision to treat themselves or not be

---

[4] As Ludwig Edelstein emphasizes ([5], p. 329), healers in the skeptical, methodist tradition were especially aware of the uniqueness of each patient and had constantly to consider, "What about the patient who is putting himself and 'his all' into the hands of the physician?" The patient had to ponder (even at times be helped to ponder) the unavoidable trust placed in the hands of the healer, and the healer correspondingly had to be alert to this and to do everything necessary to make himself trustworthy.

treated at all. In our times—through informed consent, patient rights, and the like—this difference has received moral and legal support. Still, the asymmetry of the relationship does not automatically imply that it is the physician who should make decisions. It has thus become imperative for physicians to develop an understanding of the relation with patients that is quite different from that expressed in the more traditional "medical model" ([4]; [16]; [20]; [23]; [32]). In order to develop coherent, acceptable, and practical therapeutic plans, and to enable sound decisions, physicians must learn the patient's (and family's, at times even the significant others') experiences, interpretations, meanings, and values.

(ii) At the very least, this suggests that the patient's place is not, and in many ways has never been, simple passivity—despite the typical usage of "patient." The ability to alter a patient's condition and life thus does not thereby signify having *power-over*—and that is morally significant. In clinical encounters, *power-to* has most often been understood as a sort of benign *power-for* (parentalism) or at times *power-over* or *on-behalf-of* (paternalism). Increasingly, however, the physician has had to understand the *power-to-alter* as *power-with*: decision-making requires the active participation of the patient (often, the family and others in the patient's circle of intimates)—indeed, decisions are the responsibility of patients or their legal surrogates (within certain limits).[5] This, I think, is surely one reason underlying the idea that medicine is an inherently moral enterprise: to act *on behalf of* the patient, or, if nothing else to *do no harm* (i.e. "beneficence"), requires acting *with* the patient (which suggests that the traditional sense of beneficence needs to be rethought). The physician's place in the relationship is a form of *caring* and is strictly correlated with patient *trust*.[6]

(iii) Another aspect of this was strikingly evident already in ancient medicine, especially in the Hippocratic Oath ([5], pp. 6-10; [43], pp. 202-223). Reflection on this covenant, along with the recognition of the asymmetry in the ability to bring about changes in the patient, makes it evident that the relationship itself makes it possible (even seductively

---

[5] It has become more common to acknowledge that the physician must recognize that the patient (family, legal surrogate) is "the true source of authority," not the physician (who acts solely as "advisor" or "consultant" to the patient) ([27], p. 199).

[6] Even if, so far as doctor and patient are strangers, the best that may be hoped for are situations where only "temporary trust" is possible ([18], p. 52).

tempting[7]) for the physician *to take advantage of* the multiply dis-advantaged patient (family, household). Having the power-to-alter, the physician is obviously *able* to take advantage. Clearly understood by the ancient empiric and skeptical physicians, a significant moral cognizance undergirds the Oath's strong injunctions: *not* to take advantage, but rather to act on behalf of the patient.[8] For the physician *to take care of* the patient, thus, signifies the moral responsibility *to care for* the specific patient encountered in each clinical situation. It may well have been just this that prompted those ancient physicians who developed the Oath to understand the "Art" as governed by the virtues of justice (*dike*) and self-restraint (*sophrosyne*), and to take these as the core of wisdom ([5], pp. 36-37).

Medicine is an inherently moral enterprise, at the core of which is a striking moral insight. Physicians are in the nature of the case involved in a complex moral relationship with persons who, due to impairment and to the relationship itself, are uniquely vulnerable, exposed to the power of those who wield the "art"—who must themselves always act justly and with restraint.

IV. Circumstantial Understanding

Neither patient nor physician (nor other health care provider) is focused on the relationship itself as the primary theme. Just that, however, is precisely the focus of the clinical ethics consultant. This needs to be clearer.

Experienced by the impaired person, the impairment is interpreted by and has meaning for that person. Others also experience and interpret

---

[7] Consider the Laches myth related in Plato's *The Republic* (Book II): finding the ring with which he could become invisible, Laches promptly goes to the palace of the Lydian king, seduces the Queen and, with her help, kills the king and assumes the crown for himself. This tale stands dramatically opposed to that at the heart of ancient medicine, that of Aesclepius. In the one, possession of power over others is promptly used; in the other, its possession to the contrary prompts stringent restraint (*sophrosyne*) and the exercize of justice (*dike*) [47].

[8] The Oath's injunctions are quite specific: e.g., to refrain from having sex with the patient and members of the patient's family and household, and to refrain from "spreading abroad" what is learned in the intimacy of the relationship with the patient. Although expressed in this way, there is no reason to understand the Oath as limited to just these acts, as they express the more generic responsibility never the take advantage of the disadvantaged person, family, or household.

the person's condition: family members, those in the person's circle of intimates (especially close friends and associates), persons in the wider social ambiance, but also the physician, nurses, and other providers helping to care for the person. Hence, to speak of "the experience and meaning of illness," as many have done ([2]; [16]; [24]; [45]), is necessarily to face a highly complex phenomenon, as too few have recognized ([43], pp. 29-52).

Nor is this all. As Schutz has demonstrated, every situational participant experiences and interprets the encounter within his or her own biographical situation ([29 I], pp. 243-247): typifications, life-plans, undergirding moral and/or religious frameworks, etc. These encounters are also socially framed by prevailing social values, as well as by written and unwritten professional codes, governmental regulations, hospital policies, unit or departmental protocols, etc.—any or all of which may and often do contribute to "what's going on" in any specific case.

Finally, cautious probing suggests that experience and meaning are still more complicated. Again, consider only a patient and her physician. She, like every patient, is a *self* and thus is essentially a *reflexive* being. Briefly, this signifies ([39], pp. 144-164) that the patient experiences and interprets *her own* impairment. She also experiences and interprets the *physician's* experiences and interpretations, including his experiences and interpretations of her (how she thought to experience and interpret the doctor, her illness, etc.)—and both she and her physician are, in the nature of the case, aware of, though not always focally attentive to this very complexity. In a word, the relationship is complex and reflexive: minimally, each experiences and interprets the other, their respective interpretations, and at the same time within their relationship, each experiences and interprets the relationship itself ([15]; [39], pp. 199-216).

For example, the pregnant woman in the one case not only experienced her pregnancy and her developing fetus, but this experience was complexly textured by the ways in which she experienced and interpreted what her physicians (husband, and others) told her. Similarly, her physicians experienced and interpreted her "anger" *as* directed at them and *as* about "abortion." In some respects, moreover, both experienced and interpreted *the relationship itself*. Regarding diagnostic data, for instance, she said "I know they're only trying to do their best" (i.e. she interpreted the relation as "they are trying to help"); and her physician told me, "She seems to think we're being deliberately unclear" (i.e. the relation was "not going well").

To work as a clinician (whether physician or ethics consultant) is thus to be something like a detective: deliberately probing into the multiple ways in which the situational participants interrelate, variously experience and interpret one another and, within that relationship, the relationship itself—as Kierkegaard would have appreciated [15]. The involvement of the ethicist is thus a *work of circumstantial understanding* (both understanding, and being-understanding); reflection on this and other cases is a matter of *phenomenological explication* [34].

The ethicist's work is principally a matter of *enablement* or *empowerment* ([39], pp. 175-216; [43], pp. 248-250, 285), and in that sense is designed as *therapeutic*. The aim is to help the participants identify what is at issue for each person; to help each become reflectively alert to and consider their respective moral frameworks;[9] to help delineate, weigh, and imaginatively probe the available options that are most reasonable and fitting within those respective moral frameworks; and to help each attain clarity about the "stakes" so as to enable them to live with the outcomes or aftermaths of needed decisions.

Reflection also suggests that to enter into any ongoing clinical situation is inevitably to find oneself as *also* a participant—hence, to be enmeshed in the different "stakes" and decisions ingredient to and helping to define the situation. As therapeutic (enabling or empowering), the clinical ethics consultant must therefore be quite as *accountable* (and *held accountable*) as any physician or other provider ([43], pp. 27-28, 36-41). To be sure, this accountability must be *appropriate*: *for* what is and is not said and done, and *to* those whose situation it is most immediately. The idea of responsibility is thus central to ethical involvement in clinical situations: to be *responsible for* what is and is not said and done, and *responsive to* those persons whom one seeks to enable or empower.

V. Illness Meanings and Narratives

Toward the end of his *Formal and Transcendental Logic* [12], Husserl rejects the idea that there is any direct, immediate access to "the truth" ([12], p. 277). Whatever else may be said about that august topic, much less about what sorts of evidence might back up claims about truth, it is necessary to recognize that "truth" and "evidence" are inherently

---

[9] That is, to consider, as thoroughly as circumstances permit, what is "worthwhile."

contextual and strictly co-relative to the particular interests and pursuits, or to what Schutz terms the "cognitive style" of any province of meaning or activity ([29, I], pp. 25-28). Husserl remarks, "the trader in the market has his market-truth," relevant to his specific province of concerns—so too, we must add, the scientist, the physician, and the patient ([12], p. 278).

The patient experiences and interprets her own condition—which harbors complex *meaning* for her. She has her own "truth" no less than the trader or the physician. Her experiences and interpretive meanings are quite legitimate within their own context, even if she needs help in improving her understanding (perhaps in the light of a better understood sense of what is "worthwhile" to her, as well as diagnostic data, etc.). To paraphrase Husserl, simply because the physician works from within a different framework (with its appertaining meanings, evidences, methods, and truths) does not mean that the patient's views are inappropriate or should be regarded as suspect. Some physicians may still view the patient's understanding, if not also her experiences, as suspect—as some have emphasized ([3]; [16]; [20]; [22]; [32]), and as was clear of the couple in our case when they were trying to understand statistical matters regarding their fetus's probable diagnosis and future.

She has to reach an accurate and adequate understanding of her own situation and condition. That crucial task, moreover, may require that physicians and other providers (including the ethicist) help. Precisely because of the specific asymmetry of the relationship, providers have the responsibility of initiating candid conversations to ensure trust (even if only temporary), so as to help patients and their families (and significant others) understand and talk about the patient's condition. As Arthur Kleinman emphasizes, "patients order their experiences of illness—what it means to them and to significant others—as personal narratives." Thus, he continues, a "core clinical task is the empathetic interpretation" and "reconstruction" of these narratives so as to help them take account of what the clinician knows, and thus reach a more valid understanding ([16], pp. 49, 227-236). As people are compromised and often demoralized by impairment, the clinician's work is an effort to "remoralize" their circumstances—the principal means for which is "an experiential phenomenology" that requires "the first and essential, great leap into the world of the patient" ([16], p. 232), which in turn requires the physician to understand and appreciate the patient's experiences, understanding, and narratives.

The point might be expressed somewhat differently. In trying to understand the issues presented within any clinical encounter, it is critical to be mindful of the highly complex set of biographical situations and situational definitions this involves: the multiple people variously involved and their respective experiences, meanings, truths, interpretations, attitudes, relationships with other people, life-plans, values, etc.—each of which must be weighed and respected for its own sake and in its own terms even while attempting to enhance understanding of the context. The ethicist's task is to identify, weigh and deliberate, eventually to reach reasonable, defensible judgments about every aspect of the context:[10] the patient's, family's, significant others', as well as the providers' (physicians, nurses, etc.). This must be done by attending strictly to "the things themselves," the situational participants and their specific circumstances, in their own terms.

As Kleinman emphasizes, helping patients and families develop a more valid, "remoralized" narrative understanding requires "the clinician to place himself in the lived experience of the patient's illness . . . to understand (and even imaginatively perceive and feel) the illness experience as the patient understands, perceives, and feels it," and to do the same for the family ([16], p. 232). It is critical to understand the ways in which situational participants themselves talk about their respective experiences, for the moral issues they face are presented and must be worked out solely within the context of their actual occurrence.

Sick or impaired people experience, interpret, and talk about their illnesses, typically couching their narratives in their own commonsense categories (in part personal, in large part socially derived). These narratives include and are invariably shaped by the person's under-standing of the clinician's words, gestures, and conducts; similarly, the physician's work as such includes and is shaped by the patient's narrative. The relationship is thus a special form of *mutuality* ([39], pp. 199-216): the intimate, reflexive interrelating of persons within an experienced asymmetry designed to help those who are unable to help themselves.

---

[10] The ancient empirics and, later, the methodical skeptics termed this critical act *epilogismos*, which is based on interpretations of patient discourse, bodily symptoms, life-styles, etc. They vehemently rejected the dogmatic's focus on *analogismos*—causal reasoning designed to connect the inner bodily pathologies ("invisibles") with outer physical symptoms alone ("visibles"). Where the former were guided by *semeiosis*, the latter were led to *diagnosis* ([43], pp. 130-176).

VI. Phenomenological Approach in Clinical Encounters

A. *Epoché and Reduction*

As is well known, Husserl sought to gain access to and clarify the "genuine concept" of science. For this, he proposed to "put out of action all the convictions we have been accepting up to now, including all our sciences," so as to "immerse oneself in the scientific striving and doing" specific to the pursuit of science ([11], p. 7). Although the idea of genuine science at first only "floats before us as a vague generality," taking it as a clue (*Leitfaden*) for reflective consideration makes its inner "intention" or claim progressively clearer ([11], p. 9). The intention is taken as a "precursory presumption" that can guide reflection: what is it that one *claims* or *aims* to be doing when one strives to be "scientific"?

Crucial to the entire undertaking is adopting a rigorous reflective "attitude" or "orientation" (*Einstellung*) in order "to immerse oneself in" (*sich einleben in*) and thereby become reflectively cognizant of the practice specific to science—to its inherent claim or intention. This methodical shifting to a reflective orientation Husserl termed the "*epoché;*" rigorously maintaining it throughout the course of inquiry is the "*reduction*" [38]. To shift one's attention phenomenologically is also to consider clinical encounters as examples and no longer to be preoccupied with them for their own particular sakes. As regards medicine, the specific complexity of the clinical phenomenon must be rigorously appreciated: not only the people involved, their respective experiences and interpretations, etc., but more especially the specific kind of practice exhibited by medicine.

B. *Reflection as Circumstantial Understanding*

Every field of practice includes the possibility of this reflective-attentive shift. MacIntyre's understanding of "practice" is apropos:

> By a 'practice' I . . . mean any coherent and complex form of socially established cooperative human activity through which goods internal to that form of activity are realized in the course of trying to achieve those standards of excellence which are appropriate to, and partially definitive of, that form of activity, with the result that human powers to achieve excellence, and human conceptions of the ends and goods involved, are systematically extended ([19], p. 175).

Science and medicine are practices in this sense. Indeed, precisely because the practitioner experiences the field, s/he comes to recognize what MacIntyre terms the "goods" internal to it (its "standards of excellence" or "virtues") ([19], p. 175) and perhaps even periodically reflects on one or another aspect of it. On the basis of such experiences and reflections, one can (and sometimes must) stop and think about the field and one's involvement.

To reflect in this sense requires a specific *shift of focal attention* (epoché)—*away* from actual involvement in clinical cases for their own sake, *to* considering them as examples of the practice—an attentional shift that needs to be *sustained* throughout the reflective project (reduction). To stop and think about the efforts and actions specific to any field of practice is, on the basis of one's own practical experiences, to become reflectively attentive to its inherent "intention" ([38], pp. 125-141). This complex act might be termed a sort of "practical distantiation," or circumstantial understanding. It is, more simply, to reflect in the specifically phenomenological sense.

At one point Husserl observed that this method is the very same as that which "a cautiously shrewd person follows in practical life wherever it is seriously important for him to 'find out how matters actually are'" ([12], pp. 278-279). Just this type of seriousness is quite evident in every clinical encounter, where patients and physicians (and others) know how vital it is for everyone involved to know and understand what they face. To the extent that situational participants seek to understand, they engage in at least a form (perhaps only the first stages) of reflection—"stages," for they rarely go further into philosophical issues. Moreover, although their reflections are (Husserl might have noted) only the "beginning" of wisdom, it is "a wisdom we can never do without, no matter how deep we go with our theorizings" ([12], p. 279).

C. *Evidence*

Whatever may be at issue, the point is that the "seriously shrewd" person must judge and decide on the basis of sound evidence—precisely because the issues faced are vital. To apply Husserl's words, the patient no less than the physician must judge "on the basis of a giving of something itself, while continually asking what can be actually 'seen' and given faithful expression" ([12], pp. 278-279). Whether one has practical or theoretical concerns (a patient trying to understand what's going on and what to do, or a physician trying to interpret diagnostic tests), to seek sound judgment and reach right decisions requires a concerted effort to know just how matters actually stand. The matters at hand, if you will, "matter" enormously to both.

"To come to know" something is to take it to be "thus and so," to take up some modal position (*Positionalität*) ([14], pp. 223-250) toward it. For instance, one may think one knows something "fairly well," "sort of," "surely," "not at all well," or some other modality. To take a position is in effect to make a claim leading directly and ineluctably to the practical "outcomes," which appeal to some evidence that is supposed to ground or account for the position taken—the more vital the issue, the more crucial becomes the need for solid evidence to ground the decisions that must be made.

So far as decisions rest on claims that in turn appeal to evidence, relevant experiences are essential. As Husserl insists, evidence is, "in an *extremely broad sense*, an '*experiencing*' of something that is, and is thus . . . " ([11], p. 12). Evidence is not a sort of rare and special 'datum,' a magical wand or a conferral from on high having special privilege or guarantee ([12], pp. 161, 177, 180, 289). To the contrary, evidence is a matter of *relevant experience* and is essentially contextual; it is *relative to* whatever it may be that best serves as the grounds for what is claimed, relative to the specific types of experience through which the affairs in question are encountered or by means of which one is at all aware of them: evidence about fetal hydrocephaly cannot be the same as evidence for anger.

Furthermore, even if there seems to be good evidence for believing that something is this or that, the possibility of error or deception is not precluded. The *possibility of deception* is inherent in the evidence of experience, in Husserl's broad sense, "and does not annul either its fundamental character or its effect . . . " ([12], p. 156). No genuine evidence can provide "an absolute security against deceptions" or errors ([12], pp. 157, 284-289). Thus, in ethics and especially clinical ethics, there is a clear demand for addressing uncertainty, error, deception, and ambiguity—and therefore a need for the physician to have specific plans in place in the event of error or mistake.

As "experience" in the generic sense (*Erfahrung*), evidence refers to the particular ways in which some affair ("anger," "hydrocephaly," etc.) is experienced (is given or otherwise presented), or through which one is able to become aware of or to encounter that affair—and on that basis, come to know it, make claims about it, and reach appropriate decisions. *Evidenz* is strictly correlated to the modes of givenness (*Gegenbenheitsweise*), the ways in and by means of which the things known are encountered as "they themselves," as Husserl says, "in person"[11] (*leiblich*)—in the ways specific to the "things" in question.

In the case of the pregnancy, this demand for sound evidence was very prominent. A family obstetrician first referred the couple on the basis on what he saw on ultrasound and from his clinical examination. The referral center's obstetricians not only saw that first ultrasound but obtained another as well as other tests and, together with radiologists experienced in "reading" such images and laboratory findings, came to a weighted judgment that the fetus had certain severe problems. Having no experience or knowledge of such things, the couple could base their beliefs and concerns only on what they were told by the physicians ("hear-say" evidence, even if from "experts"). But the physicians' notion that the couple "seemed angry at us for mentioning 'abortion'" was not well-grounded. Indeed, it turned out not to be evident at all; that belief was quite wrong even while the claim that they were "angry" was correct. Since these matters are pre-eminently practical, leading directly to decisions with their aftermaths, it was quite essential that these respective claims and evidences be cautiously sorted out on pains of reaching mistaken or misleading decisions.

D. *Free-Phantasy Variation*

In more technical terms, to consider any clinical case as an *example* is to practice a version of what Husserl termed "free-phantasy variation" [36] to which he gave extraordinary significance. For instance, it is said to be " . . . *the fundamental form of all particular transcendental methods*," even to provide "the legitimate sense of a transcendental phenomenology" ([11], p. 72).[12] While Husserl sought to establish

---

[11] Thus, Husserl can say that evidence *"consists in the giving of something-itself,"* along with the discipline of giving them faithful linguistic expression (which he includes in his "normative principle" of evidence) ([11], p. 14).

[12] Husserl carefully distinguished his method as "a variation carried on with the freedom of pure phantasy and with the consciousness of its purely optional character—the consciousness of the 'pure' Anything Whatever," and thus as focused on each actual or possible example as exemplifying the purely possible. He thus distinguished it from *"empirical variation;"* the method is a strictly philosophical one, although there are, as I argue in the text, a number of interesting versions of the method even in empirical science and medicine ([12], pp 247-248). Earlier, he

phenomenology as a transcendental discipline, it has seemed to me that certain versions of this method are at work in a variety of areas. It might even be more accurate to say that there are a whole battery of variational methods ([35]; [37]; [39], pp. 166-180, 190-216; 242-249); [42].

One specific type became clear, for instance, in the course of reflection on certain clinical cases that seemed suggestive for the sense of "self." Reflecting on cases of autism and those involving brain-injury, for example, a crucial and distinctive ability of the self becomes prominent precisely through its absence: "possibilizing" or "thinking for the possibly otherwise"—i.e. the ability to transcend the immediate present ([39], pp. 242-249). Just this striking absence of an otherwise normal human capacity is forcefully *presented* in those cases—while, at the same time, its absence *presents* a patient as needing therapeutic attention ([39], pp. 173-180).[13]

VII. The 'Goods' of Practice

I have been considering various clinical encounters as examples of clinical ethics involvement. It is essential at this point to be much clearer about that "involvement."

---

emphasized that the method is the key for apprehending "essences." Feigning" *(Fiktion, Phantasie)* is *"the vital element of phenomenology as of every other eidetic science . . . ,"* such as geometry or pure grammatics. He was so impressed by this form of systematic imagination *(Einbildung)* that he urged philosophers to "fertilize" their imaginative abilities by means of "abundant and excellent" observations—especially, he wrote, in poetry and history ([14], p. 160).

[13] Once having apprehended this example of the method, many variations can be recognized in other types of inquiry, each with its own specific objects, and methodical and evidentiary requirements. For example, the sociologist D. Polkinghorne provides an excellent case of an empirical variation of the method ([25], pp. 618-637). It deserves mention, too, that my own reflective inquiry into variations of the method is itself a clear example of that very method—a reflexive characteristic that is rich with implications. Kurt Goldstein, I believe ([39], pp. 173-175), followed precisely this type of method in his work with brain-injured patients ([6]; [7]), as did the psychiatrist Gerhard Bosch [1] in his study of infantile autism ([39], pp. 182-198). This variational or exemplicative method can as well be found, with interesting variations, within clinical medicine. For instance, clinical diagnosis—specifically its "detective-work" (methodical differential diagnoses) and eventual selecting (with the patient) of the best for this patient, no less than the ongoing conversations patients and families—all turn out to exhibit strikingly similar features.

MacIntyre argues that it is only "in the course of trying to achieve those standards of excellence which are appropriate to, and partially definitive of, that form of activity" that one can understand and make judgments about its internal "goods" or "virtues" ([19], p. 177). The type of internal understanding gained through actual practice requires that I "subject my own attitudes, choices, preferences and tastes to the standards which currently and partially define the practice" ([19], p. 177).

Thus, one could say, the evidence for making informed judgments about clinical practice lies within that practice.[14] For MacIntyre, the "goods internal" to a practice are "virtues"—acquired human qualities the possession and exercise of which *"tends to enable us to achieve those goods which are internal to practices and the lack of which effectively prevents us from achieving any such goods"* ([19], p. 178). It seems to me that this is equivalent to Husserl's emphasis on "immersing oneself in" the ongoing striving and doing characteristic of science by "freely varying" relevant examples in order to identify and explicate its common themes.

VIII. Affiliation

Clinical encounters characteristically present a range of emotional, volitional, and valuational components, and are thus textured by uncertainties and ambiguities. Impairment typically breaks into the fabric of a person's daily life. The relation with providers is not only asymmetrical but most often between strangers. All these give the phenomena of clinical encounters their characteristically charged atmosphere.

In the case mentioned, the couple's anger, anxieties, frustration, and the like obviously textured their experiences and understanding, as was equally clear regarding the physicians. On what did the one physician base his judgment that "they are angry at us"? What is the evidence for

---

[14] This is surely true for the practice of clinical ethics. It should be emphasized that since this discipline is novel, attaining a sense of its clinical standards has unique difficulties. In part, as noted in the text, this requires a keen appreciation of the physician's clinical work ("being in the trenches"); but it also demands the rigorous appreciation of the experience of illness and impairment, the clinical standards of other providers, and practical experience working within the institutional contexts of that set of practices and experiences. It goes without saying that there is currently little by way of common understanding of the standards appertaining to clinical ethics; there is thus a pressing need to discover these and make them integral to "ethics consultations," for instance.

that, or for my own judgment about the anger? How is this at all presented or experienced, such that judgments can either be well-made or not? Clearly, things are quite complicated: how emotive qualities are experienced and interpreted, ultimately how other persons are experienced in such situations. Without going much further into these issues ([39], pp. 181-240), it is still possible to tease out some interesting matters.

Consider merely one of the themes that have become prominent in this methodical process. Every concrete instance of clinical encounters displays a remarkable array of "feelings"—strongly expressed wishes, wants, desires, aims, purposes, puzzlements, etc. These "feelings" are *evoked*, most often by the patient's condition; in the mentioned case, it is primarily the fetus's condition that evokes the range of feelings noted (but also the couple's and their physician's responses). The feelings are, moreover, *directed* to the developing fetus's present condition (diagnosis) and, as efforts "to do something about it" are suggested (therapy), these are *aimed* at its range of possible, probable, or likely futures (prognosis). Accordingly, feelings are constituted as *oriented expressions of moral life*. They are, that is, *displayed concerns* that orient the participants about how to be most responsible and responsive to the specific persons involved—regarding the fetus, how best, if at all, to help it be or become "whole," "hale," or "well" to whatever extent is possible; in the case of the mother and father, how best to help them understand and adjust to the likely outcome (whether of diagnosis, intervention, or prognosis). Whatever else "feelings" may be, thus, their clinical presentation is the concrete, practical display of *moral concerns* about doing what is believed and hoped are the right, good, and just thing for the persons involved—within the constraints imposed by the patient's particular condition, biological wherewithal, values, social circumstances, and available medical procedures and resources.

Feelings are *indices of moral life* more generally. They are not simply overt reports about a person's subjective (and presumably inaccessible) life at the moment. They are not something like literal "ex-pressions:" the "pressing-outward" of what is thought to lie essentially "inside" as purely private and available only to the one whose feelings they are. As evoked by, directed to, and aimed at the specific patient within her present distress and possible restored futures, such feelings are the experiencing person's own moral orientations toward the circumstances and decisions, choices and outcomes, actually confronted. Feelings are thus not in the least "subjective," if by that one means private ([39], pp. 48-66; [43], pp.

55-56). They are rather encountered situationally by anyone who takes the time to observe, or who makes the effort to become involved, and in this feelings are quite as objective as any of the scientific, financial, political, or other "facts" ([17], pp. 3-54).

Every clinical encounter thus invites a crucial question: what is it about just *this* particular situation that evokes, directs, and aims just *these* specific feelings and serves to orient the discussions, decisions, and actions of the situational participants in just the ways they observably do? More briefly: why was the woman angry and what does that suggest about what, for her, is "worthwhile" and "desirable"? To notice, focus on, and probe clinically presented feelings is to gain access to what moral notions are at work giving the situation its particular issues and urgency.

The fact is, of course, that in daily life we do indeed experience another person's emotive responses as well as our own, though not in the same ways. However I may realize that I am angry or sad, I clearly do not realize this in the same ways as I became aware of that couple's anger, or those physicians' dismay. While the range of possible error about my own feelings, or what I thought about the couple, will obviously differ, it is just as obviously the case that we do experience other people's emotive feelings. *How* we accomplish this is a fascinating theme ([39], pp. 181-241); *that* we do so is unquestionable.

In the case mentioned, thus, it was evident that the woman was deeply worried, angry, distressed, although it turned out that the physician was mistaken about the object of their anger (as they quickly emphasized). But how was her "worry" presented or experienced? How do we experience other persons? Max Scheler's study of such issues is helpful. He emphasizes that the other person is experienced as "an integral whole" and with considerable depth:

> For we certainly believe ourselves to be directly acquainted with another person's joy in his laughter, with his sorrow and pain in his tears, with his shame in his blushing, with his entreaty in his outstretched hands, with his love in his look of affection, with his rage in the gnashing of his teeth, with his threats in the clenching of his fist, and with the tenor of his thoughts in the sound of his words . . . . I do not merely see the other person's eyes, for example; I also see that "he is looking at me" and even that "he is looking at me as though he wished to avoid my seeing that he is looking at me." So too do I perceive that he is only pretending to feel what he does not feel at all, that he is severing the familiar bond between his experience and its natural expression, and is

substituting another expressive movement in place of the particular phenomenon implied by his experience ([28], pp. 260-261).

Accordingly, he continues, our experiences of others are not simply of their bodies (not even in special situations such as physical exams), nor just their "souls" or "selves"—these are not abstracted from each other; rather, we experience *embodied persons* ([33]; [39]; [43]):

> What we perceive are *integral wholes*, whose intuitive [i.e. perceptual] content is not immediately resolved in terms of external and internal perception . . . . The primary awareness, in ourselves, in animals and in primitives, invariably consists of *patterns of wholeness*; sensory appearances are only given in so far as they function as the basis of these patterns, or can take on the further office of signifying or representing such wholes (28], pp. 261, 264).

We experience others—the woman, her husband, the physicians, etc.—as *integral wholes*, as embodied persons. In turn, it is the latter that allows us to "see the worry" in the "furrowed brows" and "wringing hands," not to mention in her words and their paralinguistic features, in her physiognomic expressions, etc. [3]—all of which occur in the actual context and setting in which she is encountered [39].

Not only do we judge about "objective" things, but also hold various beliefs and make knowledgeable judgments about a wide range of other things too often regarded as merely "subjective," including the emotions: remembering, touching, thinking mathematically, calculating, liking, being happy, and on and on [35]. Judgments about these must also be based on evidences, for otherwise there would be no way whatever for us to make sense of that couple's worry or anger, or that physician's mistaken interpretation. We experience the anger, claim to know about it, form judgments about a wealth of affairs (e.g., the physician claimed they were angry and that this was directed to him), and do so on the basis of some mode(s) of awareness, alertness, or experience—where evidence, however, requires just certain of these modes, as they are the ones through which the affair is at all encountered as just what it is (as "it itself").

IX. Dialogue

Talking and listening to situational participants, certain matters become especially prominent and need emphasis. (1) In general, phenomenological inquiry serves to make prominent the contextual themes essential to every clinical encounter. Of course, much has been left out: the intentionality of experience, the differences among the ongoing, merely operative mental processes (*Bewusstseinserlebnisse*) and active awarenesses (*Icherlebnisse*), and inner-time consciousness, to mention but a few. A full phenomenology of the clinical event has thus been merely suggested.

(2) More specifically, making sense of the moral dimensions of cases such as the one discussed, requires primary focus on the complex, multiple relationships constitutive of each encounter—at the core of which is the physician-patient relationship. To consider but one facet of this, to engage in dialogue with the patient the clinician needs to encourage the patient (and, where they are significantly involved, others in the circle of intimates) to articulate her own narrative. In Kleinman's terms, this is a sort of "mini-ethnography" ([16], pp. 248-250, 285) that, interlacing continuously with the clinician's narrative, helps her to express what things mean for her, a narrative that may "remoralize" her experience and in that sense help to validate it.

The clinician's conduct here seems very much like what I have elsewhere termed "affiliation" ([30]; [43], pp. 315-319). The idea is that by placing oneself in the lived experience of the patient—in everyday terms, "putting yourself in her shoes"—a fundamental moral cognizance is achieved. This act does not signify that the clinician is supposed to think about the woman's situation as if it were his own: "what would you do if you were me?" is not the issue. For the clinician—whether physician, nurse, or ethicist—is precisely *not* the woman. But that is not the point anyway. It is rather that the clinician must make the effort to understand the woman's circumstances from her own perspective, as she lives and understands it—disclosed contextually through her discourse, word-choice, paralinguistic features, bodily demeanor, etc.

Phenomenologically, the patient wants the clinician not merely to understand (competence), but just as importantly, *to be understanding* (to take care of, to care for). In this respect, the basic clinical method (for physicians, nurses, or ethicists) is a specific form of *dialogue* ([41]; [44]; [45])—not merely "mini-ethnographies," as this notion fails to capture the crucial interactions characteristic of clinical encounters. It is not, that is, simply a matter of a patient telling her story, the physician then giving his interpretive version of what's going on, etc. It is rather a matter of *shared talking and listening* (asking and responding), whose point is to find out what's wrong, what can be done about it, and, fundamentally, what should be done about it [21]: among the available options, what decision leading to what action ought to be reached (including the respective aftermaths each carries with it). This is shared discourse which patient and physician pursue together in search of what the impairment means (and, at some point perhaps, ought to mean) to the one who has it (and, for the clinical ethicist, for those who profess to help), with the longer-range goal of enabling patients and physicians to fit their always-evolving narratives into the broader story (and history) Schutz terms their respective "life-plans" ([29, I], pp. 47-49, 134-154) with their respective moral frameworks.

(3) The case I've been considering is merely one among many. To consider it, not for its own sake, but as an *example*, is to practice a specific version of free-phantasy variation: practical distantiation. It is to leave the immediacy of clinical work and to enter the terrain of philosophical deliberation—where, among other issues, the concern is to delineate what such actual encounters reveal about moral life in clinical settings (and, possibly, more generally). It is, if you will, *to learn about moral life from the clinical circumstances of those who actually face these difficult situations.*

X. Review: The Clinical Event as Context

In these encounters, the clinician is concerned with the specific, unique situation itself: physician and nurse seek to help, benefit, cure, perhaps only comfort the patient and family members; the patient and family want to get better, have the problems resolved, or perhaps begin to grieve appropriately. The clinical ethicist, on the other hand, is focused on the highly individual clinical relationship itself (the encounter itself), for its own sake (if you will, *practical* distantiation itself. On the other hand, the ethicist is at any time free (*beliebig*) to shift attention so as to consider any encounter as an example ([10], pp. 413, 422; [12], pp. 247-249), where one then seeks what is "invariant" through inspection of a range of examples—if you will, practical *distantiation*. Focused reflection on a range of examples, that is, leads one to what Husserl terms the then presented "coincidence in conflict"—to "what is common" throughout the range of examples by means of "free-phantasy variation."

From this reflective inquiry, it becomes clear that every clinical encounter has certain common contextual themes. Each is, at a minimum, constituted as a complex, mutual relationship between two reflexive

persons, both of whom at once experience and interpret the constituents
of the encounter—including themselves and one another—because of
which a focus on each person's situational definition is critical. Moral
(and other) issues are embodied and expressed through a range of
feelings that are presented solely within the contexts of their actual
occurrence—which is itself a complex, ongoing mutual relatedness between
a person seeking help and one professing the ability to provide that help.
Although mutual, the clinical event is an asymmetrical relationship with
power (knowledge, skills, access to resources, social legitimation and legal
authorization) in favor of the professed helper, a relation that involves
physical and personal intimacies, often among strangers.

Accordingly, medicine is an inherently moral enterprise, highlighted by
that asymmetry. On the one hand, to be a patient is to be disadvantaged
by the very condition that brought one to the physician in the first place.
It is thus marked by various forms of unavoidable trust (on the part of
the one seeking help), and by taking-care-of and caring-for (on the part
of the helper). The asymmetry does not imply that the physician alone
does or should make unilateral decisions; having power-to-alter does not
signify either exercising power-for or power-over. It rather signifies power-
with: it requires, that is, the active, shared participation of both patient
(family, circle of intimates) and physician (and other providers). To be
sure, the nature of the asymmetry makes it possible (even tempting) for
the physician to take advantage of the multiply disadvantaged patient
(family, circle of intimates), but just that is morally prohibited—*by* the
patient's own existential vulnerability. In somewhat different terms, at the
core of the asymmetrical therapeutic relationship is a special form of
dialectic between trust and care, between having-to-be-trusting (patient)
and having-to-be-trustworthy (physician).

Every impairment is experienced and interpreted by the impaired
person, for whom it has meaning. Others also experience and interpret
that impairment: the patient's family, circle of intimates (often but not
always including the family), physicians and other providers, as well as
persons and institutions in the wider social ambiance. Encounters are thus
framed by cultural values, professional codes, governmental regulations,
hospital policies, unit or department protocols, etc. A clinical encounter
is a specific instance of a certain kind of *context* with its specific
appertaining set of multiple interrelationships, functional significances and
functional weights, as Gurwitsch astutely noted ([8], pp. 85-154; [9], pp.
175-286; [13], pp. 435-462; [39], pp. 67-109).

The involvement of the ethicist is a work of practical distantiation: reflective, circumstantial understanding ("detective work") whose primary focus is the relationship itself—the set of multiple interrelationships, significances and weights—seeking to enable or empower those whose situation it is and who must make needed decisions. The clinical ethicist is thus to be held accountable, in ways appropriate to that work: to be responsible for what is and is not said and done, and responsive to those whose situation it is (and who request the ethicist's involvement in the first place).

The narrative forms people give to their illnesses are displayed and embodied in a range of feelings that are most immediately lodged in specific images and figures, which become expressed in prevailing commonsense categories (in some part personal, in large part socially derived), and include the person's own understanding of the clinician's words, gestures, settings, and conducts. Their relationship is, while asymmetrical, a special form of *mutuality*: the encounter between reflexively, intimately related selves within an experienced asymmetry of vulnerability and power, whose exercise is under profound moral constraints.

If nothing else, the phenomenological explication of the clinical event—with its emphatic focus on the specific circumstances, experiences, interpretations, feelings and circumstances of those actually involved; on evidence; and on freely considering and varying the range of examples in order to detect and give faithful expression to the themes invariant or common to clinical encounters—forces us to re-think not only such topics as "informed consent" or "confidentiality," but more fundamentally, to appreciate the range of evident feelings that are situationally displayed as embodied expressions or presentations of the participants' moral concerns, their own sense of "what's worthwhile." Through that, interesting and important glimpses into the sense of the moral order itself are achieved.[15]

---

[15] An earlier version of this paper was presented at the annual meeting of the Society for Phenomenology and the Human Sciences, in Memphis, TN, October 18, 1991.

Bibliography

1. Bosch, Gerhard. *Infantile Autism: A Clinical and Phenomenological-Anthropological Investigation Taking Language as the Guide*, New York/Berlin: G. Springer-Verlag, 1970.
2. Cassell, Eric J. "Making and Escaping Moral Decisions," *The Hastings Center Report*, 1, 1973, pp. 53-62.
3. _____. Talking With Patients, two vol., Cambridge, MA: MIT Press, 1985.
4. _____. *The Healer's Art: A New Approach to the Doctor-Patient Relationship*, Cambridge, MA: MIT Press, 1985 [1972].
5. Edelstein, Ludwig. *Ancient Medicine*, Baltimore, MD: The Johns Hopkins University Press, 1967.
6. Goldstein, Kurt. *Aftereffects of Brain Injuries in War*, New York: Grune & Stratton, 1942.
7. _____. *Language and Language Disturbances*, New York: Grune & Stratton, 1948.
8. Gurwitsch, Aron. *The Field of Consciousness*, Pittsburgh, PA: Duquesne University Press, 1964.
9. _____. "Phenomenology of Thematics and of the Pure Ego: Studies of the Relation Between Gestalt Theory and Phenomenology," in A. Gurwitsch, *Studies in Phenomenology and Psychology*, Evanston, IL: Northwestern University Press, 1966, pp. 175-286.
10. Husserl, Edmund. *Erfahrung und Urteil*, Hamburg: Claassen Verlag, 1954.
11. _____. *Cartesian Meditations*, The Hague: Martinus Nijhoff, 1960 [1950].
12. _____. *Formal and Transcendental Logic*, The Hague: Martinus Nijhoff, 1969 [1929].
13. _____. *Logical Investigations*, two vols., New York: Humanities Press, Inc., 1970 [1900/1901; 1913].
14. _____. *Ideas Pertaining to a Pure Phenomenology and to a Phenomenological Philosophy, Book One*, The Hague: Martinus Nijhoff, 1982 [1913].
15. Kierkegaard, Soren. *The Sickness Unto Death (with Fear and Trembling)*, Princeton, NJ: Princeton University Press, 1954.
16. Kleinman, Arthur. *The Illness Narratives*, New York: Basic Books, Inc., 1988.

17. Langer, Susan. *Mind: An Essay on Human Feeling*, three vols.,
    Baltimore, MD: The Johns Hopkins University Press, Vol. I:
    1967.
18. Lenrow, Peter B. "The Work of Helping Strangers," in H.
    Rubenstein and M. H. Bloch (eds.), *Things That Matter: Influences
    on Helping Relationships*, New York: Macmillan Publishing
    Company, 1982, pp. 42-57.
19. MacIntyre, Alasdair. *After Virtue*, Notre Dame, ID: University of
    Notre Dame Press, 1981.
20. Odegaard, Charles E. *Dear Doctor*, Menlo Park, CA: The Henry J.
    Kaiser Family Foundation, 1986.
21. Pellegrino, Edmund D. "The Anatomy of Clinical Judgments: Some
    Notes on Right Reason and Right Action," in H. T. Engelhardt,
    Jr., S. F. Spicker, and B. Towers (eds.), *Clinical Judgment: A
    Critical Appraisal*, Dordrecht/Boston/London: D. Reidel Publishing
    Company, 1979, pp. 169-194.
22. _____. "Being Ill and Being Healed: Some Reflections on the
    Grounding of Medical Morality," in V. Kestenbaum (ed.), *The
    Humanity of the Ill: Phenomenological Perspectives*, Knoxville, TN:
    University of Tennessee Press, 1982, pp. 157-166.
23. _____. "The Healing Relationship: The Architectonics of Clinical
    Medicine," in E. A. Shelp (ed.), *The Clinical Encounter: The
    Moral Fabric of the Physician-Patient Relationship*, Boston/Dord-
    recht: D. Reidel Publishing Company, 1983, pp. 153-172.
24. Pellegrino, Edmund D. and Thomasma, David C. *A Philosophical
    Basis of Medical Practice*, New York/Oxford: Oxford University
    Press, 1981.
25. Polkinghorne, D. *Methodology for the Human Sciences: Systems of
    Inquiry*, Albany, NY: State University of New York at Albany
    Press, 1983.
26. Rawlinson, Mary. "Medicine's Discourse and the Practice of
    Medicine," in V. Kestenbaum (ed.), *The Humanity of the Ill:
    Phenomenological Perspectives*, Knoxville, TN: University of
    Tennessee Press, 1982, pp. 69-85.
27. Ruark, J.E., Raffin, R.A., et al. "Initiating and Withdrawing Life
    Support," *The New England Journal of Medicine* 318.1, 1988
    (January), pp. 25-30.
28. Scheler, Max. *The Nature of Sympathy*, New Haven: Yale University
    Press, 1954 [1928].

29. Schutz, Alfred and Luckmann, Thomas. *Structures of the Life-World*, two vols., Evanston, IL: Northwestern University Press, Vol. I: 1973, Vol. II: 1989.

30. Spiegelberg, Herbert. *Steppingstones Toward an Ethics for Fellow Existers: Essays 1944-1983*, The Hague: Martinus Nijhoff, 1986.

31. Thomas, W. I. *The Child in America*, New York: Alfred Knopf, 1928.

32. White, Kerr L. (ed.). *The Task of Medicine*, Menlo Park, CA: The Henry J. Kaiser Family Foundation, 1988.

33. Zaner, Richard M. *The Problem of Embodiment*, Phaenomenologica 17, The Hague: Martinus Nijhoff, 1964.

34. _____. *The Way of Phenomenology*, New York: Bobbs-Merrill, Inc., 1970.

35. _____. "Reflections on Evidence and Criticism in the Theory of Consciousness," in L. Embree (ed.), *Life-World and Consciousness: Essays for Aron Gurwitsch*, Evanston, IL: Northwestern University Press, 1972, pp. 209-230.

36. _____. "Examples and Possibles: A Criticism of Husserl's Theory of Free-Phantasy Variation," *Research in Phenomenology* III, 1973, pp. 29-43.

37. _____. "The Art of Free-Phantasy Variation in Rigorous Phenomenological Science," in F. Kersten and R. Zaner (eds.), *Phenomenology: Continuation and Criticism, Essays in Memory of Dorion Cairns*, The Hague: Martinus Nijhoff, 1973, pp. 192-219.

38. _____. "On the Sense of Method in Phenomenology," in E. Pivcevic (ed.), *Phenomenology and Philosophical Understanding*, London: Cambridge University Press, 1975, pp. 125-141.

39. _____. *The Context of Self*, Athens, OH: Ohio University Press, 1981.

40. _____. "The Disciplining of Reason's Cunning: Kurt Wolff's Surrender and Catch," *Human Studies* 4.4, 1981, pp. 365-389.

41. _____. "Chance and Morality: The Dialysis Phenomenon," in V. Kestenbaum (ed.), *The Humanity of the Ill: Phenomenological Perspectives*, Knoxville, TN: University of Tennessee Press, 1982, pp. 39-68.

42. _____. "The Logos of Psyche: Phenomenological Variations on a Theme," in S. Koch and D.E. Leary (eds.), *A Century of Psychology as Science*, New York: McGraw-Hill Book Company, 1985, pp. 618-637.

43. _____. *Ethics and the Clinical Encounter*, Englewood Cliffs, NJ: Prentice-Hall, Inc., 1988.
44. _____. "Failed or Ongoing Dialogues? Dax's Case," in L. D. Kliever (ed.), *Dax's Case*, Dallas, TX: Southern Methodist University Press, 1989, pp. 43-62.
45. _____. "Medicine and Dialogue," *The Journal of Medicine and Philosophy* 15, 1990, pp. 303-325.
46. _____. "The Phenomenon of Trust and the Patient-Physician Relationship," in E. D. Pellegrino, R. M. Veatch, and J. P. Langan (eds.), *Ethics, Trust, and the Professions: Philosophical and Cultural Aspects*, Washington, D.C.: Georgetown University Press, 1991, pp. 45-68.
47. _____. "Encountering the Other," in C. S. Campbell & A. Lustig (eds.), *Duties to Others*, Medicine and Theology Series, Boston: Kluwer Pubs., Inc., in press.

Chapter 2

Phenomenology and Cognitive Science

Osborne Wiggins
University of Louisville

Abstract: *Concepts central to phenomenology are compared with recent proposals in cognitive science. Both emphasize the role of the embodied mind in constituting a meaningful world, both emphasize the primacy of preconceptual experience, and both approaches view language, logic, and mathematics as constructed on the basis of preconceptual typifications.*

I. Introduction

In this essay I shall contend that recent claims by cognitive scientists indirectly support and extend some of the phenomenological descriptions of Edmund Husserl, Aron Gurwitsch, and Maurice Merleau-Ponty.

Cognitive science is a large and diverse field. Those views on which I shall concentrate represent only one subfield among others. I shall rely on two books. The first one is George Lakoff's *Women, Fire, and Dangerous Things: What Categories Reveal about the Mind* (1987). In this book, he both summarizes the main findings of empirical studies in cognitive science and develops a general philosophical interpretation of these findings. Lakoff maintains that these empirical findings render implausibe the reigning philosophical position that he calls "objectivism." After critically rejecting objectivism, Lakoff defends his own philosophical approach which he labels "experiential realism." The second book on which I shall draw is Mark Johnson's *The Body in the Mind: The Bodily Basis of Meaning, Imagination, and Reason* (1987). Johnson too criticizes "Objectivism" and offers his own alternative position which emphasizes the foundational roles of the body and the imagination in the constitution of a meaningful world. The similarities between Lakoff's and Johnson's views are not coincidental. They collaborated in writing the book, *Metaphors We Live By* (1980).

M. Daniel and L. Embree (eds.), Phenomenology of the Cultural Disciplines, 67–83.

I shall sketch some of Lakoff's and Johnson's views and indicate their similarities to certain phenomenological views. I shall then suggest ways in which Lakoff's experiential realism can be critically addressed from a phenomenological vantagepoint.

II. Phenomenology and Experiential Realism: Common Theses

I would like first to mention six of the philosophical theses which, in my judgment, Lakoff's and Johnson's positions share with phenomenology.

1) *Epistemological thesis of embodiment*: Any adequate philosophy of the knowing mind must conceive it as an *embodied* mind (Gurwitsch, 1979 and 1985; Merleau-Ponty, 1962).[1]

2) *Existentialist thesis of the primacy of the lifeworld*: The practical activities of everyday social life are largely preconceptual, and the structures of these preconceptual activities provide the necessary basis for all conceptualization, including natural, technical, and formal languages (Husserl, 1973; Gurwitsch, 1974; Merleau-Ponty, 1964).

3) *Phenomenal thesis of Gestalt organization*: The objects and situations encountered in the lifeworld exhibit a phenomenal structure that is best described in terms of Gestalt part-whole relationships (Gurwitsch, 1964; Merleau-Ponty, 1962).

4) *Constructivist thesis of science, logic, and mathematics*: Science, logic, and mathematics must be understood as constructed through processes of generalization, formalization, and idealization that presuppose the more basic preconceptual, embodied experiences of the lifeworld (Husserl, 1970; Gurwitsch, 1974; Merleau-Ponty 1962).

5) *Transcendental thesis of constitution*: The structure of the experienced world must be conceived as *dependent on* the structure and activities of embodied mind (Husserl, 1970; Gurwitsch, 1966; Merleau-Ponty, 1962).

6) *Ontological thesis of realism*: The experienced (constituted) world is also a real world, and this real world includes the real embodied mind that experiences and constitutes it (Merleau-Ponty, 1962 and 1964).

---

[1] The thesis I have expressed here should be compared with Mark Johnson's formulation of his position: *"any adequate account of meaning and rationality must give a central place to embodied and imaginative structures of understanding by which we grasp our world"* (1987, p. xiii).

As I proceed I shall sketch the similarities between cognitive science and phenomenology on these six theses.

III. Typifications in Phenomenology
Basic-Level Categories in Cognitive Science

In order to describe preconceptual experience phenomenologists, like Aron Gurwitsch, frequently follow Husserl in speaking of "typification" (Husserl, 1973, pp. 31-39 and 331-334). Objects, according to phenomenology, are perceived as being of certain generic kinds. As examples of preconceptual typifications Gurwitsch cites trees, automobiles, dogs, and human beings (1966, p. 394). Moreover, Gurwitsch and Merleau-Ponty contend that the typical senses that objects present to us are constituted by the actions through which we use and manipulate them (Gurwitsch, 1979, pp. 66-84; and 1985; Merleau-Ponty, 1962).

The preconceptual types described by Husserl and Gurwitsch are precisely what Lakoff calls "basic level categories" (Lakoff, pp. 31-57). Although he employs the word "category," Lakoff is referring here to *preconceptual* or *prelinguistic* rather than conceptual or linguistic categories. Roger Brown, one of the psychologists who initiated such investigations, uses as his examples of basic level categories dogs and cats (Lakoff, p. 31). From the traditional point of view of the hierarchy of classification, we think of the category of *dog* as occupying an intermediate position between more specific (subordinate) categories, like retriever, and more general (superordinate) categories, like animal. In our everyday, preconceptual experience, however, *dog* functions as a *basic level category* in the sense that we tend immediately to see objects as dogs rather than as *retrievers* or as *animals*. In other words, *the sense that objects directly and immediately present to us in our ordinary perception is this "basic level" sense rather than some more specific or general sense.* In order to arrive at the more specific or more general senses of these objects, we would have to perform mental acts in which we selectively attended to certain of their features and disregarded others. Apprehending the more specific or general senses, in other words, requires mental acts of selective *abstraction*. Prior to such acts of abstraction—and as a necessary precondition for them—we perceive objects at the "basic level" (Lakoff, pp. 12-57).

Why, according to Lakoff, do these categories function as *basic* to human perception? First, it is at this level of meaning that objects are perceived as *gestalt-wholes* (Lakoff, p. 47). At this level what is recognized is the overall shape of the object and the functional interrelationships among the object's parts. Gurwitsch and Merleau-Ponty described perceived objects as Gestalt-wholes (Gurwitsch, 1964; Merleau-Ponty, 1962). Here Gurwitsch, Merleau-Ponty, and Lakoff agree entirely.

Lakoff claims, moreover, that at the level of basic categories we possess distinctive schemas of bodily action for using and manipulating the objects. Lakoff writes,

> Our knowledge at the basic level is mainly organized around part-whole divisions. The reason is that the way an object is divided into parts determines many things. First, parts are usually correlated with functions, and hence our knowledge about functions is usually associated with knowledge about parts. Second, parts determine shape, and hence the way that an object will be perceived and imaged. Third, we usually interact with things via their parts, and hence part-whole divisions play a major role in determining what motor programs we can use to interact with an object. Thus, a handle is not just long and thin, but it can be grasped by the human hand. As Tversky and Hemenway say, "We sit on the *seat* of a chair and lean against the *back*, we remove the *peel* of a banana and eat the *pulp*" (p. 47).

This third aspect of basic level categories is illuminating from a phenomenological point of view. Gurwitsch and Merleau-Ponty contended that we perceive objects as Gestalt-wholes and that, as thus intended, they present to us the sense they have as objects for a distinctive practical *use* (Gurwitsch, 1979, pp. 66-84; Merleau-Ponty, 1962). To take one of E. Rosch's examples to make the phenomenological point, the table is perceived as a Gestalt-whole and as "something to eat on" (Lakoff, p. 51). Lakoff maintains that these two constitutive features of basic level objects are interdependent. A chair, for instance, is intended as an object with two dependent parts, a seat and a back. The reason why I perceive the object as a whole composed of these two dependent parts is that I *sit* on the seat and I *lean* against the back. In other words, there is a bodily action, sitting and leaning, that I typically perform with such objects, and this typical action is also a Gestalt-whole composed of two dependent parts. *Correlated with the typical sense of the object, then, is a typical bodily action.* My mental life constituted this typical sense of the object precisely by my bodily use of the object in this activity. When I perceive the object now, those of its Gestalt-constituents that it would present through such use are emptily and automatically co-intended by me. The theories of the cognitive scientists thus allow us to carry phenomenological analyses to their logical conclusions: objects are perceived as Gestalt-wholes because their typical senses have been constituted through typical actions. Within the field of cognitive science the views that Lakoff expounds here are revolutionary. Gurwitsch and Merleau-Ponty formulated these notions decades ago.

IV. Preconceptual Senses and Conceptual Meanings

In his book, *The Body in the Mind: The Bodily Basis of Meaning, Imagination, and Reason* (1987), Mark Johnson argues for a pre-conceptual constitution of meaning structures. Independently of phenomenology, Johnson develops what I have called above the existentialist thesis of the primacy of the lifeworld and the epistemological thesis of embodiment. In opposition to the view that he calls "objectivism," Johnson claims that not all meaning is propositional or linguistic (pp. xix-17). In order to be meaningful, propositions presuppose a large underlying sphere of prelinguistic meanings. These preconceptual meanings are constituted by what Johnson calls "image schemata." Image schemata are themselves constituted through bodily activity and perception (pp. 18-30).

Johnson characterizes image schemata as dynamic and rather general patterns of preconceptual meaning. Schemata are dynamic in that they bestow meaning on processes that involve organized sequences of steps or movements that unfold in time. These patterns of meaning are somewhat general in that a schema can bestow a similar structure on situations that differ in many of their features (pp. 28-30).

The structure of image schemata is a gestalt structure (pp. 41-64). This means that each component of a schema is an intergral part of the whole. Each component, by playing the particular role it plays within the whole, contributes to the overall structure and meaning of the whole; and the component derives some of its own meaning from its intrinsic relatedness to the other constituents that compose the whole. Johnson claims that we can theoretically discover and describe these "image-schematic gestalts" because they are repeatable patterns throughout our experience. Each one is, a pattern that can . . . contribute to the regularity, coherence, and comprehensibility of our experience and understanding. To say that a gestalt is "experientially basic," then, is to say that it constitutes a recurring level of organized unity for an organism acting in its environment. Gestalts, in the sense that I am using the term, are not unanalyzable givens or atomistic structures. They can be "analyzed" since they have parts and dimensions. But, any such attempted reduction will destroy the *unity* (the meaningful organization) that made the structure significant in the first place (p. 62).

Johnson's position thus includes what I have called above "the phenomenal thesis of gestalt organization."

Johnson concedes that propositions exhibit logical structures and even meanings that go beyond the gestalt structures of preconceptual image schemata which they presuppose. Propositions, in other words, are meaningful objectivities of a higher order; they consequently exhibit their own special patterns, rules of construction and transformation. But what Johnson is denying is that propositions and linguistic entities in general are *autonomously* meaningful or *autonomously* structured. To some extent linguistic meanings *depend* for both their meaningfulness and structure on prelinguistic image schemata.[2]

As a simple illustration Johnson describes "out" schemata (pp. 31-37). Drawing on the work of Susan Lindner, he notes that many verbs are followed by "out," e.g., *take out, spread out, throw out, pick out, leave out, shout out, draw out,* and *pass out* (p. 32). These verb forms appear in such sentences as:

> John went out of the room.
> Pump out the air.
> Let out your anger.
> Pick out the best theory.
> Drown out the music.
> Harry weasled out of the contract (p. 32).

---

[2] The claim of dependence here should not be misunderstood as a claim of reduction. Propositions cannot be reduced to the image schemata on which they nevertheless depend.

Although Johnson describes three different "out" schemata, all of the sentences cited here are based on one such schema. In this schema an entity is first contained within something and then moves outside of that thing (p. 33). The meaningfulness of these propositions depends upon the meaningfulness of a single prelinguistic schema. The schema itself is first constituted through bodily activity: we move our bodies out of a container (say, a room or a bed) to some space outside of this container. It is, then, what Merleau-Ponty would call "the body-subject" that turns out to be the structuring source of meaningfulness and a meaningful world (Merleau-Ponty, 1962).

If it is true that all of these propositions depend for their meaningfulness on an "out" schema that is constituted through bodily movement "out" of some spatial location, then the first proposition, "John went out of the room," is closest to this root meaning. And as the list of propositions proceeds, the verb forms become more "metaphorical" while still remaining based on this root meaning. "Weasling out of the contract" signifies "movement" from one "place" (viz., within the contract) to another "place" (viz., outside of the contract); but the "places" are no longer spatial locations strictly speaking, and the "movement" is not bodily movement through space. This metaphorical projection of the "out" schema illustrates one of Johnson's main theses: "schematic structures salient in most of our mundane experience . . . can be extended and elaborated metaphorically to connect up different aspects of meaning, reasoning, and speech acts" (p. 65).

In order to demonstrate these processes of metaphorical projection, Johnson first explicates the image schema of "balance." This schema, like all others, is first constituted in bodily experiences and activities. As Johnson phrases it, ". . . the *meaning* of balance begins to emerge through our *acts* of balancing and our *experience* of systemic processes and states within our bodies" (p. 75). The schema that emerges in this way includes, among its gestalt-constituents, weight and force.

Johnson then shows how other realities such as paintings and masks can be perceived as "balanced." Such perception involves a *metaphorical projection* of the gestalt-constituents of the embodied image schema onto another domain. Seeing the mask as balanced involves "the projection of structure from one domain (that of gravitational and other physical forces) onto another domain of a different kind (spatial organization in visual perception)" (p. 82). In seeing the mask it is not simply the case that the lines and shapes on each side of it are perceived as symmetrical.

It is also true that "force" and "weight" are perceived in the mask. But as Johnson remarks, "we no longer have 'weight' and 'force' in the gravitational and physical sense. Instead, we have complex *metaphorical* (but very real) experience of *visual* weight and force" (p. 80).

Johnson even views our conceptions of justice as involving a metaphorical projection of a schema of balance into the moral domain. As he writes,

> Justice itself is conceived as the regaining of a proper balance that has been upset by an unlawful action. According to some assumed calculus, the judge must assess the *weight* of the damages and require a penalty somehow equal to the damages as compensation. We have linguistically encoded manifestations of this juridical metaphor, such as "an eye-for-an-eye" and "let the punishment fit the crime" (p. 90).

This exemplifies Johnson's claim that highly abstract or even normative domains can derive their "rationality" from more basic experiences of bodily activity.

In searching for a term to refer to this founding level of preconceptual experience, Johnson appropriates the word "understanding." For him, this term signifies our most fundamental way of comporting ourselves toward the world. He provides a good summary of his position when he writes,

> . . . understanding is not only a matter of reflection, using finitary propositions, on some preexistent, already determinate experience. Rather, *understanding is the way we "have a world," the way we experience our world as a comprehensible reality.* Such understanding, therefore, involves *our whole being*—our bodily capacities and skills, our values, our moods and attitudes, our entire cultural tradition, the way in which we are bound up with a linguistic community, our aesthetic sensibilities, and so forth. In short, our understanding *is* our mode of "being in the world." It is the way in which we are meaningfully situated in our world through our bodily interactions, our cultural institutions, our linguistic tradition, and our historical context. Our more abstract reflective acts of understanding (which may involve grasping of finitary propositions) are simply an extension of our understanding in this more basic sense of "having a world" (p. 102).

In *Women, Fire, and Dangerous Things,* Lakoff appropriates and summarizes Johnson's views. As Lakoff writes, "One of Mark Johnson's basic insights is that experience is structured in a significant way prior to, and independent of, any concepts" (Lakoff, p. 271). The most basic level of experience is preconceptual. And it is the human body that structures the world at this preconceptual level. Moreover, this preconceptual structure exhibits a "proto-logic" that receives explication and elaboration at the higher levels of conceptualization. The "kinesthetic image schemas" that connect the acting body to its surrounding world are proto-logical.

An example is the schema of a "container." The human body is experienced both as a container itself and as a reality contained within larger spaces, e.g., rooms. This lived schema exhibits certain structural elements: interior, boundary, and exterior. These structural elements have their own "logic": Everything is either inside a container or outside of it—P or not P. If container A is in container B and X is in A, then X is in B. If all A's are B's and X is an A, then X is a B. Lakoff also claims that this container "logic" is the basis of the Boolean logic of classes (p. 272).

In *Experience and Judgment*, Husserl demonstrated a similar "proto-logic" that already lies embedded in preconceptual experience. Gurwitsch has characterized the project of a phenomenology of preconceptual experience as follows:

> Its task consists in disengaging the "logos" of the perceptual world, the logicality which prevails in it. Of course, logicality as here meant must not be understood in the sense of fully conceptualized—still less formalized—logic but, rather, in the same sense in which Husserl understands the a priori and the categories of the perceptual world, namely, determinateness as to style and type but absense of exactness. Since the logicality in question proves to be the germ from which logic in the proper and formal sense develops, it may be appropriately denoted as "protologic." In fact, the transition from protologic to logic proper (understood in the widest sense so as to include all mathematization, algebraization, and formalization) requires specific indealizing operations which, of course, work on the protological structures as underlying pregiven materials (Gurwitsch, 1974, pp. 29-30).

It is tempting, then, to view the detailed analyses developed by Johnson and Lakoff as explicating this protological structure. Johnson and Lakoff could thus be seen as furthering a phenomenological project. Gurwitsch himself has called for such an advancing of the phenomenological project:

Future phenomenological research will have first to complete
Husserl's work in exhaustively setting forth the protological
structures and then to account for the acts and operations of
consciousness which are involved in the transition to the logical
level in the wider sense (Gurwitsch, 1974, p. 30).

From Preconceptual to Conceptual Experience

Lakoff contends that the basic level of the bodily structuration of things
is "the level at which things are first named" (Lakoff, 1987, p. 32).
Language, in other words, draws on the preconceptual structure of the
world already constituted through bodily action. Phenomenology concurs.
Husserl has shown that any philosophy of language cannot understand
linguistic structure without first carrying out a "retrogression" to the
lifeworld where objects are prepredicatively experienced as types (Husserl,
1973, pp. 11-68). And, according to Gurwitsch, conceptualization first
arises through detaching and disengaging a typical sense from an
individual object in which it is perceptually embedded (Gurwitsch, 1966,
p. 395). General concepts are first constituted through this thematic
disengagment of preconceptual generic types. This entails that at the first
level of language the meanings of concepts will embody the preconceptual
senses of perceived things; and, as we have seen, this preconceptual sense
is correlated with bodily action. Lakoff's view, which he calls "experiential
realism," is precisely the same. Lakoff writes,

> Experientialism claims that conceptual structure is meaningful because
> it is embodied, that is, it arises from, and is tied to, our preconceptual
> bodily experiences. In short, conceptual structure exists and is understood
> because preconceptual structures exist and are understood. Conceptual
> structure takes its form in part from the nature of preconceptual
> structures." (1987, p. 267)

The Embodied Subject and the Perceived World

If, as both the phenomenologists and Lakoff claim, the perceptual
sense of objects is constituted through bodily action, then at the level of
basic categories or typifications the sense of objects will depend on what
the human body does and can do. Certain senses cannot be constituted
because there is or can be no corresponding bodily action to constitute
them (Lakoff, 1987, pp. 50-52). Objects are perceived, for example, as

chairs; but we do not perceive objects through the more general (superordinate) category of "furniture" (unless we are unfamiliar with the objects presented to us and discern only their most general purpose). We do not perceive objects as furniture because there is no generic bodily use that we make of all furniture. There is no distinctive action that we perform with regard to all furniture that would lend that category a distinctive perceptual meaning for us. We do perceive objects as chairs, however, because there is a distinctive schema of bodily action through which we use them. This entails that the sense of objects depends upon the capacities, limitations, and habits of our embodied mental lives. In terms that Gurwitsch adapts from Max Scheler, the surrounding world is "existentially relative" to our embodied, active selves. As Gurwitsch writes,

> Universally, the milieu is "existentially relative" to living beings at large; that implies the "existential relativity" of every concretely present milieu to a living being of determined organization and determined drive acquisition (Gurwitsch, 1979, p. 58).

In arguing for this "existential relativity" of the surrounding world Gurwitsch is arguing against a view of perception and knowledge that has influenced Western thought since Descartes.

Lakoff too opposes a view of perception and knowledge that permeates Modern philosophy (Lakoff, 1987, pp. 157-259). He calls this view "objectivism." I believe that we may say that one of Lakoff's central disputes with "objectivism" is that it overlooks the "existential relativity" of the experienced milieu that Gurwitsch describes. As Lakoff complains, "objectivism defines meaning independently of the nature and experience of thinking beings" (1987, p. 266). In opposition to objectivism, Lakoff explains his own position: "experiential realism characterizes meaning in terms of embodiment, that is, in terms of our collective biological capacities and our physical and social experiences as beings functioning in our environment" (p. 267). For Lakoff too, then, the meanings of things are "existentially relative" to the embodied subject who "constitutes" those meanings. The dependence of the experienced world on an experiencing subject is also captured by Rosch when she writes, "It should be emphasized that we are talking about a perceived world and not a metaphysical world without a knower" (Lakoff, 1987, p. 50).

The Origin of Mathematics

In order to advance his thesis concerning the construction of formal systems out of more basic (lifeworldly) experiences, Lakoff adopts a view of the foundations of mathematics put forward by the mathematician Saunders Mac Lane (Lakoff, 1987, pp. 361-365). Mac Lane maintains that if we conceive set theory as the foundation of mathematics, certain questions in mathematics remain unanswered. Mac Lane mentions in particular the question: Why does mathematics have the branches it has? Mac Lane contends that "the grand set-theoretic foundation" of mathematics,

> does not adequately describe which are the relevant mathematical structures to be built up from the starting point of set theory. A priori from set theory there could be very many such structures, but in fact there are a few which are dominant . . . .Some mathematical structures (natural numbers, rational numbers, real numbers, Euclidean geometry) are intended to be unique but other structures are built to have many different models: group, ring, order and partial order, linear space and module, topological space, measure space. The "Grand (Set-Theoretic) Foundation" does not provide any way in which to explain the choice of these concepts (Lakoff, 1987, p. 361)

Mac Lane proposes an answer to this question that resembles the genetic thesis of phenomenology. Mac Lane writes,

> The real nature of these structures does not lie in their often artificial construction from set theory, but in their relation to simple mathematical ideas or to basic human activities . . . mathematics is not the study of intangible Platonic worlds, but of tangible formal systems which have arisen from real human activities (Lakoff, 1987, p. 361).

Mac Lane thus seeks to correlate specific portions of mathematics with specific kinds of human activities. He devises the following list of correlated human activities and branches of mathematics:

> counting: to arithmetic and number theory
> measuring: to real numbers, calculus, analysis
> shaping: to geometry, topology
> forming (as in architecture): to symmetry, group theory
> estimating: to probability, measure theory, statistics

moving: to mechanics, calculus, dynamics
calculating: to algebra, numerical analysis
proving: to logic
puzzling: to combinatorics, number theory
grouping: to set theory, combinatorics

Having listed these different human activities Mac Lane notes their interrelationships: "These various human activities are by no means completely separate; indeed, they interact with each other in complex ways . . . .The two parts of this table should (and do) fit together by many crosslinks" (Lakoff, 1987, p. 362).

The sorts of relationships that we normally think of as relationships holding among formal mathematical terms, such as symmetry, asymmetry, or transitivity, turn out to be rooted and already exhibited in the more basic human activities that generate these mathematical formations. Sequences of human actions are already symmetrical, asymmetrical, etc. As Mac Lane says, mathematical systems "codify deeper and nonobvious properties of the originating human activities" (Lakoff, 1987, p. 362). Mathematical formations thus always retain within themselves what phenomenologists would call their "meaning-genesis": the forms of the activities that generate mathematical formations remain embedded within these higher-level formations, only now emptied of content, i.e., formalized.

Lakoff summarizes Mac Lane's position in the following way:

> Mac Lane claims that the branches of mathematics are as they are because they arise from human activities that each have a general schematic structure, made up of various substructures, or "basic ideas." These basic ideas both occur in the structure of the human activities that give rise to the various branches of mathematics (and in these branches of mathematics themselves). Mathematics describes them and their connections and interrelationships in an absolutely rigorous way (p. 362).

Mac Lane's position here resembles Husserl's in "The Origin of Geometry." Husserl claims that the "meaning-genesis" of geometry lies in the idealization and formalization of certain human activities. Husserl mentions in particular the activities of estimating, measuring, designing buildings, and surveying fields (1970, p. 376). Geometry thus had its roots in everyday practical life, not in pure theory. As Husserl writes, "in the

life of practical needs certain particularizations of shape stood out and
. . . a technical praxis always (aimed at) the production of particular
preferred shapes and the improvement of them according to certain
directions of gradualness" (1970, p. 375). Geometry arises historically
through the idealization and formalization of these practical activities; and
once developed this science is handed down to subsequent generations as
a ready-made tradition. Both Gurwitsch and Merleau-Ponty adopted this
Husserlian position (Gurwitsch, 1974; Merleau-Ponty, 1962). Gurwitsch
sought to develop it further by following Piaget in the study of how
children learn to think mathematically through performing basic human
acts (1974, pp. 132-149).

V. Experiential Realism and Phenomenology: Points of Divergence

I have sought to indicate areas of agreement between the phenomenologies of Edmund Husserl, Aron Gurwitsch, and Maurice Merleau-Ponty
and the experiential realism of George Lakoff and Mark Johnson. In
conclusion I would like only to mention some crucial points of disagreement.

Despite the primacy which they repeatedly ascribe to the active,
experiencing human body, Lakoff and Johnson present no developed
theory of the body. And since many of their explanatory concepts, such
a "image schema," are closely related to the body, these concepts remain
vague and free floating as long as they are not integrated into a theory
of the body. Husserl, Gurwitsch and Merleau-Ponty have developed
phenomenologies of the "body-subject" (Husserl, 1952; Gurwitsch, 1985;
Merleau-Ponty, 1962). Lakoff and Johnson's insights could, I submit,
receive much systematic clarification through confrontation with these
phenomenologies of the embodied mind.

Several problems stand in the way of any attempt to join phenomenology with experiential realism, however.

The positions of Lakoff and Johnson are explicitly realistic.
Merleau-Ponty's phenomenology too implies a form of realism
(Merleau-Ponty, 1962). The phenomenologies of Husserl and Gurwitsch,
however, are decisively non-realistic by being transcendental (Husserl,
1970; Gurwitsch, 1966).

Lakoff and Johnson, on the one hand, view embodied mental life as
"world-constituting." That is, they view objects and situations as disclosing
the features they do because of the structuring processes of an embodied

mind. In this sense, then, the world and all of its components and features are "mind-dependent." On the other hand, this embodied mind is a real being, existing as a part of a real world that encompasses it and transcends it. From this realistic point of view the nature of the embodied mind is dependent upon the worldly processes that have produced it and are now acting on it: real processes determine and shape the human mind. In this philosophy, then, both mind and world have a dual status. (1) The embodied mind bestows structure and meaning on the world: the mind is constituting, and the world is constituted. (2) The mind is shaped by worldly events. From this point of view the world has the features and events it has independently of mental processes. The world is causally determining, and mind is causally determined.

Husserl has offered numerous criticisms of realistic philosophies that ascribe this dual status to both mind and world (Husserl, 1970). Modern philosophy from Descartes through Kant has come to recognize and analyze more and more fully the world-constituting status of mental life and the mind-constituted status of the world. On the other hand, a persistent "natural attitude" has prompted philosophers to assume the real existence of the world, an existence independent of mental life. The world turns out, then, to be both transcendentally dependent on the mind and really independent of the mind. Husserl thinks this duality generates absurdities.

I shall not attempt here to adjudicate this complex disagreement between experiential realism and transcendental phenomenology. I wish only to mark its presence and to suggest a need for its resolution.

A second area of disagreement lies in what Husserl would call "anthropologism" (Husserl, 1987). The world-constituting mind can be viewed as a human mind. If we deem it a human mind, however, then we must view it as shaped by all those realities—biological, evolutionary, historical, and cultural—that shape the human mind. There might thus be features of reality that we humans experience in a certain manner because our neurophysiological make-up and our sociohistorical conditioning determine us to experience them in this manner. Human experience is dependent upon the contingent deliverances of human biology and history; and hence human experience is relative to the contingent constellations of biology and history. Epistemological relativism threatens to undermine any truth-claims, including, of course, the truth-claims of the theory that implies relativism. Lakoff and Johnson directly confront the problem of relativism (Lakoff, pp. 304-307; Johnson, pp. 195-202). But charges of relativism will continue as long as they view the mind as world-constituting and they deem this mind a human mind.

Early and late in his career Husserl criticized anthropologisms that imply relativism (Husserl, 1970 and 1987). His solution lay in distinguishing between essential (necessary) features of mental life and empirical (contingent) features. For Husserl, philosophy was the science of the essential; and the empirical sciences—both natural and social—were disciplines devoted to the contingent. Lakoff and Johnson make much use of findings in the empirical sciences, especially psychology; and yet they draw philosophical conclusions from these findings. Aron Gurwitsch and Merleau-Ponty also thought it possible to appropriate some of the results of empirical science -- Gestalt psychology in particular -- for use in philosophical reasoning (Gurwitsch, 1964 and 1966; Merleau-Ponty, 1962). Clearly there should be some methodological lines of connection between philosophy and empirical science. The difficulty lies in securing those connections in ways that do not result in a metaphysical anthropologism and an epistemological relativism.

The similarities between the phenomenologies of Husserl, Gurwitsch, and Merleau-Ponty and the experiential realisms of Lakoff and Johnson are manifold and profound. The differences, I think, force us to think strenuously about foundational issues in both phenomenology and experiential realism. Such thinking, if successful, could in the long run render the similarities even more evident and fruitful.

References

Gurwitsch, Aron. *The Field of Consciousness*. Duquesne University Press, Pittsburgh, 1964.

Gurwitsch, Aron. *Studies in Phenomenology and Psychology*. Northwestern University Press, Evanston, 1966.

Gurwitsch, Aron. *Phenomenology and the Theory of Science*. Edited by Lester Embree, Northwestern University Press, Evanston, 1974.

Gurwitsch, Aron. *Human Encounters in the Social World*. Edited by Alexandre Metraux. Translated by Fred Kersten. Duquesne University Press, 1979.

Gurwitsch, Aron. *Marginal Consciousness*. Edited by Lester Embree. Athens, Ohio: Ohio University Press, 1985.

Husserl, Edmund. *Ideen zu einer reinen phänomenologie und phänomenologischen philosophie, Zweites Buch: Phänomenologische untersuchungen zur konstitution.* Herausgegeben von Marly Biemel. Martinus Nijhoff, Haag, 1952.

Husserl, Edmund. *Logical Investigations: Prolegomena to Pure Logic, Volume One.* Translated by J. N. Findlay. Humanities Press, New York, 1970.

Husserl, Edmund. *The Crisis of European Sciences and Transcendental Phenomenology: An Introduction to Phenomenological Philosophy.* Translated by David Carr. Northwestern Univesity Press, Evanston, 1970.

Husserl, Edmund. *Experience and Judgment: Investigations in a Genealogy of Logic.* Revised and edited by Ludwig Landgrebe. Translated by James S. Churchill and Karl Ameriks. Northwestern University Press, Evanston, 1973.

Husserl, Edmund. *Aufsätze und Vorträge (1911-1921).* Herausgegeben von Thomas Nenon and Hans Rainer Sepp. Martinus Nijhoff Publishers, Dordrecht, 1987.

Johnson, Mark. *The Body in the Mind: The Bodily Basis of Meaning, Imagination, and Reason.* The University of Chicago Press, Chicago, 1987.

Lakoff, George, and Johnson, Mark. *Metaphors We Live By.* The University of Chicago Press, Chicago, 1980.

Lakoff, George. *Women, Fire, and Dangerous Things: What Categories Reveal about the Mind.* The University of Chicago Press, Chicago, 1987.

Merleau-Ponty, Maurice. *Phenomenology of Perception.* Translated by Colin Smith, The Humanities Press, New York, 1962.

Merleau-Ponty, Maurice. *The Primacy of Perception and Other Essays.* Edited by James Edie, Northwestern University Press, Evanston, 1964.

Chapter 3

The Body as Cultural Object/
The Body as Pan-Cultural Universal

Maxine Sheets-Johnstone

Abstract: *In addition to implicitly carrying forward a Cartesian-inspired depreciative assessment of the body, many cultural disciplines (including philosophy) have been heavily influenced by postmodern dogma which basically regards the body as little more than a cultural artifact. Received wisdom and dogma together preclude an appreciation of the body as pan-cultural universal. A consideration of early stone tools in the light of phenomenological corporeal matters of fact shows how the body is the source of fundamental meanings, a semantic template. The analogy between the two major hominid tooth forms—molars and incisors—and the major early stone tools—core tools and flake tools—is in fact obvious once animate form and the tactile-kinesthetic body—the sensorily felt body—is recognized. A consideration of the experience of eyes as windows on two worlds exemplifies a further dimension of the body as pan-cultural universal. The experience of eyes as centers of light and dark is tied to an intercorporeal semantics that is rooted in morphological/visual relationships and attested to by biologist Adolf Portmann's notion of inwardness. The experience is furthermore shown to be the basis of cultural practices and beliefs related to the creation of circular forms such as the mandala. Phenomenological attention to corporeal matters of fact as exemplified by paleoanthropological artifacts, by the experience of inwardness, and by cultural drawings of circular forms underscores the desirability of a corporeal turn, an acknowledgment of animate form and of the tactile-kinesthetic experiences that consistently undergird hominid life.*

M. Daniel and L. Embree (eds.), Phenomenology of the Cultural Disciplines, 85–114.

Many times in the course of thinking of ways in which I might begin this essay, my thoughts turned back to Paul Valery. While it would have been appropriate to attempt to situate the essay in the context of Schutz's work, Valery's provocative piece titled "The Problem of the Three Bodies"[1] proved in the end too magnetic. In this short piece, first published in 1943, Valery proposes that each of us in our thoughts is not two but three bodies—"at least." That Valery diverges numerically from phenomenological and existentialist accounts is precisely what is of moment.[2] He distinguishes not only the felt body from the physical body but the *seen* body from both. This "second" body comes after what Valery terms "the privileged object" that is "My Body" and before the imaginatively unified but visually disjoint and fragmented body of science, the body which, as Valery puts it, "has unity only in our thought, since we know it only for having dissected and dismembered it." In sharp distinction from the privileged object that is My Body, the second body, Valery says, "knows no pain, for it reduces pain to a mere grimace."

Valery's three-body schema invites us to consider the living body in what I call sensory-kinetic terms, and in turn to realize that the usual living body/physical body distinction can be more finely analyzed and understood, indeed, to realize that corporeal analyses can be generated on the basis of sensory-kinetic understandings. Valery's first two bodies coincide actually with what I have elsewhere described as the tactile-kinesthetic body and the visual body,[3] and his third body with what I have elsewhere described as the progressively materialized body of Western science.[4] The distinctions constitute a phenomenologically-informed insight, an insight suggestive not only of the distinctive

---

[1] The essay is actually an essay within an essay. See "Some Simple Reflections on the Body," in *Aesthetics* (Collected Works, vol. 13), translated by Ralph Manheim (New York: Pantheon, 1964), 35-40.

[2] One might well wonder whether Husserl's distinction between two bodies, and two bodies only, is a function of the German language. Having no such ostensibly complete and ready-made linguistic corporeal categories, Valery's "thoughts" may have been open to the possibility of a range of bodies.

[3] Maxine Sheets-Johnstone, *The Roots of Thinking* (Philadelphia: Temple University Press, 1990). See also Maxine Sheets-Johnstone, "Existential Fit and Evolutionary Continuities," *Synthèse* 66 (1986), 219-248.

[4] Maxine Sheets-Johnstone, "The Materialization of the Body: A History of Western Medicine, A History in Process," in *Giving the Body Its Due*, edited by Maxine Sheets-Johnstone (Albany: State University of New York Press, 1992), 132-158.

perceptual modes in which we are bodies for ourselves and for others, but of the epistemological ordering of those perceptual modes. There is to begin with a recognition of the tactile-kinesthetic mode in which I am first and foremost a body for myself. Second, there is a recognition of the visual mode in which I am a body for others and to a more limited degree, a body for myself. Third, there is a recognition of the fragmented object body of Western science.

Now the visual body that is perceptually distinct from both my tactile-kinesthetic body and the body that, with all its tubules, organs, nerve fibers, and so on, is an object of science, is a body that Valery says "goes little farther than the view of a surface." As I shall hope to show, however, that surface is a complicated tapestry on and within which two principal modes of meaning are constituted. On the one hand, our social relations, like the relations of many social animals, are anchored in our visual bodies. Because our visual bodies are part of what we are as animate forms and because animate forms are evolutionary forms of life, our visual bodies are the ground of an intercorporeal semantics whose roots run both deep and wide. It is not surprising, then, that that intercorporeal semantics is foundationally describable in terms of corporeal archetypes; thus, with respect to humans and to other culture-bearing creatures, it is a semantics that in the most fundamental sense is not culturally relative. Moreover because visual bodies are animate forms and because animate forms are evolutionarily linked in distinctive ways, it is furthermore not surprising that the intercorporeal semantics that fundamentally defines our own human social relations defines the social relations of many extant primate species and by the same token, necessarily defined the social relations of ancestral hominid species as well. On the other hand, our visual body is consistently related to our tactile-kinesthetic body, our first body. Indeed, the underside of the tapestry is interwoven in fundamental ways with meanings transferred from the tactile-kinesthetic body. The two bodies are thus coordinated and in myriad ways, as any close analysis of empathy or close observation of normally competent adults, growing infants, gymnasts, and choreographers attest. Tactile-kinesthetic concepts are in consequence open to visual elaboration; that is, concepts originally formed on the basis of the tactile-kinesthetic body are—or may be—the spawning ground of visual concepts—the concept of an edge giving rise to the concept of line, for example.

My paper deals with both of these principal modes of meaning of our "second body"—with conceptual offshoots as it were, and with an intercorporeal semantics. In both cases corporeal invariants, tactile-kinesthetic and visual, come into play. My first concern will be with conceptual offshoots. I will first describe how the tactile-kinesthetic body is a semantic template and then how, in an epistemological sense, vision learns from touch, that is, how the visual body, fashioning itself after the tactile-kinesthetic body, itself becomes a model upon which tactile-kinesthetic concepts are elaborated.[5] I will then describe how our own seen bodies are the ground of our social relations and how those relations are part of an intercorporeal semantics more ancient than we. I will interweave the two themes originally set for the 1992 Research Symposium in the process of treating both of these phenomena: the theme of what philosophical reflection on a non-philosophical discipline might bring forth and the theme of "what the 'cultural disciplines' in general might be."

I

Paleoanthropologists, archaeologists, and anthropologists have all consistently remarked on how ancestral hominids, in fashioning stone tools, made tools do the work of teeth. They speak consistently of how teeth were replaced by tools. For example, one archaeologist writes, "Seen in an evolutionary perspective, the use of hands and tools, sticks, bones, and stones, to tear, cut, and to pound and grind foodstuffs is but a simple extension of the functions performed by the jaws."[6] He goes on to describe flaking techniques and the creation of edges. What he and other evolutionary scientists do not speak of, and what they do not even stop to question is "the simple extension," that is, how the replacement came to be. Where did the notion of a tool come from? What similarity was conceived between teeth and stones? Where did the notion of an edge come from? Such questions never surface in paleoanthropology and related disciplines not simply because there is typically no interest in conceptual origins, but because there is no explicit acknowledgment of a

---

[5] This section of my paper is based on a section of "The Hermeneutics of Tool-Making: Corporeal and Topological Concepts," Chapter 2 of *The Roots of Thinking*.

[6] Jacques Bordaz, *Tools of the Old and New Stone Age* (Garden City: The Natural History Press, 1970), 8.

tactile-kinesthetic body. In the world of paleoanthropology, the body is reduced to an object in the same sensory-effacing way that pain is "reduced to a mere grimace." Unlike the surface that is no mere externality but that is undergirded by a tactile-kinesthetic life and in this sense has a living density about it, the visual body of the paleo-anthropologist has insides only in the sense of an objectified anatomy and physiology. As for the surface itself, only what is visible counts—thus, behavior—and the behavioral surface with respect to tactility, for example, is all on the side of the object. Experience in the sense of the felt character of things is consistently discounted. It is ironic, then, that while ancient stone tools are spoken of precisely in terms of tactility—they are *retouched* or not (meaning they have been manually *worked* or not)—language never opens up a vista on the tactile-kinesthetic body. The touching/touched relationship is ignored. "My Body" is ignored. Because the artifactual evidence is never grounded in tactile-kinesthetic experience, the body ends up being consistently reduced to, and treated as an object, indeed, a *cultural* object to the degree that although recognized as having evolved, it is understood only in the reflected light of its products—most notably, its stone tool-making, its cave drawings, its burial practices.[7]

There is actually a compound irony here in that paleo-anthropology—like other cultural disciplines—is itself a cultural object complete with, for example, hunting males and gathering females, and gathering females who being no longer periodically receptive are—in the memorable words of several evolutionary scientists—"continuously copul-

---

[7] This blindered understanding of the body is not unlike the postmodernist's blindered understanding of the body, for the postmodernist too sees the body only in a reflected light—the reflected light of language or of socio-political practices. Thus it too reduces the body to nothing more than a cultural object—a linguistic entity or a cultural construction inscribed with power relations. In both cases, fundamental meanings of animate form and of tactile-kinesthetic experience are overlooked. The body that is coincident with these fundamental meanings is precisely the body that is a semantic template, the body that is not a mere semiotic conveyance but is rather the very *source* of fundamental human concepts, indeed, *hominid* concepts—the concept of power, the concept of a tool, the concept of drawing, the concept of death, the concept of language itself. This body is nowhere to be found in accounts which reduce the body to a cultural object.

able."[8] The striking insight of postmodern thought—that 20th century evolutionary biology in its paleoanthropological understandings and reconstructions has been a cultural construction—unfortunately stopped short of its full potential, and this because the built-in opacity of postmodernist thought with respect to the living body precludes realization of how and in fact *why* evolutionary biology need not be so skewed.[9] An understanding of the human body as first of all a hominid body, and as such engendering pan-hominid invariants by way of animate form and tactile-kinesthetic experience, is the basis for an understanding of the body as pan-cultural universal. Moreover an understanding of the human body as a social body, and as such engendering intercorporeal invariants—again by way of animate form and tactile-kinesthetic experience—is a further basis for an understanding of the body as pan-cultural universal. In other words, evolutionarily speaking, there are corporeal matters of fact to be discovered. The hominid body, of which humans are the latest variation, has in fundamental respects not changed over the past three and a half million years. It has changed styles of living, its brain has grown, and so also has its size, but it is still bipedal; it is still weaponless; its developmental sensory scheme, beginning in tactility, has not changed. What paleoanthropologists fail to recognize is the full import of these corporeal invariants. To do so, they would need to begin asking questions about origins beyond the single, typical question they already ask, and they would furthermore need to take those other questions about origins seriously in a methodological sense. Whereas their typical question—"What was it like?," i.e., what was it like to live two

---

[8] The phrase comes originally from Frank A. Beach "Human Sexuality and Evolution," in *Reproductive Behavior*, edited by William Montagna and William A. Sadler (New York: Plenum Press, 1974), 357. But see also Donald Symons, *The Evolution of Human Sexuality* (New York: Oxford University Press, 1979), 106. For a critical discussion of the characterization, see Maxine Sheets-Johnstone, "Corporeal Archetypes and Power: Preliminary Clarifications and Considerations of Sex," *Hypatia* 7.3 (Summer 1992): 39-76. (The latter article is a version of chapter 3 of *The Roots of Power: Animate Form and Gendered Bodies* (forthcoming from Rowman and Littlefield, 1993).

[9] For a critical analysis of how the living body is put *sous rature* by postmodernism, see Maxine Sheets-Johnstone, "Corporeal Archetypes and Postmodern Theory," a paper delivered at the symposium "Philosophy of Bodymind," American Philosophical Association Pacific Division Meeting, Portland, March 1992. The paper is a version of chapter 4 of *The Roots of Power: Animate Form and Gendered Bodies*.

million years ago as *Homo habilis* or one million years ago as *Homo erectus*—sneaks wistfully into analyses of fossil evidence and not infrequently precipitates speculative scenarios, it could turn into a bona fide quest undergirded by a bona fide methodology that would elucidate the living meanings of animate form and tactile-kinesthetic experience—or in more general terms, that would elucidate what it means to be the bodies we are, and to have been the bodies from which we evolved.[10] What is needed to realize the full import of corporeal invariants is thus a shift in attitude both about animate form and tactile-kinesthetic experience and about method itself.

---

[10] For a discussion of paleoanthropological methodology and its possibilities, see Sheets-Johnstone, chapters 13 and 14 ("Methodology: The Hermeneutical Strand" and "Methodology: The Genetic Phenomenology Strand") in *The Roots of Thinking*.

It is pertinent to point out that paleoanthropologists are not above self-admonishments and -criticisms with respect to engaging in what they commonly call "story-telling," but what one well-known authority more dramatically and derisively called "theatre." See Lord Zuckerman, "Closing Remarks to Symposium," in *The Concepts of Human Evolution*, edited by Lord Zuckerman, *Symposia of the Zoological Society of London* 33 (New York: Academic Press, 1973), 451.

It is pertinent to point out too that a lack of recognition of the experiential dimensions of hominid corporeal invariants and a concentration instead on the body as mere featured surface can egregiously skew interpretations of the fossil evidence. One need only consider the status of Neandertals. Until recent times, when multicultural awareness and pluralism have become *de rigueur* and new theories have begun to upset the long protected and privileged applecart of *Homo sapiens sapiens*, Neandertals were paleoanthropological outcasts. With their prognathous features, strongly recessive chins, prominent brow ridges, and bulky frames, they were not appealing creatures, at least not to most white European male evolutionary scientists. No matter that their cranial capacity was larger than ours—that fact was either brushed quickly aside or explained away—and no matter that they buried their dead—not only the first such known practice, but a practice carried out in extraordinary ways that necessarily signify a concept of caring as well as death—they were simply not comely. In view of the facts and non-facts of the matter, it is difficult not to interpret their long disinheritance as merely a felt repugnance: "We don't want to be related to *them*!" The abhorrence is similar to the reaction of people in Darwin's time who recoiled from the thought of being related to apes. One hundred-thirty and more years later, some people are still fussy. For discussions of recent re-evaluations of Neandertals, see Bruce Bower, "New Evidence Ages Modern Europeans," *Science News* 136.25 (16 December 1989), 388; Bower, "Tracking Neanderthal Hunters," *Science News* 138.15 (13 October 1990), 235; Bower, "Neandertals' Disappearing Act," *Science News* 139.23 (8 June 1991), 360-361, 363. For an early discussion of the facts and non-facts of the matter, see C. L. Brace and M. F. Ashley Montagu, *Man's Evolution* (New York: Macmillan, 1965).

Through a sensory-kinetic examination of hominid stone tool-making, I will exemplify what that shift in attitude—and thus what a bona fide methodology—would provide in the way of knowledge.

II

Hominid teeth that mash and grind food have a specific tactile character; so too do hominid teeth that bite and scrape. If you run your tongue along the occlusal surface of your upper teeth starting at your molars and progress toward your front teeth (and I hope you, the reader, will do this, and several times over), you will discover a distinct tactile change: an irregular, bumpy surface ends and an even, thin edge begins. In very brief terms, a thick, grooved, and discontinuous array of edges gives way to a thin, even, single edge. Although not spoken of in biology as major hominid tooth forms, there is no doubt but that molars and incisors constitute fundamental hominid dental types. Actual experience shows further that these two dental types are connected with two basic kinds of eating acts: mashing and biting.

Consider now that there are two major forms of early stone tools recognized by paleoanthropologists: core tools and flake tools. Core tools are relatively thick pieces of stone. As testimonial to that thickness, they are usually held not between fingers and thumb but up against the palm of the hand. They have several protruding edges that commonly stand out in relief in the same way that the edges of molars stand out in relief: a relative jaggedness with respect to functional surface is typical of each. Flake tools are in contrast relatively thin pieces of stone. As testimonial to their thinness, flake tools are commonly pinched between finger(s) and thumb—like a razor blade. A flake tool has a single manually traceable edge, and unlike a core tool, it has more readily distinguishable sides as well as a single pronounced edge. Pinched between finger(s) and thumb, it is a lengthier, more vertically aligned object than the more squat and thick core tool. Its surfaces are moreover relatively flat like an incisor rather than rounded and irregular like a molar.

The analogy between the two major tooth forms and the two major tool forms is obvious once the tactile-kinesthetic body is recognized. The ground of the analogy is palpably evident in the tactile experience of the occlusal surface of teeth, of what is called the "dental arcade." As one well-known paleoanthropologist described the arcade—*but quite apart from any reference to tactile-kinesthetic inspection and quite apart from any reference to stone tools*—"a continuous anterior cutting blade" replaces a slicing and grinding instrument; in other words, starting at the back teeth as originally suggested, one readily observes that molars give way to incisors.[11]

Now edges either stand out in relief, as with molars and core tools, or they define contour, as with incisors and flake tools. In either case, however, whatever has an edge has power. Understandably, then, the critical character of a stone tool in the beginning was not that it have a definite, set shape but that it "have an edge." The beginnings of stone tool-making were in consequence not a matter of flaking stones in specific ways; rather, early hominids took whatever edges resulted from flaking. A concern with shape becomes evident at a later stage, namely, in Acheulian handaxes. (Acheulian handaxes were general purpose tools that were widely used in Africa, Southwest Asia, and Europe.) Where edges are not created as surface properties of an object but a single all-over edge dominates, the surfaces of the object are likely to be perceived as sides, and this because the same edge that defines simple contour *ipso facto* defines sides. Our hands are a paradigmatic instance of this relationship. In fact, where sides are not *accidents* of flaking but are created in their own right, the stone has been shaped by turning it over, the knapper working the edge first from what will ultimately be one face, then what will ultimately be the other face, then from the first, and so on. Rather than jagged edges, "evenly trimmed sides" result.[12] In effect, a three-dimensional act eventuates in a two-sided object. As with the Acheulian handaxes, the visual character of such a tool is prominent; shape in the form of an all-over contoured edge is a strikingly notable feature. Just so with our hands whose shape is perceived as an all-over contour and whose sides are readily in evidence by turning them over. Furthermore, just as the *entire* stone is a tool, so our entire hand is a tool as in gripping and striking. Analogical thinking is thus again clearly evident.[13] Acheulian handaxes are conceptually linked with hands in the

---

[11] David Pilbeam, *The Ascent of Man* (New York: Macmillan, 1972), 59.

[12] Andre Leroi-Gourhan, *Prehistoric Man* (New York: Philosophical Library, 1957), 66.

[13] There is a single recent reference in the literature—by British anthropologist K. P. Oakley—to a correspondence between Acheulian handaxes and hominid hands. The correspondence was originally suggested by German archaeologist R. R. Schmidt (See his superimposed drawing of a hand on the outline of an Acheulian handaxe in his *The Dawn of the Human Mind* [London: Sidgwick & Jackson, 1936],

same way that the earlier Oldowan tools are conceptually linked with teeth. Not that the tactile-kinesthetic character of stone tools disappeared in the course of the development of the handaxes. On the contrary, it was elaborated in visual terms as is evident from the fact that the contour of an Acheulian handaxe can be followed not only *manually*, but *visually*, that is, as a linear form. This shift toward rectilinearity added a new dimension to the tactile-kinesthetic foundations of stone tool-making. There are several points to be made in conjunction with the innovation.

To begin with and as indicated earlier, a stone made into a tool is first of all transformed by touch. The stone is given a new tactile character, one whose power is tested not by *looking* but by *feeling* along the edge created by flaking. Second, to understand the passage from an essentially tactile-kinesthetic object to an essentially tactile-kinesthetic-visual one is to understand the way in which an edge, while losing nothing of its original tactile character, comes to be seen as a line. Where vision and tactility are confused rather than understood, an understanding of sensory differences is compromised. Lines are visual translations of tactile contour. They are a semantic advance, an advance in meaning. An Acheulian handaxe is not simply the result of more refined hand-eye coordinations, as some paleoanthropologists claim;[14] it is the result of a sensory-kinetic development in perceptual meaning, a transfer of *sense* in the double sense of that term. Consideration of straight edges clarifies the nature of that transfer and exemplifies a third distinctive feature of stone tool-making as an evolving art. Straight edges are paradigmatic of synaesthesia, straightness being a visual datum, edges being a tactile one. Tactility, in other words, determines the evenness of an edge, not its straightness, as any few moments with a blindfold will attest. Thus, with respect to straight edges, vision appropriates what is originally a tactile datum and makes it its own. In the process, a new

---

96-97.) Oakley interestingly comments that "a bifacial hand axe was perhaps subconsciously visualized as representing a third hand, a hand that unlike the flesh-and-blood original had the capability to cut and skin the carcasses of the animals scavenged or hunted." He speaks of this representation as "symbolic thought," but offers no further analysis. "Emergence of Higher Thought 3.0-0.2 Ma B.P." *Philosophical Transactions of the Royal Society of London B*, Biological Series, Vol. 292 (8 May 1981): 205-211; quote from p. 208.

[14] See, for example, Milford Wolpoff, *Paleoanthropology* (New York: Alfred A. Knopf, 1980), 186.

meaning is forged. Straight edges produce straight lines, as both Euclidean constructions and projective sightings presuppose.

Viewing the evolution of stone tools from a sensory-kinetic perspective provides understandings of the way in which a hominid sensorium functions. It furthermore provides understandings of analogical thinking, the roots of which lie in a gnostic tactility-kinesthesia—"gnostic" in the original etymological sense of "knowing." Analogical thinking is both basic to hominid thinking and basically corporeal.

III

The tactile-kinesthetic invariants I have described are pan-hominid corporeal invariants: what we each separately discover in our individual mouths is essentially the same. I would like now to consider how a pan-hominid corporeal invariant is linked to an *inter*corporeal invariant, or in both finer and broader terms, how fundamental tactile-kinesthetic experiences, being the ground of fundamental visual experiences, are in turn the ground of fundamental social experiences. I will in these terms progressively elucidate how a fundamental intercorporeal semantics informs our social lives. To begin with, I will describe a corporeal archetype within that semantics, perhaps the most basic corporeal archetype insofar as it both anchors and intensifies our sense of ourselves and anchors and intensifies our sense of others.

When we close our eyes, another world comes to the fore, and in a regular cycle every day of our lives. Though illuminated from time to time by flashes or dots of light, by images, and by dreams, this world is typically described as quintessentially dark. There is more in the experience of sightless eyes however than a quintessential darkness. When we close our eyes, we exit one sensory world and enter another. With the closing off of vision, a clear-cut boundary is established between an outer world and an inner world—or in more precise sensory-kinetic terms, between a *seen* world and the *felt* tactile-kinesthetic world of my body. The purely tactile boundary felt between myself and the outside world is in fact much less clear. Indeed, tactile boundaries between ourselves and what we touch are vague and in a way we cannot clarify. Moreover we cannot perform any tactile act which would nullify our surface tactility and thus possibly intensify sight in a way commensurate with the way a lack of vision intensifies the whole of our tactile-kinesthetic body.[15] We cannot either bracket our tactile-kinesthetic body. We cannot de-actualize its presence. We are always, in Valery's words, "my body." No matter that we pass over or ignore our actual tactile-kinesthetic connections with things in the course of driving, eating, writing, walking, listening, or sitting; we are in perpetual contact with the world. In motion or at rest, being in touch is central to our aliveness. Indeed, as Aristotle long ago noted, "touch . . . is the essential mark of life"; without it, "it is impossible for an animal to be."[16] What a lack of vision does in Aristotelian terms is illuminate the heart of our soul. In closing our eyes, we become aware of sightlessness as entrance to the primordial tactile-kinesthetic world which is "my body."

I would like to ask you, the reader, to close your eyes and open your eyes, alternating between the two acts and taking time in each case to experience what is there. I ask you to do this in order to verify by your own experience the brief description I will offer.

* * * * *

Eyes are mystic circles, mystic not in an occult sense but in the sense of generating wonder, even of inspiring profound awe. Open them, and a dazzling, bustling world is present. Close them, and an opaque and dense but sparsely-populated landscape appears. Open them, and awareness not only meets with an expanse of objects but moves freely within it, springing from one focal node to another. Close them, and the field of possible attention contracts; wandering randomly within an unmarked terrain, attention illuminates only the place it happens to be. Open them, and the tactile-kinesthetic character of one's roving, active glance is hardly felt. Close them, and one's eyes are transformed into a tactile-kinesthetic playground of sensations—pressures, pinchings, flutterings, squintings: sightless eyes caught short of an object.

Eyes are mystic circles that open on otherness and open on inwardness. They are windows onto two worlds. To the degree they stay open

---

[15] Of course I can sleep, and in sleeping, dream. In this sense, I can nullify my tactile-kinesthetic body and intensify vision. The "I can" here is, however, illusory since any *actual* powers to enact sleep or dreams are fictional. The acts come—or they do not come—by themselves.

[16] Aristotle *On the Soul* 435b 16-17.

on inwardness, the initial darkness begins to dissipate. We begin seeing into the darkness. The space within becomes luminous. What was initially dark is no longer dark. It is as if the light that is normally cast upon the open eye continues inward to the point that the eye no longer either searches for one of its customary objects or illuminates afterimages. The flutterings of the eye subside. The light within is felt rather than seen.

* * * * *

When our open eyes meet the open eyes of another, the experience of eyes as mystic circles is potentially as great. This is because the eyes of others are the locus of their mystery as persons. We know in a thoroughly intuitive way that we have the possibility of experiencing them as entrance to a tactile-kinesthetic world similar to our own, a world each of us calls "my body." What is the nature of this possible experience of another? Again, I ask you to verify a brief descriptive analysis, this time by actually meeting the eyes of another person with your own.

* * * * *

The eye of the other is a circle, the *first* circle we experience. We endow the visual—the circle that we see—with a tactile-kinesthetic reality. We meet the eyes of another and immediately apprehend a rounded form leading to that inside space which is another person. Our own sense of inwardness that is immediate upon closing our eyes is the ground upon which the inwardness of the other is appresented. Correlatively, we see reflected in that same experience the nature of our own bodily being. Transferring onto our own body the visual form that we perceive before us, we apperceive the circular form as part of ourselves. We thus endow the tactile-kinesthetic with a visual reality. We appropriate *circles* as part of our own bodily being, circles not as mere geometric forms but as bodily forms resonant with the mystery of life. Our eyes, like the eyes of others, are apperceived as *circles* of light, *circles* leading to insides.

Of course the eyes we see can also be in and of themselves semantically potent. The eyes of another can stop us in their tracks; they can pierce us; they can magnetize us; they can invite us. They can do all of these things because the mystic circle is itself alive. Indeed, that it *moves* and has *tactile values* is part of its mystery. It constricts, glistens, hardens, widens, turns vacant, flits about. Though I do not actually see the circles that are *my* eyes widen in curiosity, grow vacant in boredom, flit in discomfort, constrict in fright, glisten in rapture, harden in anger, I know the *feel* of all of these mystic circles. I know their dynamics. I have a felt bodily sense of their meaning. Thus, when I see the mystic circles which are the eyes of another, I intuitively recognize their expressive character.

When I go beyond the intuited qualitative meaning of the mystic circles—their curiosity, apprehension, or boredom, for example—toward the density which is the full bodily presence of the other, I allow my own seeing eyes a greater space of vision beyond me. To the degree that I begin to fathom the mysterious interior of the other, I let go of my own inwardness. I cannot, after all, be in two places at once: either I stay centered inside the circles of my own being or I expand the boundaries of those circles toward the other. To the degree I enter into the mystic circle of the other, there is an ebbing of my own felt bodily sense and a growing sense of the mystic inwardness of the other, an inwardness that remains dark, that is not illuminated by momentary flashes of light or by images, but that is pregnant with the rich, interminable, awesome density of another being.

*   *   *   *   *

Eyes are organs of sight; but they are also organs of social relationship. They are the privileged site of our contact with others. Fundamental aspects of our intercorporeal semantics are rooted in just such aspects of ourselves, in our being the bodies we are. Because we tend to forget that an intersubjectivity is first and foremost an intercorporeality, we tend to forget that meanings are articulated by living bodies. Common linguistic and conceptual focus is in fact wrongly placed: an intersubjectivity is more properly conceived and labelled an intercorporeality. We are there for each other first of all in the flesh. An understanding of intercorporeal archetypal meanings demands that we recognize this fact. It demands secondly that we recognize the seenness of each other and the communal somatic verities that go with that seenness. Third, it demands that we recognize not merely what we *do* as forms of life, but recognize ourselves as a form of life.

With further respect to an intercorporeal semantics, we tend to forget that the perceived world is already alive with significations, that it is not dead and inert until we christen things with names, or indeed, as if we christen things into *being* by giving them a name. As the previous experiences I hope show, animate form is already meaningful. Our own eyes are meaningful in and of themselves. They are archetypally meaningful as windows onto two worlds, as centers of light and dark, as entrances to a tactile-kinesthetic world, our own and that of others. They are archetypally meaningful as circles, as I will presently show in a more developed manner. These meanings can be and are reworked—amplified or suppressed—in diverse metaphysical and epistemological ways from culture to culture. Rather than directly setting forth these ways, I will attempt to demonstrate archetypal meanings of eyes in the context of actual disciplinary practices, including the disciplinary practice of philosphy. I will do this by pursuing the notion of animate form—first with reference to its general semantic import, and then with specific reference to its morphological/visual import.

IV

Paleoanthropology (not to mention other disciplines closer to home—if not home itself) misses the semantic dimension of animate form because it fails to recognize the corporeality and intercorporeality of life as something other than mere anatomy on the one hand and mere behavior on the other, and fails as well to take seriously the actual ways in which anatomy is destiny.[17] Not dissimilar oversights are apparent in experimental primatological studies and in studies in the philosophy of mind. The consistent problem is first, that bodies—animate forms—are not acknowledged and understood and second, that descriptive analyses of what is actually there are passed over in favor of explanatory hypotheses of what is there. With respect to the latter problem—and to gloss on a comment of Joseph Campbell—"If you haven't had the experience, how can you explain what is going on?"[18] With respect to the first problem,

---

[17] Freud is reputedly the source of this notion. It is ironic then that, credited with such a rich insight, he actually left the body behind and unattended: he developed the idea that anatomy is destiny only in terms of a single bodily organ. Indeed, he never mined his initial insight that, in his own words, "The ego is first and foremost a bodily ego." (Standard Edition XIX, translated by James Strachey [London: Hogarth Press, 1955], 26. The phrase is repeated on p. 27.) The ontogenetical *corporeal* psychoanalytic ego is phenomenologically related to the phylogenetic heritage of the body as semantic template.

[18] Joseph Campbell, *The Power of Myth* (New York: Doubleday, 1988), 61: "If you haven't had the experience, how can you know what it is?"

it is clear not only that phenomenological studies are needed, but that what Husserl described in the name of "psychophysical organism" is directly related to descriptive renderings of animate form. Husserl himself consistently emphasizes the *animateness* of bodies, nonhuman as well as human, and even including works or products fashioned by bodies: they too are psycho-physical unities that "have their physical and their spiritual aspects, they are physical things that are 'animated'."[19]

Now to realize this psychophysical unity is to realize that sense-making is a built-in feature of animate life. Making sense to others and making sense of others is, in the most fundamental sense, in our bones—in archetypal forms of sensing and moving. We are indeed *rational animals*: we make sense *of* our bodies—and the bodies of other creatures—and we make sense *with* our bodies. Fundamental human beliefs and practices are testimonial to this double form of sense-making. But *all* creatures are rational in this sense; they all make sense *of* their bodies and *with* their bodies, in manifold ways and to radically different degrees. They communicate with each other—not only social primates or mammals generally, for example, but social insects such as ants and bees. They care for themselves—by licking, by scratching, by preening, and by other bodily acts. They furthermore both sense and understand their own bodies and those of others: "If I do this, then I can reach the fruit"; "If I give chase, the other will run." In particular, *any creature that must learn to move itself discovers*—in the deepest Husserlian sense—*its own kinestheses*. From a paleontological viewpoint, we could indeed ask how otherwise such animals, including we humans, could possibly have evolved. Sense-making is a corporeal fact of life. Our own intercorporeal seenness is part and parcel of this corporeal fact. Social animals are "born to see and bound to behold" each other as animate forms.[20] More than this, creatures from butterflies to mountain sheep, from coral fish to humans are patterned in morphological ways that correlate with eyes that behold. Were paleoanthropologists to balance renditions of behavior with

---

[19] Edmund Husserl, *Ideas Pertaining to a Pure Phenomenology and to a Phenomenological Philosophy, Second Book*, translated by R. Rojcewicz and A. Schuwer (Dordrecht: Kluwer Academic: 1989), 333.

[20] I borrow the phrase from Erwin W. Straus. ("Born to See, Bound to Behold: Reflections on the Function of Upright Posture in the Esthetic Attitude," in *The Philosophy of the Body*, edited by Stuart F. Spicker [Chicago: Quadrangle Books, 1970], 334-361.)

renditions of experience, that is, were they to eschew mere explanatory accounts of behavior in favor of both descriptive and explanatory renderings of experience, they would have the possibility of uncovering those fundamental morphological/visual relationships that are the corporeal foundation of our intercorporeality. In both broader and summary terms, attentiveness to what is actually there in corporeal experience would afford them the possibility of discovering those fundamental ties that bind animate form to animate sensibilities and that in so doing entwine creatures in an intercommunal life.

Though not described in quite such terms, a remarkable description of the intercorporeal import of morphological/visual relationships is presented by Swiss biologist Adolph Portmann in his book *Animal Forms and Patterns*.[21] Of particular moment is Portmann's description of *inwardness* as a basic biological character, a character common to all creatures who see one another and who mutually express what he calls "psychical processes," that is, moods and feelings by way of postures, colorations, and other bodily markers. Marjorie Grene, in an article on Portmann's biology, translated Portmann's phrase for inwardness—literally, "relation to the environment through inwardness"—by the word "centricity." "'Inwardness' alone" she said, "is too exclusively subjective and fails to convey the *relatedness* that Portmann's concept entails."[22] Because important difficulties as well as insights are to be found in Grene's essay, I will briefly indicate them and consider their adverse effect upon an understanding of corporeal archetypes and an intercorporeal semantics. I hope in this way to exemplify the hazards as well as the benefits of reflecting philosophically on a non-philosophical discipline and to show that philosophical reflections must themselves be validated.

Grene notes that Portmann's concept acknowledges a quality of life that is itself acknowledged only "at the boundary of science,"[23] that is, it acknowledges a *subject* of existence. Her general point in the beginning is that "Portmann's reflections about living things cannot be contained within the frame of Galilean science."[24] In her initial discussion of

---

[21] *Animal Forms and Patterns*, translated by Hella Czech (New York: Schocken Books, 1967).

[22] Marjorie Grene, "The Characters of Living Things, I," in *The Understanding of Nature: Essays in the Philosophy of Biology* (Dordrecht: D. Reidel, 1974), 272-73.

[23] *Ibid.*, 273.

[24] *Ibid.*, 275.

centricity, she considers a possibly strong objection to Portmann's thesis, namely, that while we humans experience centricity, and unquestionably so, how can Portmann extrapolate on the basis of our experience "to the whole living world?"[25] In answering to this objection, she glosses Portmann's concept in three ways. She first states that Portmann's concept, while embracing a notion of consciousness, is not claiming that consciousness may be predicated of *all* creatures, but rather that it is "one style . . . of centricity," and that "sentience" may be more generally extended to all animals.[26] She then states that sentience is "the inner expression of centricity as such" and that in "reaching out toward an environment and taking in from it," a creature is a dynamic *individual*—a subject—thus altogether different from something inorganic.[27] Finally, she states that in claiming centricity is a basic character of living things, she, like Portmann, is not "extrapolating . . . to a vast and remote past, but only trying to pin down with a fitting phrase a description of a common quality of our present experienced world." In other words, "Portmann's extrapolation" she says, "is descriptive and contemporary, rather than explanatory of an inaccessible past."[28]

Now in conjunction with the rather odd notion that Portmann's biological descriptions have no historical significance, Portmann merely describing what is before us here and now, Grene makes the further peculiar, even cryptic, comment that in *describing* "the living forms we see before us and around us here and now," that we at the same time "try in imagination to lessen the intensity of centricity in its aspect of inwardness."[29] What is queer is to find both a denial of historical interest in what is actually before us and an attempt to minimize it. Indeed, the denial and the attempt together appear backward steps among the forward ones that Portmann seems to be taking; they are a perplexing refusal of what Portmann as an evolutionary biologist would seem clearly to be affirming. What is at stake such that it is necessary to rein in experience in this way? Further, why is an attempt to lessen what we experience tied to the attempt to detach ourselves from the past? A

---

[25] *Ibid.* 274.

[26] *Ibid.*

[27] *Ibid.*

[28] *Ibid.*, 275.

[29] *Ibid.*

reticence to extrapolate, most particularly a reticence to aver something of those hominids who were our ancestors, who invented stone tools, who conceived of death, who conceived of themselves as sound-makers, who, by artifactual evidence, likely conceived of numbers,[30] who drew replicas of animals and other artistic forms on the walls of caves, and who, to begin with, began walking in a consistently bipedal manner, thus radically changing the morphological/visual relationship of their social bodies,[31] is a reticence to look ourselves in the eye. Either this or it means looking ourselves in the eye and seeing only gene pools. If we consider what is of central importance in paleoanthropology—that all humans are hominids but not all hominids are human—there is no justification for hedging with respect to extrapolation. Taking evolution seriously means taking our own historical past seriously, to the point that we realize that, short of avouching the truth of creationist doctrine, we humans did not arrive here *deus ex machina*; short of avouching the truth of related received wisdom, we humans did not invent all the cognitive wheels on which we run; short of avouching the truth of postmodern theory, we humans are not cultural artifacts. What is of moment beyond this historical obligation to embrace and comprehend our paleoanthropological past is that to temper what is actually there by trying to lessen its actual experienced impact is to invalidate the description. Indeed, *why not understand how the experience comes to be what it is by unbuilding inwardness and attempting to fathom its origin and impact*? Most importantly too, if *relatedness* is of seminal significance in Portmann's concept of inwardness, then certainly it is of moment in terms of *any* eyes that have the power to recognize inwardness in others, and this because *any* eyes that, in their relatedness to what is about them, can see centricity in others—and

---

[30] Ashley Montagu's remark with respect to a two-sided Acheulian handaxe is noteworthy. He states that "It is clear that each flake has been removed in order to produce the cutting edges and point of the tool with the minimum number of strokes; for if one examines this tool carefully, one may readily perceive that no more flakes have been removed than were minimally necessary to produce the desired result." "Toolmaking, Hunting, and the Origin of Language," in *Origins and Evolution of Language and Speech*, edited by Stevan R. Harnad, Horst D. Steklis, and Jane B. Lancaster, *Annals of the New York Academy of Sciences* 280 (1976), 271. For a phenomenological analysis of the origin of counting, see Sheets-Johnstone, *The Roots of Thinking*, chapter 3.

[31] For a discussion of the import of these radical changes, see Sheets-Johnstone, *The Roots of Thinking*, particularly chapters 3, 4, 5, and 7.

Portmann's examples are many—cannot be denied their intercorporeal understandings.

When we look at Portmann's work itself, we see that he emphasizes over and over again both visual powers and the morphological characters—what he calls the "organs"—that go with them. These "organs" may be patterns such as ocelli, they may be appendages, or, as suggested earlier, they may be postures or colorations, but regardless of their specific character, all of them act upon the eye of the beholder; they are "organs of social relationship."[32] Innumerable instances in the literature on nonhuman primates and other nonhuman social animals support Portmann's insight; they clearly suggest intercorporeal experiences of inwardness, recognitions of another as an Other. The experiences are implicit in instances in which one creature understands what it is to be seen by another, for example, or what it is to be fixed by the gaze of another.[33] On the one hand, these recognitions underscore the fact that there are nonhuman animals who understand, just as we do, that vision itself has power, and indeed, that, in a Foucauldian sense, on the other side of any optics of power, there is *the power of optics* itself.[34] On the other hand, they indicate that there are nonhuman animals who understand, just as we do, that morphological aspects of animate form are expressive. Surely, then, given these understandings of morphology and vision, it is no great leap to acknowledge that in the world of animate form, *morphological organs may be eyes themselves.* Indeed, what are the eyes of another for my beholding eye if not mystic circles leading to a body I do not feel but whose depths I can fathom as a density of being? Eyes are indeed morphological organs in exactly the sense Portmann specifies. They are doors opening onto an experience in which inwardness is adumbrated, an experience that is grounded in the sensory-

---

[32] Portmann, *Animal Forms and Patterns*, 197. It is important to note what Portmann himself emphasizes, namely, that his concentration on *visual* form should not make us forgetful of "how great the social importance of stimuli of touch may also be in animals; nor how powerfully scents and sounds may act on ourselves as well as on animals." *Animal Forms and Patterns*, 185.

[33] Perhaps the most immediately telling examples are those in which one animal deceives another by enacting a behavior within the species's normal repertoire but for other than "normal" purposes. See A. Whiten and R. W. Byrne, "Tactical Deception in Primates," *Behavioral and Brain Sciences* 11 (1988): 233-273.

[34] For a discussion of this relationship, see Sheets-Johnstone, *The Roots of Power: Animate Form and Gendered Bodies*, especially Chapter 1.

kinetic nature and potentialities of vision itself. In precisely this sense, they are archetypal aspects of animate form. While we might have a shallow and momentary glimpse of the power of vision and its capacity to lead us to an experience of inwardness as we merely *watch* an animal carry on its activities—buzzing from flower to flower, building a dam, chasing a ball—that faint and fleeting experience of the power of vision is apprehended—arrested, as it were—when we actually meet the eyes of another animal with our own and dwell in those mystic circles which speak to us of the being of another. Inwardness is reflected back to us by those archetypal organs we call eyes, eyes that are not simply receptor organs but morphological aspects of animate form.

The eye indeed is a mystic circle. To put this biological fact of experience in much closer historical perspective, and to flesh it out further in the direction of a phenomenologically-informed philosophical anthropology, I would like to consider the eye as it has been cross-culturally understood at two extremes: the mystic circle which is the evil eye and the mystic circle which is the reverential or sacred eye. The former can be traced back to pre-Semitic Sumerian cuneiform texts.[35] The latter has been symbolized cross-culturally for millennia and is readily exemplified by the mandala.[36] In fact, I have space here only to consider

---

[35] It can also be traced back to the Book of Proverbs in which one reads: "Eat thou not the bread of him that hath an evil eye." (Proverb 23.6)

[36] Although in general, and as mandala scholars Jose and Miriam Arguelles point out, "literature concerning the Mandala is not extensive," cross-cultural evidence demonstrating the universality of the mandala is not lacking. (Jose and Miriam Arguelles, *Mandala* [Berkeley: Shambala, 1972], 20.) Mandalas are circular drawings common to Navajo Indians, for example, as well as to Buddhists. Moreover the Aztec stone calendar was drawn in the form of a mandala. In addition, there are ancient architectural constructions that have a notably circular form. Stonehenge is a well-known example. The rounded barrows believed to have been constructed by King Sil (or Zel) in England during the Bronze Age are further cases in point. With respect to these burial or treasure barrows, the "Great Round" that is Silbury Hill is an extraordinary formation. (Regarding "The Great Round," see Erich Neumann, *The Great Mother*, translated by Ralph Manheim [New York: Pantheon Books, 1955], Bollingen Series, Vol. 47). The significance of *its* rounded form, as Michael Dames describes it, has striking parallels with the psychocosmological significance of mandalas as explained in the present text. (Michael Dames *The Silbury Treasure* [London: Thames and Hudson, 1976]. Clearly what *is* lacking is not cross-cultural evidence demonstrating the universality of the mandala but a phenomenologically worked out concept of the mandala. The Arguelles's say as much when they write that "most of [the literature] deals with the Mandala as a sacred art form of the Orient, and although some thinkers—such as Eliade and

the mandala, and that briefly. I leave the evil eye for a future time.[37]

V

Jung tells us that the Sanskrit word "mandala" means circle and that where used in rituals, it is "an instrument of contemplation," an instrument that "is meant to aid concentration by narrowing down the psychic field of vision and restricting it to the centre." He goes on to say that the circles which describe the mandala "are meant to shut out the outside and hold the inside together."[38] The initial question is, how did mandalas originate? Where did the *concept* of a mandala come from? If we look to the body as a semantic template, then a spatio-teleological similarity is immediately apparent. The mystic circle that is the mandala is spatially patterned on that original mystic circle that is the eye. The teleological correspondence is borne out in the fact that the eye is both the original "instrument of contemplation" and the instrument par excellence that aids our concentration by focusing our attention. In contemplation, the eye shuts out the outside and holds the inside together—by literally closing itself, for example, by intentionally avoiding other eyes or other objects, or by glazing itself, thus keeping the world of things at bay. Mandalas are circular forms morphologically and teleologically modeled on the image of eyes.

The similarity is thus not a mere surface similarity. Moreover the mandala is generally interpreted as a symbol of the self and of the cosmos. It is in fact an aesthetic instantiation of *the correspondence* between self and world, microcosm and macrocosm. It is, in effect, a "psychocosmogram."[39] To understand how the circular form that is the

---

Jung—have related the Mandala to other cultures and traditions, no one has developed a concept of its universality to any extent." As the present paper will show, what is needed is a phenomenological analysis that recognizes and elucidates the psychophysical unity of the mandala.

[37] An initial analysis of the evil eye was given as part of a paper presented at a panel session titled "The Corporeal Turn," Society for Phenomenology and Existential Philosophy, Boston, October 1992.

[38] Carl G. Jung, *The Archetypes and the Collective Unconscious*, 2nd ed., translated by R. F. C. Hull, Bollingen Series XX (Princeton: Princeton University Press, 1968), 356.

[39] Giuseppe Tucci, *The Theory and Practice of the Mandala*, translated by Alan H. Brodrick (London: Rider & Company, 1961), 25.

mandala and that is patterned on the eye came to stand for the self and the world is first of all to experience eyes as channels through which a dual light is cast, that is, as windows on two worlds, as openings on a luminosity both within and without. What needs clarification is how those dual experiential possibilities of eye-light take on the specific symbolic import that they do in the mandala.

That the language in which the power of the eye is described is consistently anchored in images of light is readily tied to the original experience of the eye as an organ of light. From the fact that I open my eyes and it is light comes the possibility of my being enlightened. The archetypal power of the eye to inform and edify is thus linked to clarification, to elucidation, and so on. That the metaphoric language in which the powers of the mandala are described is also consistently anchored in images of light is an indication of a further symbolic extension from original experience. This further symbolic structuring is evident in the equation of light with consciousness. There is, in other words, a cognate relationship among light, consciousness, mandala, and eye, eye being the root form. Thus, the language of consciousness is also the language of light. Furthermore, like the center of the eye that reflects light and like a luminous consciousness that is a center of light, so also the center of the mandala radiates light. Of striking interest in this context is Husserl's language in describing the Ego. "It is the *center*," he writes, "whence all conscious life *emits rays and receives them*" (italics added).[40] It is emphatically clear from Husserl's descriptive analyses that Objects as well as the Ego radiate light, and that the epistemological relationship between Ego and Object is similarly structured in images of light. Husserl speaks, for example, of "two-fold radiations, running ahead and running back: from the center outward, through the acts toward their Objects, and again returning rays, coming from the Objects back toward the center in manifold changing phenomenological characters."[41] Over and over again, the original source of those two-fold radiations is described as a center, a center that Husserl says is analogous to the body as center of all sensory awareness.[42]

Now before continuing with the descriptive analysis of mandalas as a symbolic extension of the archetypal power of eyes, it is apposite at this

---

[40] *Ideas II*, 112.

[41] *Ibid.*

[42] *Ibid.*

juncture to raise some critical questions. Do we really experience rays, as Husserl says? Do we really experience the Ego, and as a center which emits and receives rays? If we do not, then we may ask, why use this language? or perhaps better, why are we drawn to using this language? I would answer precisely in terms of the fundamental *but unelucidated* relationship of eyes to light and to consciousness. The archetypal power of eyes to shed light, to lead us to self- and other-understandings, is not the power of receptor organs. It is in fact a power realized in bracketing, for it is a power to see into the constitutive form or essence of a thing. Precisely insofar as it is structured in light, the relationship Husserl describes between Ego and Object is akin to seeing into the dark, to grasping the quintessential nature of things, to grasping "inwardness." Husserl's descriptive languaging of the experience of understanding another individual further exemplifies the correspondence. Of comprehending another, he writes, for example, "[I] look into his depths"; "I see deeply into his motivations."[43] What Husserl is doing in his phenomenological analyses of the Ego and the Object is elucidating the light, that is, elucidating the archetypal power of eyes to illuminate the dark. In broader terms, the quest of phenomenology is to throw light on the light—to know knowing, to understand understanding. What this suggests is not only that the very idea of phenomenology is grounded in the experience of inwardness, in the experience of illuminating the dark, *but that the original of that experience has not yet been brought to light and described*. In other words, while the aim of phenomenology has been to arrive at and elucidate inwardness, it has done so without elucidating that original archetypal experience by which experiences of oneself and of the world are structured in images of light. In still other words, phenomenology itself is grounded in bodily experience. Though genetically unilluminated, inwardness is the eidos of the whole of Husserl's phenomenological undertaking.

Together with phenomenology itself, Husserl's languaging both of the relationship of Ego and Object and of the Ego itself as a center of light—a center of "conscious life"—sheds considerable light on why a mandala is a basically circular spatial form that cross-culturally symbolizes

---

43 *Ibid.*, 310, 341.

the psyche.[44] Circular forms made in the image of the eye circumscribe a symbolic Ego in terms of inwardness. But they also circumscribe within the whole of their compass a privileged place: *a center*, a unique point from which light emanates. Of the fact that mystics place themselves at the center of a mandala, a recognized authority on Indian mandalas writes that "Man places in the centre of himself the recondite principle of life, the divine seed, the mysterious essence. He has the vague intuition of a light that burns within him and which spreads out and is diffused. In this light his whole personality is concentrated and it develops around that light."[45] If man *places* the principle of life or mysterious essence at the center of his being, however, the principle or essence cannot have been *found* to be there, that is, it cannot have been actually experienced. On the other hand, if man *feels* a light within himself, a light that, while concentrated, diffuses itself throughout his being, animate essence must be a corporeally experienced fact of life. On this account, a centering of oneself inside the circle of a mandala is not the result of a seemingly gratuitous act. It is a symbolic elaboration of a bona fide felt experience, the experience of eyes as openings onto a world in which light is felt rather than seen. This experience and the experience of eyes as circles leading to inwardness together appear closely related to the dual dimensions Husserl singles out as descriptive of spirit or animate presence: "spirits are the subjects that accomplish *cogitationes*"; spirit is

---

[44] See Jung, *The Archetypes and the Collective Unconscious*, and Campbell, *The Power of Myth*, for drawings and graphic incorporations of mandalas. In light of the evidence—the drawings and the graphic incorporations—and of the extraordinary cognate relationships outlined in the present paper, it is puzzling to find analyses of the corporeal origin of the mandala lacking and indeed to find the question of why the mandala is first and foremost a circle rather than a square or a cone, for example, entirely omitted. A pervasive cultural inattention to the body and to bodily experience would seem to explain the omissions. Tucci, for example, casts experience in the role of follower rather than leader in the generation of the concept of a mandala. He speaks of the mandala as a geometric projection of the world, and though he explicitly states that he is not concerned with its origin, he nevertheless emphasizes its "worldly" genesis, i.e., the mandala is a pictorial representation of cosmic processes. In fact, Tucci explicitly states that "experience . . . suggested certain analogies" with the drawn figure *after* it was conceived and drawn. The mandala thus appears to be tied to experience only after the fact and only in the most general sense. (Tucci, *Theory and Practice of the Mandala*, 23-26; quote from p. 25.)

[45] *Ibid.*, 25-26.

"the *fullness* of the person."[46] Indeed, inwardness is to *subjecthood* and to the mystery of being at the center as a radiant center is to *fullness* and to the mystery of the center of being. Experiences of inwardness and of a radiant center are archetypal experiences of light as spirit—or psyche. Symbolically instantiated in the drawing of a mandala, they become an archetypal human act aimed at self-understanding.[47] *Drawing* the circular form is symbolic of wholeness.

Again, phenomenology itself as well as Husserl's languaging of the relationship between Ego and Object sheds considerable light on why a mandala is a basically circular form that cross-culturally symbolizes the cosmos as well as the self. To put oneself in the center of the mandala is to be at the very hub of the universe, centered rather than spinning along on the outer edge; to put oneself in the center is to be "at the still point of the turning world,"[48] at the unmoving eye of the storm, calm, quiet, unjostled, unperturbed by all that is whirling about one in three-dimensional space. Thus to be both *inside* and *at the center* of the mystic circle of the mandala is to have the potential of understanding at a cosmic level everything that is going on about one. No longer being whirled along at the spinning edge, one has the possibility of apperceiving the whole and with it, an illumination of the spirit—the animating essence—of the cosmos itself. Putting oneself at the center is thus akin to the phenomenological epoche—to bracketing the everyday fact-world the better to see clearly into its nature, to accomplishing those *cogita-*

---

[46] *Ideas II*, 292, 293.

[47] The act is dimly prefigured each time we close our eyes to sleep. As an actual journey inward, it is presaged in a psychological sense by the world we find awakened in the darkness of our fantasies and dreams. Like the eye itself, the eye that is the mandala leads to the I, to the self, to the subject; so also it leads to the *fullness* of myself as person, to my potential for wholeness, to the mandala that is my body. (See Tucci, *The Theory and Practice of the Mandala*, specifically chapter 5: "The Mandala in the Human Body.") Note also that Jung's "self-reflections," carried out over seven years and forming the basis of his analytic psychology, document the symbolic connections between creative act and inwardness. Through "active imagination," Jung actively generated and entered into a fantasy world through which he charted the unconscious and its archetypal forms. See, for example, his *The Archetypes and the Collective Unconscious*. In this illustrated work, Jung discusses mandalas and their significance.

[48] The line is from T. S. Eliot's *Four Quartets* ("Burnt Norton," IV) (New York: Harcourt, Brace and Co. 1943).

tiones that comprehend it. Eastern meditational practices bear out the kinship. "Just sitting" produces insight.[49]

Now this archetypal mode of self- and world-understanding is at odds with typical 20th century Western modes of understanding, modes in which eyes have lost touch with their archetypal power to see into the dark. They have become merely observant eyes. They are eyes that are busy gathering information, measuring, quantifying, inspecting, surveying, looking, watching. They are eyes that are perpetually on the move, and on the edge rather than at the center. They are eyes of a piece with bodies that are mere culturally-inscribed surfaces. They are eyes that have lost sight of their potential to see into the nature of things. To see into the nature of things requires a sense of their inwardness. It requires a sense of the life of the thing that one is looking at. In the words of a long misunderstood but ultimately Nobel-recognized cytogeneticist (whose work was on a most lowly form of plant life—corn, the plant equivalent of fruit flies), it requires "a feeling for the organism."[50] A feeling for the organism. A feeling for stone tools. A feeling for geometry. A feeling for perception. A feeling for the body. With particular respect to the object, the relationship Husserl describes between Ego and Object is precisely akin to "a feeling for the organism." Eyes that lack a feeling for the organism no longer bring with them a resonant and open sensibility. They are mere receptor organs. Poised in their sockets, they look at what is before them and duly record its properties; they watch what they see and duly record its reactions. All of those optics of power of which Foucault writes are generated on the basis of just such factual and fact-seeking eyes. Such eyes have all the living juice squeezed out of them and cannot fathom being inside. They cannot see into the nature of things; they cannot see into darkness, neither their own nor that of the object before them.

The two distinct sets of eyes return us directly to the themes of the symposium. To work through the body as animate form, as psychophysical organism, attending to experience and providing corporeal analyses of

[49] See Shigenori Nagatomo's "An Analysis of Dogen's 'Casting Off Body and Mind'," *International Philosophical Quarterly* 27.3 (September 1987): 227-242. See also his "An Eastern Concept of the Body: Yuasa's Body-Scheme," in *Giving the Body Its Due*, edited by Maxine Sheets-Johnstone, 48-68.

[50] See Evelyn Fox Keller's *A Feeling for the Organism: The Life and Work of Barbara McClintock* (New York: W. H. Freeman & Co., 1983).

same, is to come to deeper and fuller understandings of the ties that bind us in a common evolutionary heritage.[51] Facets of animate form have the potential of leading us to corporeal invariants, to pan-cultural universals, to archetypal meanings, to fundamental human self- and world-understandings. What philosophical reflection on non-philosophical disciplines has shown is that what is required is a corporeal turn, that is, an acknowledgment of animate form and of the tactile-kinesthetic experiences that consistently undergird the lives of living creatures. What the cultural disciplines might be is foreshadowed in this turn. An appreciation of "my body" is not only rarely apparent in Western biology, anthropology, paleoanthropology, and psychology. It is rarely apparent in Western philosophy. Were people in all of these disciplines disposed in the context of their investigations to consider kinesthesia, for example, they would discover the intimate connection between tactility and movement and with it, fundamental distinctions between the tactile-

----

[51] Our common evolutionary heritage binds us primatologically as well as cross-culturally, and in ways strongly suggestive of the theme of inwardness. At least two chimpanzees, when given the experimental opportunity, placed objects *in* a container, in preference to placing them *on* something or *under* something. Moreover, after sniffing and licking a chalk-made circle, both put themselves *inside* it—the one chimpanzee at one moment sitting in it, and at another moment rolling about in it and making sweeping motions on the floor with her arms. The other chimpanzee "suddenly jump[ed] into the middle of the circle, rubbing all around herself (in a circle) with the back of her hands," then sat down, then rubbed again. (David Premack, "Symbols Inside and Outside of Language," in *The Role of Speech in Language*, edited by James F. Kavanagh and James E. Cutting [Cambridge: MIT Press, 1975], 45-61; see in particular pp. 48-51.)

The actions of the chimpanzees strongly recall evidence from developmental psycholinguistics. The first preposition a child learns as both locative state and locative act is the preposition "in" and its derivatives, "inside," and "being inside." This linguistic fact is related in substantive ways to an appreciation of the body as a semantic template. Bodily experiences dispose all of us as infants toward a knowledge of "in." From our first acts of suckling to being put in a crib or other container, from being enclosed inside arms to being inside houses or other shelters, from being put inside wrappings to putting our arms inside sleeves, we all have had (and we continue to have) multiple experiences of *in*, insides, and being inside. Moreover though we think of ourselves only as being born *into* the world, we all came from insides, miraculous insides that protected us by shutting out the outside and holding our insides together. In effect, all humans and in fact all gestated creatures were once inside the mandala which is the womb. In a Jungian psychoanalytic sense, that experience, though no longer remembered, may resonate within our collective unconscious as an archetypal experience of in, of being inside, of inwardness.

kinesthetic and visual body. But they would in turn discover too that these fundamental distinctions lead ultimately to convergences. The lithic elaboration of edges into lines and the archetypal power of eyes to see into the dark are exemplary instances of the rich and substantive conjunctions of the tactile-kinesthetic and visual bodies.

A corporeal turn would furthermore foster an appreciation of cultural *inter*disciplinary studies. When culturally oriented academicians view humans (or other creatures) simply on a behavioral level, in particular, as a visual specimen from which information is to be gathered and on which reports are to be made, they invariably begin by separating out their particular "interest," be it economic, anthropological, religious, sociological, or psychological. Thus, not only is experience neglected or transposed to what is measurable, but the near exclusive focus on behavior compartmentalizes knowledge. The problem is that we do not *live* in this parcelled-out manner. Our sociology is not separate from our psychology; our anthropology is not separate from our medical practice. Disciplinary fragmentation exacerbates the neglect of experience. It reduces knowledge about living creatures to discrete pieces of information about them. Living creatures fail to be recognized as the "persistent wholes" that they are.[52] The mission of a philosophy of the cultural disciplines should thus be to "interdisciplinize"—to draw together—as well as to "universalize." The mission of a philosophy of the cultural disciplines should in this sense recall that original Socratic philosophy that knew no bounds, that persisted in its investigations and followed every query wherever it led.

Phenomenology is critically positioned to carry out just this philosophy. Phenomenologists who describe what is actually there in experience follow the paths of experience where they lead and in so doing have the possibility of relating their descriptive analyses to diverse fields of study, tying together fragments of disciplinary information into coherent understandings. Particularly with respect to analyses of fundamental *bodily* experiences, a phenomenologist is critically positioned to show how fundamental cultural practices and beliefs, even those stretching back to stone tool-making, are in fact founded upon the pan-cultural universal that is the living hominid body. Even further, and again, following the paths of experience where they lead, a phenomenologist is critically

---

[52] The phrase "persistent wholes" is J. S. Haldane's. See his *The Philosophical Basis of Biology* (New York: Doubleday, Doran and Co. 1931), 13.

positioned to show how the living hominid body, being something more than a particular piece of anatomy or behavior, is, in its intercorporeal semantics, strikingly similar to the bodies of other primates. Through just such phenomenological studies of animate form, human nature would be properly anchored in a natural history. Husserl's writings call us consistently to the task of forging this natural understanding of human nature. They consistently invoke our continuity with nonhuman animals and thus our anchorage in the natural world. It is indeed ironic that paleoanthropology as presently practiced should consistently ignore this task and that precisely *a suspension of the natural attitude* should allow us the possibility of seeing with the clearest of eyes into the darkness of our own natural history.

Chapter 4

Connectionism and Phenomenology

Tom Nenon
Memphis State University

Abstract: *After a brief introduction to Connectionism and some general remarks on the relationship between Phenomenology and the empirical sciences, this paper presents examples of the way that both Phenomenology and Connectionism can benefit from insights derived from the other approach.*

There is of course no discrete object clearly denoted either by the term "Phenomenology" or by the term "Connectionism." Each of these terms can be better thought of as referring to a theoretical movement centering around specific themes, approaches, and figures. For not only does each of these movements as such lack clearly delineated contours, but each is also characterized by a fair amount of disagreement among friends and foes alike over what should be taken to be its essential constituting features. This is probably even more true for Connectionism than for Phenomenology, partially because it is a fairly recent newcomer to the intellectual scene, but also because it has emerged from an inter-disciplinary background which unites researchers with very different agenda. Interestingly enough, however, this should be less of a source of embarrassment for someone who thinks of herself as a connectionist than for many who consider themselves "phenomenologists" according to some of the classic formulations of the project since one result of reflection about connectionist theories is that it leads one to think of concepts more as general types than as sets of necessary and sufficient conditions for being a certain kind of thing. I will return to that theme at the end of Part I.

First, however, I would like to begin by stating that when I use the term "Phenomenology" in this essay, I will be referring in a general way to an approach oriented upon the work of Edmund Husserl, without

M. Daniel and L. Embree (eds.), Phenomenology of the Cultural Disciplines, 115–133.

trying to make the case here that there is anything more paradigmatically "phenomenological" about Husserl than other figures, several of whom (e.g., Schütz, Gurwitsch, or Merleau-Ponty) might provide even more fruitful points of intersection with Connectionism than Husserl's work does. I should also state explicitly that the picture I will be painting will be set out in very broad strokes, sometimes resulting very consciously in more of a caricature (though I hope a not completely misleading one) of Husserl and his positions than a precise and detailed portrait, since my intentions in this paper are directed more toward a few systematic issues than to the interpretation of certain texts or doctrines.

I

Accordingly, I would like to proceed by taking the notion of Phenomenology for granted for a moment and turning to Connectionism. What is Connectionism and when did it emerge? According to a few of the more philosophically oriented writers on Connectionism,[1] it is described as a "new approach" that has emerged within the interdisciplinary enterprise known as "cognitive science." Ironically, this new approach calls into question some of the most basic assumptions that constituted this interdisciplinary field in the first place. For what originally brought together researchers from such diverse fields as computer science, neurology, mathematics, psychology, linguistics, and philosophy were two assumptions: first of all, the view that the computer provides a useful or perhaps even the best model for how to think of intelligence or cognition in general, and consequently of human cognition in particular; and secondly, the assumption that by computers one means things like the kinds of machines that had become predominant by the middle of the 50's and prevail up until today—high-powered calculators that operate

---

[1] See here, for instance, William Bechtel and Adele Abrahamsen, *Connectionism and the Mind: An Introduction to Parallel Processing in Networks* (Cambridge, Mass. and London: Basil Blackwell, 1991); John Tienson's Introduction and William Bechtel's survey of connectionism in: Terence Horgan and John Tienson (edd.), "Connectionism and the Philosophy of Mind," *Southern Journal of Philosophy*, 26 (1988) Spindel Conference Supplementary Issue; and Teinson's Introduction along with the revised version of Bechtel's paper in: Tienson and Horgan (edd.), *Connectionism and the Philosophy of Mind* (Dordrecht/Boston/London: Kluwer, 1991). For an overview of the current philosophical debate on Connectionism, see in addition to these three volumes: William Ramsey, Stephen Stich, and David Rumelhart (edd.) *Philosophy and Connectionist Theory* (Hillsdale: Erlbaum, 1991).

according to programmable rules in order to process data and store information, and then on the basis of these rules and stored information to process other data and perform specific tasks for which the machines have been programmed in the first place.[2] It follows from this model of the mind as a computer, the "classical" model of a computer, that cognition is nothing other than "rule-governed symbol manipulation."[3]

This characterization points to two things: a) the formal character of the operations, the fact that these programs function analogously to mathematical calculus or propositional logic, i.e., apply to a wide range of possible subject matter precisely because the performance of the operation is indifferent to the particular data to be processed; and b) the susceptibility of these manipulated entities, be they thought of as data, information strings, or the physical states of the machine, to being interpreted in terms of something else that they represent, i.e., as representations of beliefs and desires or things in the real world—a point which is especially relevant when applying the computer model to the mental life of humans and asking what it means for beliefs and desires to "represent" the world. I will say a little more about this in the fourth part of my paper.

When applied to the mental life of human beings, the assumption is that we should think of beliefs and desires as resembling symbols whose formal relations and patterns of derivation would be determined according to the rules of some mental "program."[4] But perhaps even more important is the notion of a rule as a rigid and programmable operation that is to be performed upon the symbol in an explicitly specified way under precisely specified conditions. Convenient models of such rules are found in algebra and other forms of discrete mathematics, and in formal logic.[5] Thus for a tradition that had long conceived of thinking as a kind

---

[2] An intriguing critique of current cognitive psychology and the computer model of the mind from a Continental perspective has recently been advanced by Fred Evans, *Psychology and Nihilism: A Geneological Critique of the Computational Model of Mind*, (Albany: State University of New York Press, 1992). Chapter 5 includes an explicit treatment of connectionism, which is followed in the next chapter by a critique of cognitive psychology that draws on Merleau-Ponty's notion of the "body-subject."

[3] Tienson 1991, 1.

[4] See here especially Bechtel and Abrahamsen, 8-14.

[5] Tienson 1991: "Connectionist mathematics is the mathematics of dynamical systems; its equations look like equations in a physics text book. The mathematics of the classical picture is discrete mathematics. Its formulae look comfortingly (to the

of reasoning, and reasoning itself in mathematical and formal logical terms, it seemed quite natural to think of machines as duplicating what made mental processes a form of intelligence. Hence the name "artificial intelligence," usually shortened to "AI" and applied not only to the man-made systems of machines and programs, but to a whole approach governed by the model of reasoning as symbol manipulation according to formulae or algorithms both in man and machines.[6] Cognitive science emerged as a distinct field by gathering together researchers from various disciplines around this model of intelligence or cognition as symbol manipulation according to pre-set, programmed rules.

Within about the last six or seven years, however, there has emerged a new way of thinking about these issues, one that can be seen both as a competitor within the field of cognitive science or "classic AI," and in another sense as a possible successor to the project of cognitive science as a whole. It is a competitor in the sense that it offers a competing model of information processing that could change the way systems and machines are constructed to perform cognitive tasks, especially those that were difficult or impossible to accomplish with classic computers. If successful, it would be a successor in that it could be seen as displacing the classic project of explaining all cognition in terms of rule-governed symbol manipulation. Even so, most versions of Connectionism at this point still share with classic AI some of the same basic assumptions about the possibility of drawing important conclusions about human intelligence from non-human information processing systems, although they tend to emphasize the reverse direction more strongly, i.e., they think that one can derive at least a few clues from human cognition about how to construct expert systems that would be better able to accomplish the kinds of things that human beings do well, but classic machines do not. These include such tasks as pattern recognition, especially where all of the relevant parameters are not spelled out in advance; dealing with vagueness; and handling multiple-constraint problems with indefinite outcomes.

Connectionism is a new way of thinking about things that is at least indirectly and partially inspired by the model of the human brain as a

---

philosopher) like the formulae of formal logic." (1-2)

[6] A phrase chosen here simply for the sake of alliteration. It is not intended to imply any particular stance on the recently popular question as to whether such a model reflects a specifically gendered, i.e., male way of thinking or not.

neural network. At its core, the basic idea of such a system involves interconnected units or nodes (something like neurons perhaps) that can be activated either by external stimuli, i.e., input, or by other units. What makes this network a connectionist one is that each of the units is connected to at least a few, more usually a large number of other units which it can then activate by sending out signals. This in turn leads the latter units to send out waves of signals to other units, including the ones from which they have just received signals (something like the activity of synapses), until the system ultimately "settles in" to a kind of equilibrium state. These signals can be either activating (turn on) or inhibitory (turn off) signals; the connections are thought of as weighted so that the strength with which a signal is passed on from one unit to the next depends on a factor called the *weight* of the connection, a factor which, along with the activation level of the units, is a variable.[7] The processing at each unit is determined solely by its initial state and the input from the nodes which are connected to it, and the whole system consists of nothing more than many such nodes and their connections.

So there is nothing in the system that controls the operation of the system as a whole (like the Central Processing Unit in a conventional machine), nor anything in the system that "knows" the state of the system as a whole. Instead of proceeding in a linear fashion, the systems employ parallel processing, that is they do not proceed one step at a time; rather, at any given time there will be a number of units passing along or inhibiting activation to a number of others. Furthermore, though the states that receive an interpretation—and thus, count as "represent-ations" of something outside the system—may be those of individual units, more often the activations of several units at once will be what is interpreted. Thus the term "parallel distributed processing" (PDP) has become another name for the kind of activity carried out by these systems and as an adjective for approaches oriented on such systems.[8] In general, however, the input presented to the system is interpreted as the

---

[7] An only somewhat longer and much clearer introduction to the basic model can be found in Tienson 1988, 6-13.

[8] Compare, for instance, the title of the two important volumes by Rummelhart and McClelland that describe one research group's experience in designing and employing such systems, volumes which were decisive in the emergence of connectionism as a widespread movement: D. E. Rumelhart and J. L. McLelland, *Parallel Distributed Processing* (Cambridge: MIT Press, 1986).

"problem" and the resulting settled state as a whole is what is termed the system's "solution."

A distinct and important feature of connectionist systems is the way they "remember" things. The only things physically present in the system are currently active representations. What in a classical system would be "stored beliefs" are not located at any one place in the system. Instead, the "weightings" are set so that appropriate representations can be recreated given appropriate input, but these representations are not located anywhere in the system except in the weightings as dispositions to create these representations under the proper circumstances.

For what follows, I also need to mention two other features of such systems and then allude to some of the characteristics they exhibit. Often the systems will be organized into layers by introducing connections that are asymmetrical in direction. For instance, one set of units will be capable of receiving input and passing on activations to a selection of or all of another set of units, which may themselves be interconnected, but not connected back to any of the units in the first set. These are called "feed-forward networks." Moreover, for certain tasks it appears advantageous or even necessary to set things up so that there are one or more intervening layers between the so-called input layer and the one which is interpreted as the output layer.

As a second point it is also important to note another feature that these systems exhibit as dynamical systems. In connectionist systems, even the slightest differences in initial settings can be hugely magnified during the process so that vastly different outcomes can result from the same input (the kinds of differences that "chaos theories" in mathematics and physics describe, for instance). Conversely, however, it is also possible for exactly the same input to result in exactly the same output in two different systems whose hidden layers may be configured quite differently, or which may even be lacking hidden layers, but have very different initial weightings which may have cancelled each other out in a specific case. Taken together with the "chaos factor," this means that one cannot rely on the observation of similar responses on the part of two different systems when presented with similar input in the past in order to predict that their responses to some identical input in the future will even closely resemble each other. Consequently, even though there is a micro-level determinacy about what will happen given a certain initial state of the system and a specific input, there is no way to derive strict laws about how two different systems will perform relative to each other in the future based merely on similarities observed about their performances in the past.[9]

One of the most striking features of connectionist systems is their response when presented with problems of what CogSci folks call "categorization," i.e., the recognition of something in terms of the class of things it belongs to or the subsumption under concepts, and pattern recognition. These responses do not fit into a framework of concepts in terms of necessary and sufficient conditions for belonging or failing to belong to a class. To take a simple example, consider a network where activation of a particular unit represents recognition of something as belonging to the category represented by that unit. The unit gets input from a number of different connected units, which might represent features of the object. Differences in weights from different input units can correspond to differences in importance of the various features. Whether a unit turns on might also depend upon its reaching a particular threshold, so very different combinations of input might lead to "recognizing" an object as belonging to that category. This makes it possible to recognize things not only as simply "in" or "out," but rather also "typical," "atypical," or somewhere in-between.

Not all features end up counting equally, some end up counting much more strongly than others, and there may not be any one feature or even a specific set of features that is essential in the sense that the system will consider its presence absolutely necessary in order to identify an object as being a member of the class of things that usually exhibit that feature. Thus, there is no strict essence of a thing, but rather cores of features around which individuals gather. This does not imply that a connectionist system could not be made to conform to a more rigid conceptual logic. Rather, like most human beings, they do not naturally tend to do so. That adds a flexibility for dealing with less than complete information and for equivocal situations, but introduces a degree of indeterminacy and ambiguity in the way that they are likely to classify specific objects.

---

[9] I am indebted to my colleague Terence Horgan for pointing this out to me (with the usual qualifier about eventual mistakes in presenting it accurately not being attributable to him, but to my having my weightings wrong when he presented me with what should normally represent the proper input needed to allow one to produce a correct version of it). One can get an idea (provided one has the proper initial settings) about how this bears on explanation in the human and social sciences from Terence Horgan and John Tienson, "Soft Laws," *Midwest Studies in Philosophy*, 14 (1990): 256-279.

Before, I proceed to a few remarks about the relationship between Phenomenology and Connectionism, I should say at least a word or two about learning in connectionist systems. "Learning" takes place not by changing inputs, or changing initial activations, but rather by "changing" or "shifting the weights" of the connections between various nodes for an output that is considered a mistake, until—perhaps after a large series of adjustments—a more appropriate correlation between input and output is achieved. Up until now the decision about whether the weights need to be changed and how much they should be changed has been made by an external operator who is also the interpreter of the outcome in terms of desired "real life" results. In the fourth part of this paper, I will suggest what might be a more appropriate, though certainly more complicated way for learning to be enacted, one which would follow rather naturally from a different notion of representation inspired by Phenomenology as transcendental philosophy.

II

From the standpoint of Husserlian Phenomenology, one might well ask what all of this has to do mental life of human beings. For as human beings, who are in Husserl's view subjects (more correctly, as subjects, who happen to be human beings as well), we learn of our mental life not by studying machines or analyzing mathematical systems, or even by learning how brains work, but rather by closely observing and analyzing what is given to us through the immediate awareness of consciousness and its products to itself. We know of ourselves as the bearers of mental states and we know of the mental states (*das Psychische*) of which we are the bearers directly and completely whenever we turn back away from the objects that are the immediate focus of our intentional states such as believing, desiring, loving, or hating, and focus our attention upon that which is immediately given to us as such, namely those acts of believing, desiring, etc. in which the objects present themselves to us.

The various procedures associated with the term "phenomenological reduction" are really nothing other than ways to make sure that we avoid making any commitments to the existence of the objects of such intentional acts and focus simply upon the acts themselves and what is involved in them, not as real events in the history of actually and indubitably existing individuals with these or those characteristics, this or that name, but rather as a realm of phenomena that are immediately given to us with some specific import and significance to each of us as the subject of that state. Indeed, this immediate and complete givenness of the thought to me as the bearer of this thought is what makes it mine, and the positing of the "me" as the bearer of the thought implies in itself nothing other than the fact that there is some consciousness for whom this intention is the intention that it is (and that I am this consciousness).

Thus when we talk about the mental (*das Psychische*), the mind (Husserl often uses the traditional term *"Seele"*), or consciousness, we are talking about a realm that is accessible to us through a completely different means than those through which we study the brain, not to mention the processing systems for humanly constructed expert systems. As such, it is a completely independent realm, the argument goes, and thus must be studied in a completely different manner, which is not subject to confirmation or verification by any empirical means regarding external objects, including the behavior of human beings as existing objects in the world. So or along similar lines might an argument be advanced that could be modelled on Husserl's refutation of psychologism in the "Prolegomena" to the *Logical Investigations* and his critique of naturalism in "Philosophy as a Rigorous Science." In fact, one might argue that the whole point of phenomenology is to establish an approach that would be immune to the vagaries of empirical science—so what could cognitive science have to offer to Phenomenology?

My first response would be that Connectionism is above all a way of thinking about things. The extent to which it should be accepted as a model for what happens in the life of the mind will depend upon what one thinks about mental life itself. Interest in Connectionism need not imply the belief that neurological research or the success in constructing network-based expert systems automatically tells us something about human cognition. Instead, for a Phenomenology of mental life, Connectionism offers a number of interesting questions and new approaches that cannot overturn or substitute for phenomenological evidence, but present us with interesting and new questions and ways of interpreting phenomenological *Befünde*. Since at least some of what we think of as mental life is familiar to us most directly and indubitably in our phenomenological apprehension of it (even before any systematic development of a phenomenological method as such), the Phenomenology of mental life must and will continue to play its own specific role in any inter-disciplinary approach to cognition in general, including Connectionism.[10]

But would that not imply that the Phenomenology of mental life can proceed independently, and thus need not be affected in any way by what we believe we have learned from other fields? I think not. Husserl's extensive methodological reflections concerning such techniques as phenomenological reduction and eidetic variation describe practices that are meant to guarantee the possibility of ascertaining essential connections between various contents of consciousness without any reference to empirical facts. However, even if one grants what Husserl says about the conditions necessary to establish Phenomenology as a rigorous science and about the immunity of the results of phenomenological research from challenges that originate in the empirical sciences, then it still does not necessarily follow that the concrete practice of Phenomenology will not and cannot be strongly influenced by what we think we have learned from other spheres.

If for instance the successful practice of eidetic variation depends upon one's first having surveyed all of the relevant counter-examples, then there will always remain a problem about whether the limits of the imagination of the researcher are the limits of imaginability *simpliciter*.[11] What one will take to be an adequate intuition at the moment might turn out not to have been an adequate intuition upon further inspection, and is often influenced by what one thinks one knows otherwise. So one can affirm the phenomenological program and what the attainment of any final scientific results in Phenomenology would involve and still deny that

---

[10] It is also interesting to note that Husserl's critique of psychologism actually applies more appropriately to classical AI than Connectionism. For it is classical AI that is committed to the view that human thinking not only conforms to rules such as those expressed in formal logic and mathematics, but that it is actually rule-governed, i.e., that these laws causally determine actual human cognition. For classic AI, cognition is simply rule-governed symbol manipulation according to things like the laws of logic and mathematics, so that it is a natural assumption to think that the laws of logic are causally determining factors in human thinking. In fact, Husserl's critique parallels that of connectionists who emphasize the differences between the models provided by formal logic and discrete mathematics, on the one hand, and the way that human beings actually think and speak on the other. Husserl and many connectionists in fact share the desire to distinguish carefully between the two and avoid reducing questions of logical validity to questions of how human beings think (or the other way around).

[11] Husserl's careful distinction between evidence and apodicticity in §§ 59 ff. of *Formal and Transcendental Logic* suggests something very close to this point.

any individual involved in the actual practice of Phenomenology can ever be certain that he or she has realized Phenomenology completely with regard to any specific result. One can think that one has good phenomenological evidence for what is actually a rather shaky claim or not see an essential connection between things, and one reason for this can be that one has allowed oneself to be misled by models that have been imported from another sphere. Moreover, the more complicated the phenomenon at issue, the more susceptible we are to mistakes.

Husserl's favorite examples of rather basic mathematical truths can be misleading in this regard, for the sphere of the mathematical is purposely a very restricted one, made up of clearly definable entities pertaining only to quantities as such. In the realm of the mental, I would maintain, things are not as clear—in spite of what Husserl maintains in a number of his earlier writings about the complete and direct givenness of mental states (*das Psychische*) as such. Thus the kind of paradigmatic examples we use in when reflecting upon mental life can have a great effect upon what the results of our investigation will be, and these examples will most often be strongly affected by what we think we know about the non-mental sphere and intimately, perhaps even unavoidably related to questions concerning the physical basis or instantiation of mental events (for instance, the relationship between sensuous fields as we find them in phenomenological reflection and what we believe about the nature of various bodily organs).

In sum, then, even if empirical science is accorded no place in the pure formulation of the phenomenological program, that does not mean that what we learn (or at any point in time, think we have learned) from non-phenomenological science cannot have a profound effect upon the practice of Phenomenology.[12]

Finally, I might add that the way we think about the workings of neural networks is especially relevant in view of Husserl's recognition that

---

[12] One should not overlook the fact that phenomenology never has proceeded in a vacuum free of the influence of other sciences. Our common ways of thinking of the mind both outside of and within phenomenology display the traces of Christian theology (the mind as a self-subsisting entity capable of willing and knowing all on its own), modern mechanistic thinking (the mind as the operator of the body as a machine), and biology (concepts like "sensory stimuli" or "input from the senses"), and computer science (the very notion of "input" itself). It is thus not so much a matter of importing models into phenomenology from outside, as being inspired by and adopting models that are more adequate to the phenomena at issue.

human beings as persons, as centers of motivation, are founded in bodies, and thus that the mental will depend for its existence on a non-mental, natural stratum in human beings.[13] This does not involves a reduction of the mental to the physical, but it does mean that a truly plausible account of the mental must at least be compatible with what we think we know about bodies and brains. If our views about these two realms do not square up, that does not necessarily mean that it is our view of the mental that must give way. In some cases that may mean that we should go back and reexamine what we think we know about brains and bodies. But in any case, it points to a closer connection between theories about cognition as processes in machines and brains, on the one hand, and the theories about mental life, on the other, than many who consider themselves phenomenologists might want to recognize.

In the following section, I suggest how Connectionism can indeed help phenomenologists better conceive of one phenomenon, namely dispositions or—to use Husserl's term—"habitualities," that has traditionally posed some problems for Phenomenology and most other approaches to a philosophy of mind as well.

III

Finding an ontological home for dispositions, in particular for cognitive dispositions that seem to function as the effective background for explicitly held beliefs and for actions, has been a problem in analytic philosophy of mind and to a certain extent for Husserl as well. We certainly seem to need to find a place for them, since they perform a lot of work in any plausible account of why and how we do the things we do. We need something to explain what motivates many of the mental states we are consciously and actively aware of, but the things that we need to complete the explanation are often states that we are not directly aware of. Thus, one posits a realm populated by unconscious beliefs and desires, which as repeated and consistent sources of certain beliefs or actions are seen as dispositions—or to use Husserl's term, habitualities. Many of our directly observable actions, conscious decisions, and explicit beliefs seem to follow patterns that can be explained in terms of other

---

[13] Cf. here for instance *Experience and Judgment*, Paragraph 8, where Husserl explains that the tendency toward our naturalistic conception of the mind has its justification in the fact that everything worldly has its place in the spatio-temporal sphere.

beliefs and desires which we hold, but which we are not explicitly and thematically aware of prior to the action, decision, or belief that issues from them.

Under the assumption that mental states—whether strictly determined or not—are motivated by other mental states, we find ourselves compelled to posit something, namely dispositions or habitualities, to fill the gap. But even so, they present a problem. For they do not fit easily into our paradigmatic examples of mental states. The way that we most directly and certainly know of the existence of the mental is that there are many beliefs and desires we ourselves have and are directly aware of. The ones that philosophy—and especially Husserl—have typically focussed on are those that also exhibit a propositional structure, that predicate something of something else. To use Husserl's words, the prototypical examples of beliefs as mental events are "judgments" (if we focus on belief for second) of which we are immediately aware. However, we know from *Experience and Judgment* for example that Husserl does not consider judgments the fundamental building blocks of mental life, since he maintains that predicative judgments—as well as decisions and actions—are in turn founded upon the prepredicative experience of the individual objects that predicative judgments are about (§§ 4-6), and that underlying this realm is the more comprehensive and fundamental sphere of the life-world as the ultimate horizon against which any experience of individual objects is at all possible.

Nevertheless, even the prepredicative apprehension of such individuals implicitly bears within itself structures resembling those of the judgments which later arise out of it, since it contains within itself the tendency towards a certain framework of explication and possesses a specific doxic character, that makes it characterizable as a species of belief.[14] In this regard each background belief seems something more like a structured image, with various aspects that are not explicitly spelled out but are nevertheless implicitly contained within it—in Cog-Sci language: it is syntactically structured. But in spite of this syntactic structure, and in spite of the important role that this level of the experience of individual objects plays in mental life as the "substrate" not only for predicative judgments, but also for decisions and actions, we are normally not directly aware of it, Husserl indicates, at least not completely and directly. He says for example that at the basis of all judgment, decision, and action

---

[14] See §§ 10, 12, and 16-21.

as mental events which we are directly aware of, lies the experience of something "that is a substrate with simple, sensually graspable qualities to which there is always a pathway of possible interpretation." (§ 12)[15]

Why "interpretation"? Precisely because it must be posited if we are going to explain those things that are founded upon it, but it does not share with the paradigmatically mental the characteristic of being completely and directly accessible to reflection.[16] So if we divide up the world into the mental as that which is directly accessible to consciousness and the non-mental as that which is not, then it turns out that we would have to assign dispositions to the realm of the non-mental[17]; but when we consider the role they play in motivating consciousness and the syntactic

---

[15] It appears that Husserl is actually talking about two different kinds of phenomena here, one involving the implicit directedness towards the individual objects given in sense perception that Husserl is positing as the basis for all predicative judgments, and the other being the results of experience as a whole which underlie our actions. Both share the quality of not being the direct objects of consciousness, but each in different ways. The case that I am making for dispositions applies more appropriately for the second sense of habitualities listed above, but a case could be made for pattern recognitions in a prepredicative grasp of objects as they present themselves to us in experience as well.

[16] In another paper, I have tried to exhibit the dilemma in which Husserl finds himself when he establishes a direct connection between the will and one's actions. In this case, it seems that Husserl is committed to the possibility of mental states which may not necessarily be accessible to the agent through reflection, in spite of his contention in other passages that the mental is always at least potentially directly given to consciousness itself ("Husserl on Willing and Acting," *Man and World* 24 (1992): 301-309. That paper suggested (even though it did not assert) that Husserl should have abandoned the second claim. In this paper, what I am suggesting is rather that Husserl can maintain the second claim as long as he is willing to countenance a class of objects (like dispositions) that are not mental in the paradigmatic sense of those states that are envisaged in the second thesis, but only in virtue of their role in producing those states—"protomental states" one might call them.

[17] In "Leaping to Conclusions: Connectionism, Consciousness, and the Computational Mind" (Tienson and Horgan 1991, 444-459), Dan Lloyd uses this dichotomy to assign dispositions and the unconscious generally as non-mental ("non-cognitive" is his term) almost automatically to the realm of the physical underpinnings of a cognitive system, thereby providing a causal link between the physical and the mental (the cognitive and the non-cognitive) that he sees as a confirmation of identity theory. To me, Lloyd seems to proceed too hastily in his conclusion that this must be biological since it is not paradigmatically cognitive. It seems to me that one could take a different route, positing a constellation in a physical apparatus that is not properly describable in physical terms but only in terms of what it leads to in the cognitive realm, thus a proto-cognitive realm that, when described in terms of its function, is closer to the cognitive than the merely physical.

structure they exhibit, then they appear to be closer to the mental than the physical. Husserl's solution to this dilemma seems to be most often that he thinks of them as sedimented beliefs (e.g., the 4th Cartesian Meditation § 31 or *Formal and Transcendental Logic* §§ 42 ff.) that could can be retrieved and made conscious again. Thus habitualities and dispositions become a kind of active memory, non-conscious representations that can in principle always been retrieved again and brought to direct consciousness.[18]

Now this seems problematic to me in two respects: first of all, it is not clear to me that it is indeed the case that sheer reflection alone is ever capable of resurrecting all of the beliefs that are supposed to lie at the bottom of these dispositions.[19] Perhaps we can in some cases, but if not in all of them, then the dilemma I just outlined still presents itself for a number of them. And secondly, even if we could make all of them explicit, it is not clear that the best way to think of them is as sedimented belief states, since it does not necessarily follow that they therefore must have ever had the status of conscious beliefs just because we can now raise them to the level of consciousness.

Here is where I think connectionist models come in and are actually quite compatible with Husserl's accounts of passive synthesis as governed by structures of association. For as I mentioned in Section I of this

---

[18] This is not the only approach Husserl takes. In his analysis of the prepredicative realm in *Experience and Judgment* (§§ 15-28), for instance, Husserl suggests that at the most basic level, our cognitive life is permeated by "tendencies" that derive from our past experiences apart from and prior to predicative judgment. To these tendencies or "interests," which constitute our "horizonal consciousness," Husserl also gives the name "habitualities." Interestingly enough, however, the phenomenological warrant Husserl provides for positing such tendencies is not that they are directly discernable, but that they can be discerned through an analysis of the role they play in the constitution of objects for us in cognition. In one sense, then, they are even more problematic than habitualities thought of in terms of stored judgments, since it is not clear in which sense they ever were present as such to be stored and retrieved. One would have to read "sedimented" no longer simply as "removed from the realm of active attention," but instead as "brought about or caused" by past experience so that it would be a kind of belief which at no time ever was consciously entertained. I think that this approach is phenomenologically appropriate as a description of a large portion of our mental life, but that does not mean the problems concerning its ontological status are any less problematic than those connected with dispositions thought of as stored-away judgments.

[19] One might take Husserl's move from mere "phenomenological description" to analyses of "intentional implications" to be an attempt to come to terms with these kinds of problems.

paper, Connectionism provides a model for explaining active representations in terms not only of other active representations stored within the system (remembered or sedimented beliefs), but rather in terms of weightings as dispositions to produce certain representation under the proper circumstances. The "experience" of a cognitive system then would consist not in a practically infinite number of stored beliefs, but in the weightings it has adopted to bring about appropriate beliefs under the specific conditions. The upshot of all of this is simply that one should not dismiss things like tendencies and dispositions within the mental realm simply because they do not seem to fit neatly into the region either of the mental, the physical, or the ideal. One must also be careful to avoid the distortion of the phenomena by all too hastily concluding that they must be kinds of stored beliefs that are kinds of "remembered" judgments upon which one no longer focusses directly. Connectionism suggests that there may be sedimentations of experience that are not now and never were judgments in any sense, but rather exist only as a tendency to process new information in certain ways, as functions within mental life whose existence we become aware of only in their functioning.

Admittedly here, these very sketchy remarks leave out much that would be necessary for this to become a serious hypothesis about the nature of dispositions. But thinking about the problem in these connectionist terms does point a way out of the dilemma I outlined above. It would provide Husserl with a way of asserting both the accessibility of all genuinely mental events in a paradigmatic sense, while at the same time being able to posit other states that may not pass this test, even though they do play a role in mental life, proto-mental states perhaps, that would not have to be thought of simply as stored beliefs (which, I might add, present an even greater problem for the traditional approach if we extend our focus beyond the strictly theoretical realm and include dispositions for actions as well).

IV

In the preceding section, I have indicated in a general way how the connectionist model might provide resources for reconceiving a set of problems within Phenomenology. In conclusion, I would like to suggest that cognitive scientists could learn something from Transcendental Phenomenology about how to conceive of representation more adequately, something that might suggest a better way to construct systems that someday would be capable of learning on their own and be less dependent upon the machinations of an outside observer.

For most researchers in Cognitive Science and many of those attracted to Connectionism, representation is still thought of as a relationship between states of the system (activations of the units, in the case of Connectionism) and objects in world, both being distinguishable as distinct entities. The problem which then presents itself is how tell what states represent which objects and why. The problem is similar to the one that presents itself to Descartes. How do I know that my ideas represent anything in the world if all I ever have access to is my ideas and never the world? This seems to hold for cognitive systems other than human beings as well, even though the problem there is somewhat different, since we have no reason to believe that they have any self-conscious states in which they ask whether there are is anything else other than those states which are immediately accessible to them. But even for an outside observer (which is a very different perspective) the problem still poses itself about the states of the system as representations. What does that even mean for a machine to "represent the world," for example? And until we know that, how can we attribute "semantic import" to these states?

Here I think a few themes borrowed from Phenomenology's confrontation with Descartes' problem could help. What is it that knowers need to do in order to be knowers from a phenomenological standpoint? Surely, at least from the standpoint of Transcendental Phenomenology, it cannot involve transcending the bounds of possible experience. Rather it involves a structure of intention and fulfillment in which those things count as genuine objects that fulfill the intentions directed toward them. Applied to a cognitive system, that would mean that it would not have to be constructed such that it could somehow jump out of the system, nor that there would have to be isomorphic resemblances between parts of the system and objects in the world, but merely that the system would have to generate something that could count as expectations regarding future input. What would give such a system "semantic import" would not be its relationship to an external observer, nor the way that it maps things completely outside the system, but rather the relationship it establishes to future input such that this input could serve as feedback about the correctness or incorrectness of the response generated in response to past input.[20]

This suggestion does not entail that cognitive scientists all necessarily have to embrace a version of transcendental philosophy (although I think they should, and not only for the sake of their machines and models) in order to accomplish this, since adopting a transcendental standpoint involves far more than what I have just suggested. It would, however, point the way to some other practical advances. Once one begins to think of representation no longer as the problem of how to compare parts of the system to things outside of it, one might also be able better to conceive what learning would be and how this could become part of the architecture. (Although for the purposes of us outside the machine who want to know what it is doing and gain information from it, we will certainly want to continue to make some such comparisons or assign interpretations from the outside until part of its output is put into a conventional language we can understand.) Learning would then simply involve shifting the weights whenever the received input did not correspond to what was expected under such circumstances. Representation would be the representation of some specific input under certain conditions, and learning would be the system's self-adjustment so that the two eventually come to coincide more closely. This would remove the necessity for an external operator to constantly readjust the machine according to a learning algorithm outside the system in the way that this is currently done using techniques such as back-propogation.[21] The point holds generally for any cognitive system, whether designed in a classical way or according to a connectionist model, but at least as far as learning is concerned, the prospects for a connectionist system are rosier, since the flexibility of a connectionist system should in principle be better able to accommodate the inherent vagueness of such terms as "under the proper circumstances" or "similar input" which will certainly never able to be completely eliminated. In this case also, what I am offering in this paper

---

[20] In a number of places, e.g., in §8 of *Experience and Judgment*, Husserl explicitly identifies anticipation as one of the key features of consciousness as intentional.

[21] If this is indeed theoretically possible, this would also have a direct bearing on Evans's criticism that connectionists systems are just as essentially as insensitive to context as traditional AI has been (121-22), since the changes would not be enacted by an outside manipulator, but by the system and its interaction with its environment, according to which would change itself in response to discrepancies between its "expectations" and the input actually submitted to it from the outside.

is only a very vague outline of how one might begin to think of these problems. My hope is that these remarks will at least serve as an indication that there is indeed interesting and significant work to be done at the intersection between these two approaches, work from which each could benefit.

Chapter 5

The Other Culture

J. N. Mohanty
Temple University

Abstract: *I question the idea of the purely "my own" world, and so have found no way of formulating the problem of constitution of the other culture analogously to what Husserl does in the Fifth Meditation. The other is a part of my world—as much as the strange, the unfamiliar and the unintelligible are.*

I

How is the sense "other culture" constituted? That is the question I will be reflecting upon in this essay.

I start with the assumption that phenomenological constitution is constitution of *sense*, not of the thing. So when in the Fifth Cartesian Meditation, Husserl undertakes to solve the problem of the constitution of the alter ego, he was—on my view—concerned with the constitution of the sense "alter ego" but not with the constitution of the alter ego itself. The charge of solipsism therefore is misplaced. He, i.e., Husserl, knew there are other egos. He wondered how transcendental phenomenology, with its thesis that all meanings are constituted within the experiences of the transcendentally purified life of the thinking ego, could have room for the sense "other ego." The extreme methodological solipsism practised in the Fifth Meditation, along with the reduction to one's sphere of ownness, is undertaken only to be able to formulate this problem in its most radical form and, from that radical solipsistic position, he tries to build up step by step the genesis of the sense "other ego." Understandably, many who otherwise admire Husserl's work have questioned the success of this project. But before one wants to judge the success or the failure of Husserl's execution of the task, we must be clear as to what the task was. For example, Husserl never set upon

M. Daniel and L. Embree (eds.), Phenomenology of the Cultural Disciplines, 135–146.

himself the task of proving that there are other egos, his task was never
to overcome solipsism (for he was never a solipsist). It was also not his
self-appointed task to decide how we *know* that there is another ego
over there inhabiting that body.

Since my reflections with regard to the other culture will begin with
asking if a Husserlian meditation on this theme, analogously to the Fifth
Cartesian Meditation, is at all possible, I should spend a little more time
distinguishing between the various problems about "other minds" which
philosophers have been concerned with. I would like to distinguish
between three levels of problems with regard to other minds. Husserl,
I maintain, was concerned only with one of them, certainly not with the
other two.

First of all, many philosophers have been concerned with the
question, how do I know, for example, that the other person—he or she
or you—has such and such mental state (pain, anxiety, fear, for example).
Answers to this question range from the theory of analogical inference
to the theory of direct empathy. But this question already presupposes
that we know the other to be another ego with its own inner experiences, mental states or intentional acts. There is therefore another
philosophical question, which many philosophers have asked: how do we
at all know that that body over there has a mind, an inner life, like
mine, that it is not a mere body with no inner life, a painted wax figure
for example? This question presupposes that I have a mind, that I am
an ego and not a mere body. It asks, how do I know that the
other—that body in front of me—is also an ego like me and not a mere
body? It thus presupposes that I have the concept of ego, and that I also
have the concept of the other ego. It only wonders, on what grounds the
latter concept is applied to this body over there.

Both these questions are questions about truth of certain cognitive
claims I may make. The first question is concerned with the cognitive
claim made by me when I say "He is in pain." The second question is
concerned with the truth or falsity of the cognitive claim I make when
I say "That is only a mannequin, not a real person" or when I say "That
computer is not a mere metal box, but has a mind of its own, it thinks,
believes, questions, remembers, supposes, etc."

There is still another level of questioning: how does the ascription
of the predicate "ego" to the other *at all make sense*? The point of the
question may be clarified thus: if my concept of the "ego" is from my
own case, then it would appear as if it belongs to the very concept of

ego that it is mine, in which case to ascribe ego to an other would involve a contradiction. From the transcendental egological stance which Husserl adopts, it is indeed necessary to explain how I could at all speak of the other ego. How am I able to transfer the predicate "ego" from my own case, where it has its original home, to the other—more specifically, to the appearing body over there? Whatever may be Husserl's account in the Fifth Meditation, that is not my present concern. I assume, he shows, in his own terms, starting from the reduced sphere of my ownness, how I can step-by-step constitute the sense "other ego," and *meaningfully* (not necessarily truly, for the other body in my perceptual space may be hallucination, an apparition, a mannequin or a wax figure) ascribe ego to the other.

Can we pose a similar problem, at the level of philosophical abstraction, with regard to the other culture, and ask how is the *sense* "other culture" constituted for a member of the home culture? Most social scientists, when they raise methodological questions regarding knowing other cultures—thereby asking questions pertaining to the insider-outsider situation or Kenneth Pike's etic-emic distinction[1]—are really asking questions regarding the epistemological basis of their sciences. In general, they are asking, how can the investigator, belonging as he does to his own home culture, forge an access to the native's world—his language, beliefs, thoughts and desires, in fine, to his "conceptual framework." The concern is analogous, on an individual level, to the skeptical worry, how can I know what is transpiring in his mind? Just as the last worry presupposes that I already have available to me the sense "other mind," so does the social scientist's epistemological concern presuppose that he has already available to him the sense "other culture." Just as in the former case, there is the transcendental question "How is the sense 'alter ego' constituted?," so in the latter case one may want to press a transcendental question "How is the sense "other culture" constituted?"

---

[1] For these debates, see Hedland, Pike and Harris, *Emics and Etics. The Insider/Outsider Debate* (Frontiers of Anthropology vol. 7). Newbury Park: Sage Publications, 1990.

II

Are the two questions really analogous? Or, do they only seem to be alike, that is to say, alike only verbally? As a matter of fact, closer examination reveals important disanalogies between the two questions.

The solipsistic ego is arrived at by a special reduction Husserl instituted in the Fifth Meditation—namely, the reduction to the sphere of one's ownness. In this reduction, one removes from one's experience all that refers to other egos, so all experiences of material objects and all cultural predicates, all intersubjectively constituted predicates. One thereby reduces one's experience to experience of one's living body with its sensations, kinaesthesias, also one's internally experienced ego-pole. It is only after such a methodological step has been carried out, that the genuinely transcendental problem regarding the genesis of the sense "other ego" can be formulated: how could my experiences, as so reduced, nevertheless provide motivations for predicating "other ego" of something that appears within that reduced sphere? Husserl's solution to this problem is well-known, and need not be recapitulated here.

Can an analogous step be taken with regard to the genesis of the sense "other culture"? Such a step would require me, us rather, to remove from our solipsistic cultural experience all components that derive from commerce with other cultures. But how can I ever be sure that some components of my culture are not derived from other sources? In other words, a corresponding reduction to the sphere of my, or our, ownness cannot be carried out at this level. Only the myth of purity of a culture may mislead one to believe that one can have such a sphere of one's ownness at this level.

The disanalogy can be pressed still further. At the level of the ego and its other, there is a real otherness, a discontinuity, a discreteness, which is hard to deny. It is, in other words, hard to deny that A and B, two different persons, have two different egos—howsoever their contents may overlap. A reduction to the sphere of ownness is therefore a meaningful project. You just strip away whatever contents you may owe to others conceptually. For example, that my experience contains perceptions of a physical Nature is undeniable, but perception of a physical Nature implies possible intersubjectivity. Striping such contents away nevertheless would leave my ego standing in its own sphere of ownness, which does not contain anything it conceptually owes to others. The fact of my ego's self-contained purity, at this level of discourse, is undeniable. The problem is then to overcome this isolation, this sovereignty, and to reach out to a meaningful alterity.

With regard to a culture, such a claim to sovereignty within its own domain is precisely what is deeply questionable. There is a surface level otherness—one speaks of German culture and Indian culture as being different—which is always questionable beyond a certain limit. If the identity of a culture consists in its unique historical development, what guarantee is there, as we go back to historical and prehistorical origins, that there are not discernible common ancestors and mingling of diverse routes of influence?

Are C_1 and C_2 different cultures? When is a C a "foreign" culture? Shall we say, as Husserl seems to have held, C_1 and C_2 are "mutually" foreign, when they have different "generative" histories? But what assures, in any given case, that C_1 and C_2 do not, at some point in their generative histories , have a common stem, a common path before branching out? Surely, that C_1 and C_2 are different linguistic groups does not determine the answer. How different languages should be—up to the point of being "untranslatable"—so that they can be regarded as being mutually "foreign"? What could "untranslatability" mean in view of Davidson's radical critique? Shall we say that there is just no absolutely "foreign" culture, but rather that there are only different degrees of foreignness? But the idea of degrees of foreignness itself is ambiguous, depending upon what cultural trait one is focusing upon. A culture that is "more" foreign with regard to a trait α may indeed be "less" foreign with regard to another trait ß.

Undoubtedly, one speaks, and is justified in speaking, of the home culture, as one speaks of the native language. But such locutions, despite their legitimacy within limits, should not blind us to the following *disruptive* considerations. In the first place, the home culture may contain—and one cannot be too sure that it does not—elements from a foreign culture (foreign words in the home language, for example). The home culture, in the second place, is not itself a monolithic structure. It rather contains strata, that are "foreign" to each other. Consider what is called Indian, or even Hindu culture. It contains beliefs, rituals, practices—not to speak of languages, art forms, musics—which are "foreign" to some natives. Thus the other belongs to it, and is not simply outside. You can avoid this consequence by looking for a subculture that is homogenous. You never find it.

These remarks are meant only to make the point that the transcendental constitution problem as it was formulated by Husserl with regard to the "other ego" cannot simply be stated with regard to the "other culture." But one may nevertheless have a problem somewhat like it. Without being able to reduce to the sphere of one's ownness, i.e., to one's own culture in its purity, one may still ask, how does one come to apprehend a foreign culture as such? Let us try to go back again to the beginnings.

III

In course of our travels in foreign lands we come across a group of natives engaged in ritualistic behavior. Since the transcendental constitutive problem with regard to other egos has been, *ex hypothesis*, solved, we perceive them as other egos, i.e., as having mental experiences like ours, intentional acts for which their world is presented, as in our case, with *their* meanings. The question that we ask ourselves is *what* do they mean by their actions and speeches. If, in our own case, behaving in a certain manner is a sign for having a certain intentional experience, does the same observed behavior in their case too, carry the same significance? Or, as Turnbull observed in the case of the Ilk in Uganda just the reverse may be the case. What do we take them to be experiencing, when we observe the man watch a child "with eager anticipation as it crawled towards the fire, then burst into gay and happy laughter as it plunged a skinny hand into the coals . . . .[Then] a mother would glow with pleasure to hear such joy occasioned by her offspring, and pull it tenderly out of fire."[2] Fraser's *The Golden Bough* abounds in such examples.

With this, we are back in the epistemological problem—indeed the first of the three questions I listed at the beginning. I did not plan to answer *that* question, which is of so vital interest to the social scientists. I would therefore return to the constitutive question, despite the failure to take a transcendental stance with regard to the other culture. How is the idea of *"foreign* culture" constituted?

One meaning of "constitution" can be explained thus: to exhibit the constitution of a concept ϕ is to show what are the sorts of intentional experiences in which objects instantiating ϕ are originarily presented.

[2] Colin Turnbull, *The Mountain People*. New York: Simon and Schuster, 1972.

Thus, to exhibit the constitution of the concept "material object" is to identify the type of intentional experiences which originarily present something as a material object. It is thus that one says, in phenomenology, that material objects—not these things, but their sense as material object—are constituted in outer perceptions, not, to be sure, in one single outer perception, but in a concatenation of them. To return to our question, we may construe it, then, as: how do we recognize a culture as a foreign culture? How is a foreign culture given as a *foreign* one?

There is no problem in recognizing that they, the natives, have a culture. To perceive them as other egos is to perceive them as having intentional experiences like mine, as conferring meanings on their intersubjectively shared world, to perceive them as dying and new ones as being born, to take them as having a generative history of their own as we have ours; thus they have their own culture. But what experiences on our part serve to present their culture as being a "foreign" culture? There is a spurious notion of "foreign" which, if my arguments are correct, we need to reject. In that sense of "totally foreign" (and its correlate "purely ours"), we only *presumptively* apply those concepts to their culture (and to ours, respectively). That is the best we can say. Even these presumptive ascriptions need to be correctly evaluated.

There is a large common framework, within which—and only within which—differences show themselves. First of all, the different cultural worlds all belong to the same Nature; they, rather their *territories*, are parts of the same spatio-temporal system. We and they perceive the same nature—plants, trees, animals, rocks, mountains, waters, and heavenly bodies, although they may not be ascribing to them the same meanings as we do. Secondly, we, as well as they, are embodied, and our bodies are human bodies—similarly structured. Third, it would be a fairly reasonable assumption that our mental lives of some level have the same structure, even if possibly our different cultures allow for differences at some other level. Fourth, our biological needs and the basic human drives are at some level the same. Given these over all shared framework, it is only reasonable to expect that no culture can be totally different from ours. Consequently, as Husserl writes, "Everything that is so foreign, so unintelligible, has a core of familiarity" (*Unbekanntheiten im still der Bekanntheiten*) (Hua XV, 432). The completely foreign is still familiar inasmuch as there are there spatial things (*Raumdinge*), men and animals, villages, landscapes, etc. (*Ibid.*, 430). They are "unfamiliarities in the style of familiarities" (*Unbekanntheiten im still der Bekanntheiten*), they are understood as "possibilities for experience" (ibid, 430).

Taken together with various remarks I made earlier (to the effect that within the home-culture, there is also an other), the above implies that the contrast between the home and familiar culture and the foreign, the unfamiliar is a permanent feature of *every* "world" (ibid, 431), one does not have to go out to meet the other, the unfamiliar, the unintelligible, the strange, the unknown. One meets it within one's own home world.

The other, the alien, is the limit to the understood and the known. The native is "foreign" because, and in so, far as he is different, he is different because, and in so far as he is not understood. Even when the social scientist, or the empathetic traveller understands the native, this understanding can overcome the foreignness (*Fremdheit*), *only when* it is based on *mutual communication*. More often than not, the attempt to understand the other is, in such cases, one-sided. The scientist "observes" and "interprets" the native. At most, there is an informant who "translates" the native's speech for the scientist. Only when this one-way track of "making sense" of the native is overcome by the "mutuality" of "making sense" of each other, the foreignness is overcome. A common world, mutually shared, thereby begins to constitute itself.

In view of the fact that there is the other's homeworld which is different from mine, which I do not fully understand, and also in view of the fact that even within my home-world not everyone has the same access to all its dimensions ("science," "religion," "music" etc.)—how is it possible to say that we all experience—in some sense of "experiencing"—the one and the same world? Husserl asks this question repeatedly in the intersubjectivity papers. What does the identical world and the constitution of it mean, how is the subject as subject "for" this world constituted? ("Was besagt da identische Welt and Konstitution derselben, was characterisiert die Subjekte als Subjekte 'für' diese Welt?") (Hua XV, 228). The question, for him, amounts to asking:

> . . . to what extent and how far I can take over, through understanding, their (i.e., of the strangers') experiential structures, and so can progress towards a synthesis of their homeworld with mine? How do I arrive at, and I must, a comprehensive consistency?

> (. . . inwiefern ich und wie weit ich ihre (die Fremden). Erfahrungsgestalten in Nachverstehen *übernehmen*, also zu einer Synthesis

ihrer Heimwelt mit der meinen fortschreiten kann. Wie komme ich, und muss ich kommen, zu einer übergreifenden Einstimmigkeit?" (p. 234).)

The question is not, Husserl continues, who, in the immanence of his experience, has a priority, but how "it is with the community of both total-experiences in their possible or yet-to-be-instituted synthesis of consistency." ("es mit der Gemeinschaft der beidseitigen Gesamter-fahrungen in ihrer möglichen oder herzustellenden Synthesis der Einstimmigkeit steht.") (p. 234). This is a task Husserl assigns to a *transcendental aesthetic*, a transcendental *"Empiriographie"* which projects the idea of "an experience—and experiential world-structure of all mankind." ("eine allmenschheitlich Erfahrungs-und Erfahrungweltstruktur") which is to serve as the "norm of the critique of relatively consistent experiential worlds and meaning-worlds of any community of humans." ("Norm der Kritik der relativ einstimmigen Erfahrungswelten und Meinungswelten irgendwelcher Menschheiten") (p. 235). This is a transcendental task. This task requires thinking not alone about how my homeworld is constituted, and then the eidetic notion of homeworld in general, but also the "problem of critique of a homeworld against the horizon of foreign home-worlds viz. the critique of a universal experience which shall set up a unity—i.e., a true world—which is to synthetically combine all home-worlds." ("Problem der Kritik einer Heimwelt im Horizont fremder Heimwelten, bzw. der Kritik einer universalen Erfahrung, die über alle synthetisch zu verbindenden Heimwelten Einheit herstellen soll, bzw. eine wahre Welt herstellen.") (p. 235). One of the interesting concepts Husserl introduces in this connection is the idea of a *"Zwischenheimatliche,"* and inevitably the problem of the possibility of an infinity of home worlds "in mediated connection, in relation to the problem of the constitution of one endless Nature out of the endless home-world natures." ("in mittelbarem Konnex in Beziehung auf das Problem der Konstitution einer unendlichen Natur aus den unendlichen heimweltlichen Naturen.") (p. 236). Note that the question is not regarding the factual existence of such worlds but about the essential possibilities.

IV. Summary

I have questioned the ideas of the purely "my own" world and the "purely" "foreign." By questioning these ideas, I have found no way of *formulating* the problem of constitution of the other culture analogously to what Husserl does in the 5th. Meditation. I have found no way of expelling the other from within my own world. The "foreign," then, is that which I do not understand. But understanding and failure to understand, the familiar and the strange, have their place within every world. It is not simply one-sidedly knowing the other, but "mutual" communication which removes "strangeness." The idea of one world for all is constituted through such communication, and may serve as *a norm* for critiquing one's home-world.

A Note on Davidson

As against Davidson, I will press for an irreducibly intentional element in the idea of culture. As for language, while the auditor interprets the speaker, the speaker does not merely produce noises but intends to mean something: thus there is a complicated web, and also layers of intentions. I agree largely with Davidson's principle of charity as it applies to an interpreter, but I see the danger of overemphasizing it in an extreme version. There is something called "understanding false sentences." To understand a sentence is not *eo ipso* so to interpret it that it comes out true, it is to know under what possible conditions it will be true or false. Thus while understanding a foreign culture (and so interpreting it), I will do my best to make its beliefs come out true, I must also recognize the limits of such an effort. The other's intention, and the other's insistence opposing me—if he is a bilingual—will remind me that I am not at liberty to interpret him to fit my belief-system. It may also be that I am to change my belief system in order to fit his. It is at this point that Davidson's physicalism helps him. Since the mental is the physical, the other's utterances are all that are there until I, or some one else, or even the speaker himself, interprets them. Interpretation *constitutes* the mental. For me, the mental (and the intentional) is *sui generis*. The interpreter's efforts aim at grasping, by approximation, the speaker's intention, the latter is its goal, and the speaker, were he aware of the interpreter's work, may, despite recognizing his generosity, oppose him. It is through such conflict of points of view that an agreed meaning may emerge. As a result, I take the question of *truth* out of the theory of interpretation. To understand the other is to grasp his intention, to confer on his speech/action a meaning which, were he bilingual, he would ideally accept—even if for me the implied beliefs (ascribed to him) were false.

To try to make them, even in large measure and not wholesale, *true*, would be to "rationalize" them in a manner that would tend to abolish the other's own mode of thinking. While saying all this, I would still insist that there would be, in the other's world, beliefs which I would not hesitate to share, and which would provide me with the foothold I need to be able to proceed to the task of interpreting the more "foreign" areas. What I want is that the principle of charity must be supplemented by the principle of empathy.

The fact is, interpretation is, theoretically and in principle, a two-way, or perhaps, a many-way track. If A, B, and C are from three different cultures, then it may be that: A interprets B, while B is also interpreting A. At the same time, B is also interpreting A's interpretation of B. A, not satisfied simply with interpreting B, also interprets B's interpretation of A. As for C, he interprets, not A and B separately, but also A's interpretation of B and B's interpretation of A, and the other higher level interpretations obtaining between A and B.

This complex situation obliterates the priority accorded to one's home language (culture, world). The other is translating mine to his, while I am translating his into mine. In and through this complicated many-layered work, we discover points of agreement as well as of difference—also an increasingly accumulating vocabulary in which to state them. The idea of "overlapping of noemata" seeks to capture this in a Husserlian conceptual framework. Note that my scheme above allows for noema of noema i.e., higher order noemata.

Lessons from Hegel and Kant

What is often forgotten by philosophers is that the tension amongst cultures is not simply a question of the authenticity of interpretations. There is also the practical conflict arising out of power and domination. Thus it is necessary to recall, in this context, the significance of Hegel's dialectic of master and slave for a culture's achieving self-consciousness through extracting recognition from the other. At the same time, we need also to counterbalance this Hegelian insight by the Kantian insight that a truly ethical community is not a political or cultural unit, but a pluralistic mankind founded upon mutual recognition and common commitment to the principles of perpetual peace. What the Kantian idea suggests is that while at the political level there is conflict of cultures, and even the interpreter is not free from the will to dominate, there is an ethical obligation which one person has for any other—no matter to which culture he belongs—which knows no constraints. The idea of an ethical community, then, is the truly universal, transcendental community—which respects political differences and cultural otherness without setting up insuperable barriers between our own and the foreign.

Chapter 6

Cultural Logics and the Search for National Identities

Algis Mickunas
Ohio University

Abstract: *The current appearance of the search for national identities may be understood in contexts of broader cultural structures. One such structure is Western Modernity, and provides a background which plays a multiple role in the search for identity: in its secular form it provides an opposition to national/ethnic groups of specific religious type; in its technical form it provides means for economic transformation or military enhancement; in its homogenizing universality, it sets a tone for postmodern movements and anti-modern oppositions; in its individuating form it becomes an attraction for freedom and liberation. This essay explores the relationships of Western modernity to various nationalistic and ethnic movements and, in a final analysis, raises a question concerning the constitution of nationality*

As if it were on cue, recent sociological, cultural, ethnic, psychological, and even communication theory literatures are focusing on identity. Various hypotheses are offered to account not only for this focus, but also to decipher what such identity would be.[1] A decade ago some sociologists contended that the problems of national identities resulted either from modernization or from some domination of one ethnic group by another and a partition or an occupation of the lands of one group's forefathers by another. Thus nationalisms and their identities have only

---

[1] Michel de Certeau, *Heterologies: Discourse on the Other*, translated by Brian Massumi (Minneapolis: University of Minnesota Press, 1986); Jack Goody, *The Logic of Writing and the Organization of Society* (Cambridge: Cambridge University Press, 1986); and Mike Featherstone, (editor) *Global Culture: Nationalism, Globalization and Modernity* (Newbury Park: Sage Publications, 1990).

M. Daniel and L. Embree (eds.), Phenomenology of the Cultural Disciplines, 147–170.

a temporary significance and, from a broader and more complex cultural outlook, are abnormal.[2]

Even if these sociological evaluations are true, they fail to account for the phenomena of the emerging quests for national identity. Are they the results of unique ethnic groups, and if so what are the claims of such groups to their identity? If it is their specific culture, then one has to show what constitutes cultural identity. The claims of Eastern Europeans reacting to the breakdown of the Soviet Empire point out that much of Eastern Europe belongs to the Western culture of Enlightenment in contrast to the Russian Byzantine autocracy that extended into the world-salvific fervor of communist domination. Here one would pit the difference between the modern Western Enlightenment with its democracy against cultural dogmatic autocracy. Yet how does one square this with the claims of Westerners that Soviet communism was a conjunction of Western political and scientific Enlightenments? Given this context, the uniqueness of Eastern European culture is not yet evident. Furthermore, if the quest for national identity arises under the conditions of one ethnic group's domination of another's land and culture, then nationalism would be equal to a group's ethnic self-identity. This presumes an ethnic purity that leads not only to the problems of philological and archaeological tracing of the original culture, but, given the typical mixture of peoples in a given region, to a myth of an original ethnic stock. This problem was apparent in the German effort to trace the pure Aryan race, which was constantly brought up short by scholars such as Eric Voegelin.

In the face of these concerns, this essay proposes to tackle the issues in terms of some broader morphologies of awareness and their symbolic designs, specifically as they relate to the constitution of the process designated as modernization. This requires a brief survey of some researchers who have raised questions of ethnicity and national identity within the contexts of symbolic systems and their possibilities either to accept and enhance or to reject and retard recent Western processes of modernization. Care must be taken not to impose on cultural phenomena some univocal symbolic design drawn from another culture. In this sense, claims by various researchers will not be regarded as final, but will be interrogated with respect to their own limitations. Thus, the Western

---

[2] Anthony Smith, *Nationalism in the Twentieth Century* (New York: New York University Press, 1979).

concept of modernization will have to be treated with care, since in some cultural contexts it may assume a symbolic significance that differs from its own enlightened purposes and rationality.

Given these concerns, it is possible to trace a transcendental logic for an eidetic articulation of broader symbolic designs of awareness within whose complex contexts one could make sense of the searches either for national or ethnic identity, and their relationships to modernization and even globalization. The best method for this venture is the *epoché* of the human sciences; it allows for an identification of invariants, their relationships, and their presence in various contexts of symbolic design.[3] In addition, it falls within the parameters of the *epoché* that if a particular explanatory thesis is offered, it will be treated equally as an aspect of a symbolic design of awareness.

I. Logics of Cultural Awareness

There is a tendency among scholars of cultural awareness to offer an encompassing, yet radically "clean" models of symbolic design. Such models range from binary all the way to quaternary structures. Sorokin's research presents an example of the binary type that separates cultural awareness into two major dimensions: ideational and sensate, each assuming different variations within cultural parameters, such as idealism vs. empiricism in the modern West and pure transcendence vs. polluting rescendence in Hinduism. Nonetheless, for him the Oriental awareness is basically ideational, although the Confucian mode of awareness allowed for mixtures. The same is the case with Islam, apart from a couple of periods that allowed the sensate awareness to have its say.[4] The ideational emphasis is a hindrance to dynamic tensions and resultantly to modernization. The West, in contrast, is deemed to be dynamic owing to a constant shift between the ideational and sensate awareness. The shifts are regarded sequentially; as soon as one modality of consciousness is

---

[3] T. Seebohm, "Die Begründung der Hermeneutik Diltheys in Husserls Transzendentaler Phaenomenologie," in *Dilthey und die Philosophie der Gegenwart*, edited by E. W. Orth, (Freiburg: Verlag Karl Alber, 1985), 98.

[4] Pitrim A. Sorokin, *Social and Cultural Dynamics* (New York: American Book Company, 1937), Vol. I & II.

exhausted, it switches to the other and thus redynamizes the cultural process.

It is peculiar that Sorokin would regard the latter process to be normal, while other cultural designs as abnormal. The distinct features of other cultures are modes of deviation from the dynamics of the West. It seems that Sorokin overstates his case by accepting a modern awareness of the West as the standard. Even the classical Greeks maintained a very close intertwining between these two modes of awareness. This can as well be said of various Medieval periods. Thus it is apparent that some Oriental and Mid-Eastern modes of awareness are similar to some of the Western modes. An important question is which among these modifications would yield frameworks for processes in search of national identities and their ability to incorporate the modernizing trends?

While Sorokin seems to be restricted to a mode of awareness most appropriate to Western modernity, he leaves a basic issue concerning awareness unattended. What is it in the Western consciousness that prompts the shifts from one to another mode of symbolic design, and what is there elsewhere that would prevent such oscillations? The problem is one of the constitution of crises. And this problem must point to a possibility that if the Western culture is more dynamic, then it might be that it has no means to absorb crises and may thus be led to a loss of its own distinctness. Whether such a loss may appear with modernization is a question still to be addressed.

Another researcher who follows a binary, although somewhat more elaborate structure of symbolic design, is Dumont. He claims that complex cultures are distinguished by their dominant ideologies, composed of a network of symbols that legitimate the relationship of an individual to society in a hierarchical pattern.[5] For him, the traditions in which the individual is subject to a society are normal; the modern type in which the individual is prior to, and is the origin of culture, is abnormal. The reason for this claim is pragmatic: normalcy belongs to the type which is best suited in preserving continuity and stability. As a result, an all-encompassing ideology, which subordinates all power, must be at the top of the hierarchy, while the lower, the praxis level, must keep power and ideology separate. This separation varies: it may appear with a distinction between priest and ruler, e.g., in India, or church and state, e.g., already

---

[5] Louis Dumont, *Essays on Individualism: Modern Ideology in Anthropological Perspective* (Chicago: The University of Chicago Press, 1986).

in the medieval West. Even if the members of the lower level threaten disruption, they would be readily integrated at the higher level. Lacking such an arrangement, the modern types of culture face a constant threat of dissolution that leads to totalitarianism as an artificial means of recouping structure. In this sense the modern Eurocentric conceptions are, for him, deficient; the standard type would be India. It is to be noted that his comparison of Western modernity with traditional India allows him to critique *only* Western modernity.

A more complex analysis is offered by Weber. He shows that complex cultures do not require ideal integration; rather their very continuity depends on struggles among diverse groups, unions of groups, social organizations, and ideologies. Coherence may be found only at the level of culturally organized life, yet what holds complex cultures together are the modes of managing the struggles. Such modes provide benefits and are partially enforced against those who obtain less benefits. The latter constitute a source of dynamic change. Yet such a change is more complex than a mere uprising. For Weber important transformations, such as breakthroughs into modernity, require a conjunction of elements that show up very seldom. One requires some form of attainable ethos addressed to restless groups with their unique imagination of the future and practical interests, the legal system, and community and kinship structures.[6] The Occident is a tension and a balance between the spiritual, the organized church, and the monastic community, and between a unique voluntarism of the feudal state, with a limited covenant and an autonomous bureaucracy possessing a general power of political coercion. But in the development toward modernity, this is the source of Western abnormality. Once it reaches maturity, it loses the positioning of various, even if unequal parts, and becomes dominated by a homogeneous logic of rationalization; it becomes abnormal, an iron cage.[7] If normal is less rational, then an opposition to modernity will appear in the guises of the antimodern, the postmodern, and the archaic. The current appearance of nationalisms tends toward archaisms that are ethnocentric and bear traces of sacrality, tensed against modernizing secularity.

---

[6] Max Weber, *The Protestant Ethic and the Spirit of Capitalism* (New York: Charles Scribner's Sons, 1958); *The Religion of China: Confucianism and Taoism* (New York: The Free Press, 1952).

[7] Max Weber, *Protestant Ethic*, 182f.

A critique of Weber is offered by Eisenstadt.[8] He argues that Weber's thesis is inadequate to encompass the complex cultural designs because this thesis is basically sociological and fails to establish an all encompassing theoretical format for the study of all cultures. For him, the latter are processes that relieve continuously the tension between the transcendent and the mundane regions. In this sense, he also operates with a binary structure. Cultures exhibit various solutions to this tension: a contemplation of an (other-worldly) transcendence that is characteristic of Hinduism and Buddhism; mundane ethical action, apparent in Confucianism and secularism; and the nexus between transcending and this-worldly action. Thus in Islam there is a strong separation of military and political practice from an ordered transcending orientation. Islam can accept modernization in military and other practical spheres. A different form of this nexus can be found in medieval Europe and early modernity, present in the interconnection of two salvations: the Faustian and the Promethean or the this-worldly and the other-worldly salvations.

Cultural symbolisms are invested with a power to transform a social order. Yet the extent to which such a transformation occurs depends on the direction taken by a social order, framed by the tension between the transcending and the secular domains. Thus the social order may move toward the transcending and contemplative, or to the this-worldly, toward their separation, or their nexus. Regarded socially, such moves depend on successive, symbolic separation of peripheral from central activities, and vertical hierchization of social strata. Given this symbolic arrangement there appear various possible alliances that can modify a culture to yield a solution of this tension: socially, elites can form alliances among themselves, e.g., the political and the religious; religious elites may ally with some segments of a gullible population, and ruling elites can join the bureaucratic technocracy, etc. How would these alliances address modernization's problematic, primarily with respect to the tension between the transcending and the worldly? Modernizations do not seem to offer anything that is transcending. Indeed, the latter might appear as postmodern factors in modernity and ethnic archaizations in nationalism.

The most complex articulation of cultural symbolic designs of awareness is offered by Gebser.[9] Among historical and current cultures,

---

[8] S. N. Eisenstadt, *Revolution and the Transformation of Societies: A Comparative Study of Civilizations* (New York: The Free Press, 1978), 322f.

[9] Jean Gebser *The Everpresent Origin*, translated by N. Barstad and A. Mickunas

he traces at least five structures of awareness: the archaic, the magical, the mythological, the mental, and finally the integral. For him, all cultures contain these structures, although what provides a cultural uniqueness is the predominance of one structure over the others. Such a structure plays a dominant role for interpreting other structures. Thus in a mythological structure, mind or rationality may play a subordinate calculative role in practical affairs, while magic in rational culture may turn out to be purely technological: while designed rationally, military and political organizations are deemed necessary to protect and enhance the vital interests.

For Gebser, as for Weber, modernization is rationalization and homogenization of all phenomena, with an added presumption of rational universality. This modernization process, specifically in its reading of rationality as instrumental, i.e., technical, tends to abolish cultural differences and national identities. This produces, for Gebser, and, we may add, for Husserl, vast cultural crises. The latter emerge on the background of one mode of awareness, such as quantitative reductionism, into the exclusion of other modes. For Gebser, such crises appear in the cultural practices of reversion to archaizations, magical ritualizations, New Agisms, as efforts to achieve authenticity and identity. One could say that the search for authenticity of Heidegger and his deconstructionist followers is premised on this reversion.

Gebser's investigations overcome the binary logic of the researchers mentioned above, although his thesis is in partial accord with Weber's and Husserl's assessment of Western modern rationality. His essential contribution for our issue is not only the tracing of multiple symbolic designs, but preeminently in his showing what modalities of awareness appear within modernizing processes when the latter become homogenizing and reductionist. Given the current modernizing processes, it is necessary to address modernity as a context within which the quests for ethnic and national identity are located.

II. Modernization

Modernizations, whether ancient or current, possess complex cultural designs. Such designs may maintain invariants that clash and even move in opposite directions. Modernizations postulate individualism and its

---

(Athens: Ohio University Press, 1984), Part 1, Chapter 3, § 2.

separate claims; they invent history as a context for articulating all significance as human achievement; they tend to postulate various forms of rationality, all regarded as impersonal and universal, each vying with others to be the final logic or master discourse, proposing a universality and an equivalent application of normative and legal rules to all social members; then there is humanization in the sense of devising rules and techniques for the benefit of humanity. Concurrently, such tendencies reveal a general homogenization and fragmentation of cultural life. In Western modernization the former assumes a specific form that allows it to be extremely virulent in an economic domain and in the production of technical power comprising main sources of dynamic that fuel a conception of development and globalization. This form is the homogenization of the environment as qualitatively indifferent matter that can be made into desired products. For Gebser this is magical awareness; instrumental rationality can change anything into anything—as if by magic. This leads back to the question whether cultural distinctions and diverse modernizations can be maintained within this globalizing energy with commodity production and exchange being its common denominator.

At the theoretical level, the conditions for homogenization are afforded by the shift from presentational thinking—Being is present—to representational thinking—standards for Being derive from representations as inherent in, or constructed by, the subject. One condition for homogenization, at one level of representational thinking, is a choice of quantitative language, as most appropriate for the representation of reality. Yet it is this choice (implicitly granting primacy to will over reason) that allows representational, modern thought to constitute within itself its own postmodernity. Quantitative language cannot retain its representational character and turns out to be, in principle, signitive.[10] Signification removes the subject from the moorings of vertical intentionality and permits both the transformation of representations at will and the quantitative homogenization that designates all events as equivalent and thus transformable one into the other. The subject becomes completely detached from the world and is posited as a source of signitive meanings. It can float and become nomadic. The latter term—as the background of various "discoveries" by postmodern writers—is, as a matter of fact, an inherent aspect of Western modernity.

---

[10] Elisabeth Ströker, *Philosophical Investigations of Space*, translated by A. Mickunas (Athens: Ohio University Press, 1987), Part II, Chapter 2, § 4.

What lends this shift from presentation to representation, and then to signification, its dynamics, its flux? We must find an account for its sudden explosion that promises, in literal terms, globalizing Macdonalization of cultural diversities, provoking Jihadic opposition by fervent nationalisms, ethnocentrisms, and their attendant archaizations. First, the choice of a specific signitive language is not innocent. Since modern signification has no vertical nexus, it must be technical in order to access the world, i.e., definitory of the object and at the same time providing the rules for constructing the object. As Husserl had noted, mathematization is a technique.[11] If one were to assume the classical Platonism where mathematics is metaphysical presence, then the modern conception of mathematics as signitive construction of the world could be called, with Volkmann-Schluck, metaphysics gone wild.[12] This is to say, there being neither presence of the world nor its representation, one can signify the world at will. In this sense, the mathematical, signitive reason, turns out to be instrumental at the service of will. The latter is the modern nomad that can resignify all events in accordance with its wants and arbitrariness, leading to the well known primacy of power for the sake of power.

The logic of power is its self-incrementation. Unchecked by the vertical presence of the world—the intentionality that requires respect for qualitative differences imbedded in symbolic designs—the metaphysics of will has no other object apart from increasing power.[13] This is the magic of modernity that subtends reason and, resultantly, constitutes the ground of postmodernity within modernity. All events can be signified at will, and produce instrumentally through human signification. Yet at another level this constitutes a catalyst for the modern emphasis on individuality and for democratization. It is to be noted that this modern level is invoked by nationalistic and ethnocentric movements, each demanding "freedom" to decide its own life and destiny—while using modern technical means to obtain it in a holy war for the sacred lands of mythical ancestors.

Given this background of postmodernity within modernity, the current proliferation of postmodern theses can make some sense. Postmoderns

---

[11] Edmund Husserl, *The Crisis of European Sciences and Transcendental Phenomenology*, translated by D. Carr (Evanston: Northwestern University Press, 1970), 23ff.

[12] Karl-Heinz Volkmann-Schluck, *Einfuehrung in das Philosophische Denken* (Frankfurt am Main: Vittorio Klostermann, 1965), 59ff.

[13] Karl-Heinz Volkmann-Schluck, *Politisches Philosophie* (Frankfurt am Main: Vittorio Klostermann, 1974).

conceive symbolic designs of awareness and reason to be fragmentary, bricolagous, decentered, multi-layered and multi-perspectival, with a stress on, and incognizant acceptance of modern, individual uprootedness. The writings of postmodernists (which in the main confuse modern social conditions of fragmentation and the lack of encompassing discourse with postmodernism as a hypothesis of radical cultural transformation) tend to embrace demands for unhindered self-expression, calls to authenticity, and acceptance of inconsistencies of occultic movements. On the whole, these works fail to distinguish between antimodernity and postmodernity. The former rejects all that is affirmed in modernity, including representational and signitive modes of thought, and is very much immersed in the disconnected presence. It is a negative, an oppositional movement that repudiates everything. It deconstructs all modern trends, proclaiming at the same time, that there never were recognizable trends. Counter to postmodernity, it also rejects signification.

Postmodernity might equally propose to reject signification, but it does so by claiming that modern significations misrepresented and marginalized different cultural designs of awareness.[14] Postmodernists, including Foucault, admit that signification is modern, yet Foucault asserts all discourse to be power.[15] But this is precisely what is already present at one level of the modern metaphysics of the magic of will and technical power. It can be argued that antimodernism also remains within the modern system upon which it is merely parasitic. It has nothing to deconstruct—since of itself it offers no tangible thesis. It is similar to Hegel's charge against Fichte's movement to be free from, but not for something. If deconstructionism were to claim that after becoming free from all theses in order then to master all culture in its own way unconditionally, then it would accept the autonomous will of modernity with its nomadic power. The various movements, from postmodernities through antimodernities, function on a background of their constantly attacked, yet always enjoyed, egalitarian culture of democracy with its specific mode of modernization.

---

[14] Gilles Deluze, *L'Anti-Oepide* (Paris: 1972).

[15] Michel Foucault, *Power/Knowledge: Selected Interviews and Other Writings 1972-1977*, edited by C. Gordon, translated by C. Gordon, K. Marshall, J. Mepham and K. Soper (New York: Pantheon Books, 1980).

III. Identity and Modern Globalization

Modernizing globalization purports to institute a universal and leveling cultural design that seems to threaten cultural differences and, within their context, national identities. The previously mentioned scholars have a variety of answers to this issue, especially in the current talk of modernization. Although there are dynamic processes, such as economic capitalization and technical improvements, secularization, democratization, and even laxity in human relationships, a fear may arise not only of homogenization, but above all of total dissolution of any guide posts for human life. Sorokin's binary and sequential recurrences of symbolic designs yield no clues concerning modernization that are secular, developmental, and simultaneously sensate-material and quantitatively ideational. After all, homogenization of the cultural environment combines mathematical ideality and atomistic materiality. Moreover, it is difficult to find even generalities within his framework that would distinguish among cultural designs. He would add very little to the current quest both for cultural and national identities and the efforts to rediscover or to invent such identities. In the context of modern dissolution of distances and a creation of interdependencies, if these units are to be identified, they might lose the identities which were possible when they maintained greater isolation from each other.

Eisenstadt's binary logic of transcendence and secularity, which allows for their mixture, fails to recognize that modern secularism contains its own transcendence—mathematical ideality and atomistic materiality—without ceasing to be this-worldly. In addition, his thesis offers no clues concerning the dynamics of transformations, such as Gebserian magical-vital awareness. Gebser allows for a multi-layered and multi-directional analyses and a possibility of crises that integrate postmodernity and even antimodernity. It was noted elsewhere that the freedom of modern signitive symbolizations allow for separation of institutional spheres.[16] Despite the superficial occultistic show, the contemporary West does not follow the solution of the transcendence-secularity tension by a "mixed" model, but is, in the main, secular. The efforts to weave sacrality into practical life have failed even in the Soviet Union's play with Marxism as one form of sacrality.

---

[16] S.Eisenstadt, *Revolution and the Transformation of Societies: A Comparative Study of Civilizations* (New York: The Free Press, 1978), 322ff.

The modern West is more akin to the pragmatic attitude of traditional China. If there is a transcending dimension in the modern West, it is globalizing homogenization and not other-worldliness. In this sense, the current West would have no direction and would thus be chaotic. Even if we presume that earlier cultural designs were composed of binary logic, we cannot regard such structures to be inherent in the contemporary modernistic processes. The latter compel current cultures toward incessant readjustments and crises. Indeed, Gebser suggests that such a binary logic only constitutes a mythological context dominated by polar tensions; it cannot be adequate for modern dualisms of reason and matter, since such dualisms are integrated in a technological magic of production of both objects and subjects.

Weber is keenly aware of, and totally focused upon, what he regards as a universalizing spread of the modernistic West. Since this spread stems from a particular culture, it posts a threat to the identities of other cultures. This spread could be called a logic of partial universality claiming to be the sole universal cultural design and rationality. It is to be noted that this form of rationality—the instrumental—is not even identical with the classical rationality embodied in the concept of *theoria*. Yet this universality, according to Weber, might affect social organizations of other cultures for the worse by imposing a rigid bureaucratic uniformity which, in its all pervasiveness, becomes culturally anonymous and inaccessible. For Weber this situation might compel cultural groups, even in the modern West, to fall prey to new prophets or to aim at a rebirth of archaic notions.[17] If there is postmodernity in Weber, it must be a resurgence of traditional symbolic designs. Such resurgence has been and is attempted around the globe across various efforts to infuse the globalizing modernity with modes of transcending archaization.

It would be too simplistic to think, with Eisenstadt, that a current secular modernity could be disrupted by the transcendent. Such efforts, in various guises of fundamentalism, cannot avoid the globalizing language of Western modernity. "Among fundamentalist circles in Iran, Egypt, and elsewhere, a new Islamic political language is emerging, which owes an unacknowledged debt to the westernizers and secularists of the past

---

[17] Weber, *The Protestant Ethic*, 182.

century and their foreign sources, as well as to prophetic and classical Islam."[18]

If modernizing is both a leveling and a fragmenting, what ought a culture have in order to achieve national identity and integrity? One could presume a perduring linguistic community, expressed in literatures deemed to be *our own*, as a saving grace of cultures and nationalities. If this is to be successful, it must be shown how the literatures could absorb and reshape the impact of modern globalization. Perhaps the Hellenic culture may provide a clue. The Greeks could easily absorb various myths from surrounding cultures, although they subjected them to Greek interpretation which operated not on a basis of archaization or their clashing ideologies, but on a shared number of institutions that included dialogical flux and mythological pluralism.[19] The institutional context, allowing for protagonal and antagonal interpretations and public dispute even what the sacred, did not propose any one mythological content, but could incorporate, in a unique way, any content. Yet in current modern globalization, the cultures absorbing modernization cannot merely treat it as an ideological content to be either contested or compared; it comes in forms such that in order to understand its language one has to conflate it with the modern bearer of the language: mass media technology.

Despite its various drawbacks, the work of Dumont may be partially suited to explicate modernizing globalization and the maintenance of cultural and national identities. In his view, each culture undergoing modernization develops its specific modernistic ideology, stressing individualism and historicism, by unifying it with archaic myths and magic rituals. What is restrictive in this conception are the examples, borrowed almost exclusively from the modern West, such as England and France, and some of the border-line Europeans, such as Germans and Russians. Even if it were possible to extend his claims to other cultural designs, they contain various options that might not square with modernity. If there were to occur a pseudo-unity of modern individualism and archaic tradition, totalitarianism might well be its result, abolishing modernization, although using its technical means to obtain archaic ends. The synthesis

---

[18] Bernard Lewis, *The Political Language of Islam* (Chicago: The University of Chicago Press, 1988), 116.

[19] Julien Ries, "Gedanken zur Hellenisierung der orientalischen Kulte," in Hartmut Zinzer, Hrsg., *Der Untergang von Religionen* (Berlin: Dietrich Reimer Verlag, 1986), 51ff.

is at the center of the problem, since it wants to abolish the distinction between the individual and the traditional unity, leading toward a dissolution of individuality (the case of totalitarian states) or a dissolution of social cohesion in favor of individualism (the case of Calvinism and its branches).

Despite the pitfalls cultures, for Dumont, could maintain their identities in modern globalization by establishing their local solutions. This option has not been appreciated by the modern West because the West tends to regard its particular universality as all encompassing. If there is a solution for Dumont, it would have to be premised on an acceptance of modern globalizing within the parameters of its particular universality, with material or practical values. Yet at the level of symbolic designs of cultures, the indigenous mythologies will remain distinct and engaged in oppositional confrontations and mythological evangelisms. This would mean that the symbolic designs of various cultures and their national identities would have to abandon their claims to be the universal civilization—as is the claim by various fundamentalisms such as Islam—and to accept a position of being a particular universal. What emerges in this context is a struggle between the claims of two particular universalities, the globalizing practical and the localized mythological, each purporting to encompass the other. Can the symbolic designs of local cultures subsume the globalizing practical under its own parameters? What could be anticipated is an increase in revitalization of mythologies and their efforts to make themselves part of a universal discourse and confrontation.[20] The current case is Azabaijan, Armenia, and Turkey, accepting modernization, and yet taking sides within the parameters of mythological confrontations supplied with technical means to do battle.

This combination poses unique problems for specific nations. The contemporary West faces an issue of its own uniqueness. Its tendency toward modernistic globalization, dependent on the primacy of secularization and—apart from resurgence of fundamentalistic claims—the historical equalizing of all mythologies, artistic styles and experimentations, feminism, the self-critical intellectual milieu, the demand for tolerance of ethnic diversities and life styles, the non-foundationalisms, etc., pose important questions: for whom, and to what extent, is the current West *traditional*? Does it have any basis or structural identity? Although the

---

[20] Jo Ann Chirico, "Humanity, Globalization, and Worldwide Religious Resurgence: A Theoretical Exploration," *Sociological Analysis* 46 (1985): 219-242.

West's particular universality, its praxis, has pushed for hegemony, it seems to have exhausted the origin of its uniqueness; one can join any cult or tribe, but without any commitment; one may become a universal nomad, citizen of the world, gracious but ridiculous, like the meetings of the United Nations—completely ineffective and nondescript.[21] Indeed, Fujimora drew no overly surprising conclusion by pronouncing the *the end of history*, although those who inhabit this end will, according to him, live bland lives.

Various fundamentalisms, notably the Islamic, Judaic, Christian, and of late, the Hindu, are not yet willing to put secularization and thus humanity first. They cannot accept themselves as parts of the world; they deem their mythologies to be self-sufficient in the sense that their specific texts are universal and constitute the very imminence of the world. Yet the very tensions between the fundamentalisms and secular modernity reveal that in the age of praxis globalization, self sufficiency is impossible. An effort to establish a particular mythology as the sole universal culture diminishes its ability even to attain its own local ends. This is equally well demonstrated by Marxian eschotology and its exclusive purpose of the constantly postponed universality.

IV. Modernization—Archaization—Postmodernity

In the contemporary setting, nationalisms must be regarded in relation to the globalization of modernity that compels cultural encounters.[22] The modernizing impetus, with the above articulated trends toward individuation, etc., institute a mode of thinking and behavior that, compared with the archaic mode, appears to be fragmented, abstract, and even artificial. This is affirmed by the Taoist wholeness in contrast to Confucian moves toward moralizings and political and social constructs. At the archaic level of a tradition that is subject to modernization, the all-encompassing, local, undifferentiated hierarchies of mythical rituals, and even medico-magical practices, comprise an eternal presence. At this level nationalisms, with their wholeness and unity, arise with their overemphasis on power rituals, mythical pronouncements, and even blood-race mysticisms. It is an identity

---

[21] Anne Roiphe, *Generation Without Memory: A Jewish Journey in Christian America* (New York: Linden Press, 1981), 214.

[22] Anthony D.S. Smith, *Nationalism in the Twentieth Century* (New York: New York University Press, 1979).

in unity, where the pronouncements are identical with the voices of the forefathers, the flag is the nation, the ritual is the divinity, and the sacral sayings are curative.

For Gebser, such currently ritualized practices are premised on magical awareness. It includes ethnic and nationalistic movements. This nationalism, even under the guise of broader cultures, has its sacral sites to die and to kill for. They are the concentrated sites radiating all significance. Thus even the emerging Hindu, Jewish, Islamic, and Christian fundamentalisms are ready to claim those sites. If there is a mosque sitting on a sacred site of Rama's birth, the mosque is to be torn down to yield room for the temple of Rama. Such are the efforts to reclaim an archaic identity of a Hindu, a Jewish, a Christian, or an Islamic nation in face of modernizing globalization. At the same time, each archaization claims its legitimacy for a sacral site (or land) on the grounds either of mythological ancestors or divinities. This archaization, manifest in its virulent form in the regions of former Soviet Union and Eastern Europe, appears in the claims by each ethnic nationality to the right of a particular place that was "ours" by dint of archeological readings of excavated potshards.

This is quite different from the modernizing nationalisms, such as that of the U.S. with its strong fragmentation and minor ritualization (except during national crises where certain archaic forms are emphasized). The emergence of archaisms with their mythical expressions in modernizing processes are designed to restore the undivided unity that became fragmented by modernity. It is to be noted that restorations are premised on the disrupted unity of a people, and its dynamism and virulence are generated by a modernizing presence, required as the point of difference and oppression. The Soviet insistence on scientific praxis and its capacity to outmodernize modernity comprised not only a homogenization both of national and ethnic identities, but became coupled with an unacceptable imposition of Russian culture. The discussions that dominate the nationalities and ethnicities of this region fluctuate between a wish to reject Marxian modernism since it was imposed by a more or less inferior culture and the wish to reject that culture while maintaining the modernizing impetus. Russian nationalists maintain a similar attitude with respect to Marxism, with some additional clashes between the Russian slavophilic soul and its orthodox mythology, on the one hand, and the Marxian Judaic mythology mixed with Western materialistic decadence, on the other.

It has been noted that modernity contains its antimodernism. While the latter is parasitic on modernity, it cannot form any ties with nationalistic archaizations, since it does not contain any unifying principle and cannot offer ritualistic resistance against modernization. Although our previous discussion painted postmodernity as an undisciplined process of accepting a modern arbitrary signification, it can become self-conscious of its own mature ability to establish connections among various analytical fragments of modernity not by positing an encompassing discourse, but by tracing resonances among institutions and disciplines of modernity.[23] At this level of postmodern awareness, it may be possible to regard the postmodern awareness of modernity as a nexus among various facets, the discrete components, of modernization and nationalistic archaization. Such a bridge comprises diverse resonances that can find connections unnoticed by modern analysis. Thus one can find legalisms in scientific language, hidden valuations in positivistic philosophies, salvific rhetoric in secular politics, and capitalist economies in the fundamentalist promises.

Given these relationships, we can enunciate the current factors that appear within and in face of Western modernity: there is the modern-antimodern crisis, requiring their mutual implication and mutual exclusion; there arise the archaic moves as nationalistic and/or ethnic searches for identity that may be the inner disruptions of modern homogenization, its inner threats that constantly promise a protective enclave of sacral unity and archaic identity against the winds of self-reliance and autonomy. Then there are the partial mythical universalities with claims to salvific transcendence. The antimodernity, as we saw, rejects the presence in archaization, the representational subject, and even signitive processes. As a result, postmodern awareness takes the modern facets and traces their resonances and intersections, without positing a necessary continuity among them. It is of note that each facet resonates with and incorporates differently the facets of other modes of consciousness into systems of shifting networkings which, in their openness, make incomplete sense. Thus the archaic level, while surpassed and fragmented, is traceable in the specific hierarchy of modernizations, at least in symbolic designs. For a particular hierarchy to be valid within

---

[23] Hans Bertens, "Die Postmoderne und ihr Verhaeltnis zum Modernismus," in *Die Unvollendetervernunft: Moderne versus Postmoderne* hrsg. Dietmar Kamper und Willem van Reijen, (Frankfurt am Main: Suhrkamp, 1987), 46-98.

modernity, it must either be posited as a structural function of modernity, itself capable of becoming global, or made impotent and merely symbolic, as in the case of the existing European monarchies.

Obviously, some archaizations may include cultural claims to universality, such as virulent moves that promote return to some mythological unity. They may comprise a totalitarian temptation, a holy war against secularization, proclaiming the latter to be demonic, the Great Satan, and thus evil. This type of temptation may be culturally enhancing for the believers, but it turns to a modern nomadism for a purely abstract conquest of evil whose secular members will be regarded as total enemies. Although this nomadic trend may be latent in postmodern writings, specifically in the claims that it is a collection of experiences, none being subject to the controls of the others, it does achieve its own culture by creating spontaneous, even if unexpected, resonances across most diverse domains. Despite the modernistic separations, one finds attunements that defy modern ideological divisions. In the face of vast differences in ideology between fundamentalists and feminists, there may be an attunement concerning types of art. In brief, what strictly rationalistic and technocratic modernity rends asunder can become congenial and even conjugal. Nonetheless, the postmodern awareness will not survive if such of its factors as archaizations remain purely symbolic without establishing political discourses. Being stuck in one's own absolute posture of uniqueness and mythological truth, no resonances will appear. While this retards human cross-cultural relationships, the more dangerous retardation of such relationships and of modernizations occurs if there is a unity between the modernistic and the archaic.

This is to say, a culture that enters modernization and organizes itself in archaic ways becomes a threat to others and destructive to itself. At one level this was exemplified by the Soviet Union. It would be equally dangerous to others and self if an archaic mode of culture becomes capable of adapting vast modernistic means to its own purposes as happened with Nazi Germany.[24] This course would also appear if a monotheistic fundamentalism were to establish a political purpose: every modernistic means, including nuclear holocaust would be legitimated to "save" the world. Traces of this tendency are manifest in the current

---

[24] Jeffrey Herf, *Reactionary Modernism* (Cambridge: Cambridge University Press, 1986).

West under the designation of "political correctness." In this, and also in the other three modifications, national identity fails.

Archaizations, all the way to ethnicity, are insufficient for the maintenance of national identities for other reasons too. A contemporary appeal to various fundamentalisms as archaic modes of retaining one's national identity disallows the latter, since current fundamentalisms transgress national and even cross some levels of cultural boundary. At such levels they become incompatible, as is the case with modern jurisprudence and any archaization appealing to divine or even naturalistic laws. In turn, various nationalities, whose origins are distinct from the origins of the fundamentalist texts, cannot claim to go back to their archaic roots by appealing to such texts. Other grounds must, therefore, be found for the constitution of nationality, specifically in the context of modern globalization. It may be plausible for postmodern culture to provide the resonances among various aspects of a nation, including the archaic, modernizing, major mythologies, ethnic claims, and globalizing requirements. Yet such resonances remain floating, exploratory, bereft of a stabilizing capacity and reliable practical judgments. It must be borne by something which allows it play-space. The task cannot be taken up by modernistic culture of the West which is overly flat and possessing homogeneous, formalistic, and rule bound instrumental and magical rationality. Its technical achievements tend to subdue, dull, exhaust, and level the identities of multi-faceted cultural designs. Genuine archaic rituals are always provincial, stuck in the nebulous and mysterious sacrality of ancient places and texts. Adherents of such texts and the continuous repetition of one's own former greatness have only revenge for those who intruded—if it only had not been *for them*. Thus we still face the issue of national identity in the context of globalization that tends to subsume cultural differences and ethnocentric and nationalistic claims.

V. Nation and Culture

Can we say provisionally that nationalism is best qualified to resist the globalization of modernity? First, it behooves us to differentiate nation from ethnicity or the state. Gleaning from a variety of works analyzing this question, it is possible to limit the view of nation to the following: culturally, it cannot be homogeneous, but must be an imagined community with institutions that sanction not only a diversity of activities but above all the destinies of individuals and groups. It may possess an actively shared language that lends itself to creation of a spontaneous collective consensus for mutual living. The sharing of public institutions for a common purpose constitutes a basis of mutual respect among the members of such a community.[25] The nation is thus an invention, even if in some instances it is formed by ethnic imposition—as was the case of the Soviet Union or Yugoslavia. Nonetheless, even in these cases the nation was premised on a conception of cultural identity with strong or weak celebrations of its founding: great achievements, sacrifices, monuments, artworks, and psychological designs built to exhibit and enhance all these. Indeed these cultural productions are imaginatory variations of the continuous inventions of a nation.

It would seem that the current turmoil, the political rhetoric—in the face of the lack of imaginatory creativity on the right—are bland efforts to continue the invention of the *American nation.* Yet comparative cultural studies suggest that a replacement of the cultural sphere by political rhetoric, leading to a conflation of politics and culture, comprises the greatest danger to others and one's own efforts to constitute and to revive a nation. This conflation might create an appearance that political rhetoric is capable of dynamizing nationalism, yet it becomes a contentless separation of the population on an ideologically fragmented battleground that reduces both the ideologies and the population to the modernistic level of practical interests. These, clearly, abolish national culture in favor of globalization. Neither corporations nor the workers find any allegiance to any nation. Zenith is pleased by the cheap labor in Taiwan, while Zenith workers who have lost jobs are more than happy to get jobs with Nissan. Here one becomes a subject not of a nation but of a global corporation whose loyalties are to global markets; the homogeneity of making and technical magic are predominant.

While nationalisms that conflate culture and politics, and even conflate nationalism with archaization, may lead to the abolition of the nation and to an institution of totalitarianism, democracies have an advantage insofar as they disconnect culture from politics. In this way, cultural symbolic designs resist—and even oppose—the efforts of political and ideological encroachments. If such enchroachments occur, various results follow. First,

---

[25] Anthony Smith, *Theories of Nationalism* (New York: Harper and Row, 1971); O. G. Anderson, *Imagined Communities: Reflections on the Origin and Spread of Nationalism* (London: Verson, 1983).

ideology becomes identical with cultural expression and thus ceases to animate an imaginatory continuity of a nation. Second, nations consisting of diverse ethnicities will become fragmented by the ideological stifling of free ethnic contributions to the nation. Third, the moment of presence of a local community will be moved toward an arbitrary ideological signification that does not even have a representational value to such a community. Fourth, such a conflation might be seen by diverse ethnic communities as an imposition of an alien ethnicity within a nation and thus a rejection of the nation. Fifth, the resultant separatist ethnic movements will call for their own "nations."

It seems, then, that nations function optimally in a democratic context such that political ideologies are spread across groups with diverse interests that allow each group to find opposing ideological connections at diverse levels. The same distribution appears across most diverse cultural orientations, disallowing a conflation of the political ideologies and cultural designs, and preventing a pure division into homogeneous majority and minority tendencies. The recent attempt at such conflation, with claims to majoritarian morality, has been exhibited by the rhetoric of Dan Quayle and the so-called Moral Majority.

In a current setting, nationalisms confront modernizations and archaizations. During the last century in Germany and Eastern Europe, and in the contemporary mid-East and the republics of the former Soviet Union, nationalism tends toward archaic cultural identity. In contrast, the American and French, and to a lesser extent British, have been culturally modernizing.[26] To repeat, the distinction between modernizing and archaic nationalisms lies in the claims that nationality derives from individuals having rights (modern) and individuals as bearers of the national spirit of an original people.[27]

Nationalisms turn to archaisms when the imagined nation is still in its formative stage, arising from ethnic or tribal units. This is the case in various African efforts to form nations. One finds a difficult diversity that forms an archaic-modernizing with conflicting tendencies in India. This conflict appeared in various guises between Hindus and Muslims, and in the heated controversy stemming, for example, from the television

---

[26] Yehoshua Arieli, *Individualism and Nationalism in American Ideology* (Cambridge: Harvard University Press, 1964).

[27] Louis Dumont, *Essays on Individualism* (Chicago: The University of Chicago Press, 1986).

series on the Mahabaratta. It is of note that some archaisms, such as fascism, may well be antimodern with respect to modernistic cosmopolitanism. There may appear postmodern nationalisms—in the case of Vaclav Havel.

If we trace the phenomena of the more established nationalisms, we shall find that nationalism has exhibited, on a variety of cultural levels of modernization, symbolic designs that appeal to diverse groups with various positive and negative intensities. Such nationalisms may be associated solely with the *tolerance* of varieties. This means that nationalistic culture is grounded, in modernity, in the possibility of everyone partaking and actively participating in a chosen variety of cultural work. In this sense nationalism of this type is more resilient than a major monistic cultural movement which has a much lesser chance of survival. It could be well argued that in the current setting, the resiliency of modernizing nationalisms will confront the fundamentalist absolutisms of archaic type, although each, in its particular universality, will claim to be *the* universal culture.

This does not imply that the relationship between modern nationalism and monistic mythologies does not contain modifications in their relationships. Monistic mythologies too have undergone modernizing modifications. Thus, an effort in the U.S. to promote a fundamentalist for the highest office would create a fusion of the American nation and monistically rigid mythology; that is, the nation would be absorbed into mythology. But modernist nationalism can maintain peaceful coexistence with a tolerant mythology and a mutuality of diversity, which is apparent in mainstream Protestanism. Even in case of conflicts, the latter is willing to accept resolutions on the grounds of mutually established rules. No doubt, modernist nationalism can find affinities with monistic fundamentalisms either by using them in cases of national defence or by treating them as one among other cultural claims. Yet in some cases, monistic fundamentalisms may tend to rule over their nationalisms in the face of secular modernization, as is the case in Islam. In other cases the degree of their influence depends on the extent to which modernist national culture, with its inherent postmodern capacity to resonate, has become generally pervasive, as is the case in the U.S. It goes without saying that archaic nationalism, joined with some typical monistic fundamentalism, would oppose, in principle, the modernistic tolerance of mythologies. All this depends on institutions that have, within them, the capacity to resignify themselves in a postmodern sense.

It seems then that a maintenance of complete modernistic nationalisms requires the endurance of a multiplicity of cultural designs promoted very strongly by the presence of postmodern culture. Given the exclusion of its virulent stage of political correctness, postmodernism creates psycho-social and conscious resonances that connect quite diverse cultural phenomena without positing rules of integration and enforced unity. Myths can resonate with anarchistic modes, jazz can become classical, and classicism can be the paradise of advertisers. Rock can become a cult of sensuality and resonate with mythical fever. Technocracy can become magic, and the latter can become political rhetorical theatre. It allows an emergence of multiple discourses that find connections at levels previously unsuspected. This leads to a language in constant and yet recognizable transformations—not to antimodern and empty deconstructivism, but, to speak with Merleau-Ponty, to coherent deconstruction.

VI. Postscript

The globalizing mode of modern consciousness, while stemming from a particular cultural universality, is deemed to be the most pervasive framework of praxis. Its reductionistic tendency toward functional efficiency and its technical reification, amoral position and presumed neutrality would disregard nationalistic cultural diversity. Even if there were an appearance of deference for cultural differences, it would be coded either for efficiency in communication or as an enhancement of production and exchange of commodities. All relevant cultural factors would be simulacra, comprising neither nationalistic culture nor an independent area of symbolic designs capable of influencing the structure of the globalizing process.

The question of nationalistic fervor within monistically designed cultures such as Islam, Christianity, Judaism, Hinduism, in their archaic and absolutist stances, leads also to a leveling of cultural designs to specific prescriptions excluding ethnic and even nationalistic identities, unless such identities become conflated with the monistic tendencies. If such a conflation were to include the political, then cultural designs would be subject to unyielding theocratic and secular powers. To speak in terms of national identity, an Islamic republic of Iran would be the same as an Islamic republic of Lybia. While these monistic movements are one form of archaization within modernizing contexts, other forms of archaization, leading to the ethnic and racial origins, provide identities that assume a sacral spatio-temporal loci of ancestral blood and land. Their exclusiveness will tend to split any presumed national imagery and create, in the area of symbolic designs, all sorts of animosity and division. Those emergent nationalisms, that fall within these parameters, will not create an imaginary nation sufficiently pliable for survival.

The imagery of cultural nationalism in a modernizing context is best served by postmodern sensitivities capable of traversing various institutions, cultural practices, and archaizations, finding among them unsuspected resonances in ways that access the same old things in novel modes. Such accesses do not get absorbed into one or another cultural mode, as would be the case with structuralism, but comprise a non-reductive mutual enrichment. In this sense, while the process of globalizing may proliferate communicative means, such means are also accessible to national cultures and their mutual dialogue, not as a universal discourse, but as an establishment of resonances among differences that open common concerns beyond the globalizing praxis. Such concerns can begin at the national-global intersections—the environment, peace, human rights, nutrition, health—by allowing each national culture to articulate such concerns by finding, in turn, other concerns that may resonate with totally different questions of diverse national cultures. Once again, such postmodern resonances cannot be absorbed fully without trace into one or another national culture; each encounter leaves a residuum of commonality and difference, and can thus be counted upon to form integration without imposed or reductive wholeness. No claim is made concerning a protracted ability of the postmodern aspect of modernizing cultures to have the staying power and critical stamina for a continuous discovery and maintenance of requisite intra-cultural and cross-cultural resonances. Yet in the current context of modernizing globalism, it is the sole facet that emerged with modernity capable of fulfilling this function of integral differentiation.[28]

---

[28] Anthony Smith, *Nationalisms in the Twentieth Century*, 84; Jean Gebser, *Ursprung und Gegenwart* (Stutgart: Deutsche Verlags-anstalt, 1966).

Chapter 7

Philosophy and Ecological Crisis

Ullrich Melle
K. U. Leuven (Belgium)

Abstract: *Humankind is caught in an escalating spiral of ecological, social and cultural destruction, of material and spiritual deprivation, and of a cancerous homogenization and "uglification" of the human life-worlds. Cancer and addiction are the universal metaphors of our time—cancerous, deadly growth and addiction to a suicidal and alienating life-style. Philosophy can not remain indifferent to this unprecedented ecological and cultural crisis. However, as Heidegger says, "it is in the very nature of philosophy never to make things easier but only more difficult." A philosophy of the ecological crisis has to sail between the Scylla of easy answers and authoritative solutions on the one hand and the Charybdis of increasing the already prevailing confusion and disorientation on the other.*

It is not easy nowadays not to be deeply pessimistic about the human condition and the future prospects for the human race. Humankind seems to be caught in an escalating spiral of ecological and social destruction. The ecology of the global human household is out of balance and the situation is rapidly deteriorating. The approaching end of the millennium only heightens widespread apocalyptic feelings and forebodings. We seem to live in an apocalyptic situation: the last call is being issued to change our ways radically, the point of no return has almost been reached, the day of judgment is near. We shall soon know whether self-extermination has been the true telos of the history of our kind, whether the evolutionary experiment with self-consciousness has been a failure and the human species a horrible mistake of evolution.

In many parts of the world it is not apocalypse soon, but apocalypse now. Indeed, a number of countries and regions may live already in a

M. Daniel and L. Embree (eds.), Phenomenology of the Cultural Disciplines, 171–191.

post-apocalyptic situation, a situation of never-ending doom without any further hope of deliverance. The islands of wealth are shrinking by the day, and the malignant tumour of poverty, misery, hunger, social disintegration and human degradation is spreading rapidly. Never before in human history have there been so many desperately poor people. While the capitalist world-economy is less and less able to satisfy the basic survival needs of food, clothing and shelter of almost a quarter of the world-population, it is more and more driven by the luxury consumption of the rich and the wealthy: "by the culture of the drive-in and the duty-free built around the car and the airplane," as the Canadian political economist Pierre Chossudovsky calls it.[1]

The small economy of abundance in one part of the world and the vast economy of misery and starvation in the rest are both ecologically destructive and unsustainable. The have-not's are destroying their priceless natural environment because of their poverty, but what if they were to develop and industrialize according to the American-European-Japanese model? In his highly acclaimed book "Erdpolitik"(Earthpolitics) Ernst von Weizsäcker asserts we would face an "ecological inferno" if consumption per head were to reach American-European-Japanese levels in the rest of the world.[2]

The excessive material growth of the anthropogenic system—in the economy of abundance mainly in the form of the expansion of the material culture, in the economy of misery and starvation primarily in the form of population growth—is leading to the extinction of other forms of life, the depletion of natural resources and the general pollution of the biosphere. The global human household as a whole is devouring an ever increasing amount of natural resources and, as an unavoidable consequence, is producing an ever increasing amount of waste.

The various destructive tendencies are interacting and reinforcing each other. The resulting exterministic dynamic is like an avalanche which is crashing downwards ever faster and in which we are all being rolled. There is a growing gulf between the ever more visible signs of social and ecological destruction on the one hand and actual policies on the other.

---

[1] Pierre Chossudovsky, "Comment éviter la mondialisation de la pauvreté?" in: *Le Monde Diplomatique* (September 1991), 3.

[2] Ernest von Weizsäcker, *Erdpolitik. Ökologische Realpolitik an der Schwelle zum Jahrhundert der Umwelt* (Wissenschaftliche Buchgesellschaft: Darmstadt, 2. aktualisierte Auflage 1990), 123.

According to the Club of Rome, "traditional structures, governments and institutions don't have the problems in their present size under control."[3] Sitting on mountains of recorded data, of scientific knowledge and technological expertise, of historical and social experience, there exists nevertheless a kind of individual and collective helplessness to stop the escalating spiral of destruction, the rapid increase of ecological and social entropy.

The complete and final breakdown of real-existing Socialism, the terrible and undeniable truth about its inhumanity, its ecological destructiveness and its economic incompetence, which have been fully revealed retrospectively in the aftermath of this breakdown, has radically discredited the socialist-revolutionary alternative. There are no social liberation movements any longer, there is no revolutionary strategy and no revolutionary hope anymore. Instead there are blind and violent revolts, pillage and looting, large-scale organized crime, barbaric strife and wars fuelled by nationalistic and religious fanaticism, much of it being the nihilistic response of the dispossessed and frustrated victims of the existing world-order which is defined by the iron logic of productivity and profitability. And there is the swelling tide of migrating people and refugees which are being uprooted and driven away from their home-regions by famine, poverty and war. In the meantime the West is reorganizing and restructuring its military capabilities and strategies in order to be able to defend Western civilization with barbaric means against the rising tide of barbarism.

This, of course, is only a very rough sketch of a global human household which finds itself locked in an exterministic dynamic which, if not broken, threatens to culminate in the not too distant future in an ecological holocaust and a global civil war. Time is running out quickly. According to Sandra Postel in the latest report of the World-Watch-Institute, "the nineties will be a decisive decade for the planet and its inhabitants."[4] And Lester Brown in his concluding article of this report states: "Either we turn things around quickly or the self-reinforcing

---

[3] Alexander King, Bertrand Schneider, *Die globale Revolution. Ein Bericht des Club of Rome* (Spiegel Verlag: Hamburg, 1991), 74.

[4] Sandra Postel, "Denial in the Decisive Decade," in: *State of the World 1992. A Worldwatch Institute Report on Progress toward a Sustainable Society*, by Lester Brown et el. (W.W. Norton & Co.: New York/London, 1992), 3.

internal dynamic of the deterioration-and-decline scenario will take over."[5]
According to him "there is no precedent for the change in prospect."[6] He
goes on to compare the required "Environmental Revolution" with the
agricultural and industrial revolution "as one of the great economic and
social transformations in human history."[7]

Deep divisions of opinion exist about the character and direction of
this so-called environmental or ecological revolution. Involved are
fundamental questions about the kind of world we want to live in, the
kind of society we find equitable and appropriate to human needs and
aspirations, about the telos of human life and human history, about our
place in nature and the relationship between non-human nature and
human culture.

Until quite recently in the history of our species there used to be
mainly non-human nature on earth. Human settlement and human culture
was surrounded by vast areas of wild nature. Human culture was too
marginal in the biosphere to have any noticeable influence on the
fundamental parameters of nature like the climate and the weather. This
primal, pre-human nature had developed over hundreds of thousands of
years to an ecological climax state. It was an extremely rich, thick,
intricate and complex web of life. Today there are only shrinking islands
of this primal nature left. And even these islands are affected in different
ways by the material reproduction of the global human household,
particularly its evaporations. As Bill McKibben in his moving elegy on
"The End of Nature" remarks succinctly: "By changing the weather, we
make every spot on earth man-made and artificial."[8] There is today no
natural nature anymore, but only anthropogenic or cultural nature. Nature
as a whole has become part of the human household. It is no longer the
huge and undisturbed backdrop of the cultural activities of the human
species. Not only will our pollution reach Antarctica, not only will the
ozone-layer and the weather over Antarctica be influenced by human
activity, but the preserved wilderness of Antarctica will only be granted
by humans on the ground of human interests and ideals.

---

[5] Lester Brown, "Launching the Environmental Revolution," in: *State of the
World 1992. A Worldwatch Institute Report on Progress toward a Sustainable Society*,
by Lester Brown et el. (W.W. Norton & Co.: New York/London, 1992), 174.

[6] *Ibid.*

[7] *Ibid.*

[8] Bill McKibben, *The End of Nature* (Penguin Books: London, 1990), 54.

The radical transformation of pre-human nature in the course of human expansion is dramatically reducing the richness and diversity of pre-human nature. The extinction-rate of plant- and animal-species today is 10,000 times higher than in pre-human nature. According to the German zoologist Bernhard Verbeek this is a catastrophe "that has not yet happened on our planet since the beginning of life."[9] The thick and intricate web of life has been torn and the resulting nature is much less stable and reliable. For its reproduction it becomes more and more dependent on human manipulation and management.

With the fantastic—critics would rather say horrific—possibilities which the new bio-technologies, in particular the DNA-recombinations begin to offer, a completely new stage in the human transformation, the humanization of nature, appears on the horizon. Through genetic engineering and manipulation new organisms can be tailor-made. As the life-forms of the old nature are dying out at an accelerating rate, as the biological productivity of the old nature is declining, we may be able to create in our laboratories a new nature which is more resilient and more productive. McKibben remarks, "just as the clouds of carbon dioxide threaten to heat the atmosphere and perhaps starve us, we are figuring out a new method of dominating the earth, a method more thorough, and therefore more promising, than coal and oil and natural gas. It is not certain that genetic engineering and 'macro-management' of the world's resources will provide a new cornucopia, but is certainly seems probable. We are a talented species."[10]

The probability of such a new technospheric cornucopia may be open to serious doubt. What is all the more likely that this will be the direction in which a major part of the scientific-technocratic, political and bureaucratic establishment will look for an answer to the pressing global problems of ecological breakdown, resource-depletion, waste-disposal and hunger. There seems to be no other alternative then to press on in the hope that that which brought us into the present crisis will eventually lead us out of it. We need more economic growth, further technological break-throughs, more and more industries and bigger bureaucracies in order to fight the afflictions brought about by previous economic growth, previous technological and industrial development. In the words of Brian

---

[9] Bernard Verbeek, *Die Anthropologie der Umweltzerstörung. Die Evolution und der Schatten der Zukunft* (Wissenschaftliche Buchgesellschaft: Darmstadt, 1990), 69.

[10] Bill McKibben, 152.

Tokar, a writer on Green politics and thought: "The further the earth's ecosystems, our health and our personal lives are degraded by technological progress, the more our civilization becomes dependent upon technological solutions to try to manipulate its way out of the mess that has been created."[11]

Edward Goldsmith, the editor of the renowned British journal *The Ecologist*, radically opposes this "more-of-the-same" strategy. He sees a radical and irreconcilable opposition between the technosphere and the biosphere, so that "the ethic of perpetual technospheric expansion is in reality no more than an ethic of biospheric destruction."[12] According to Goldsmith, humankind can only survive and the true needs of humans can only be satisfied as long as humans are an integral part of Gaia, fulfilling their assigned role in the Gaian hierarchy. The Gaian ecosphere is a highly co-operative, self-regulating, self-sustaining enterprise, which maximizes its own stability and realizes the optimal conditions of living for all of its natural parts, the human species included. There exists then an identity of interests between humanity and Gaia. For Goldsmith "it is the fundamental flaw of the world-view of modernism to ignore this perennial truth."[13] Only as vernacular man/woman in endemic societies—this is Goldsmith's radical and controversial claim—are humans supportive and supported members of Gaian hierarchy.

The project of technospheric transformation and expansion is grounded on what Goldsmith calls the world-view of modernism, the great misinterpretation, which postulates a thrifty nature that first has to be made productive by human science, technology and industry before it yields the benefits required for cultural development. The result of this great misinterpretation, Goldsmith maintains, was the creation of a technospheric surrogate-world, which is radically at variance with human nature. "As economic development proceeds, man is thereby condemned to living in a world to which he is ever less well adapted biologically,

---

[11] Brian Tokar, "Social Ecology, Deep Ecology and the Future of the Green Political Thought," in: *The Ecologist* 10.4/5 (1988), 140.

[12] Edward Goldsmith, "Towards a Biospheric Ethic," in: *The Ecologist* 19.4/5 (1989), 74.

[13] Edward Goldsmith, "The Way: An Ecological Worldview," in: *The Ecologist* 10.4/5 (1988), 168.

socially, ecologically and cognitively, and also aesthetically and spiritually."[14]

The implications of Goldsmith's view are obviously far-reaching. We have to return to the situation where human societies are in nature again, where humans are plain citizens again besides and together with a myriad of other life-forms in Gaian community and where non-human nature is the vast backdrop again for human life. We as a species have to fit in again, we have to submit ourselves to the laws and constraints of pregiven, self-organizing nature and its own evolution. This would require, it seems, a manifold reduction of the world-population and an almost complete dismantling of the industrial technosphere. According to the radical German ecologist Rudolf Bahro, who equally thinks that what he calls the industrial-capitalistic system is incurably exterministic we can only survive on earth "with a subsistence-economy of voluntary simplicity and of frugal beauty"[15] and if we limit our numbers. "This contractive way of life"[16] means "that we must give up most of our large-scale material transformations, that we stop tourism, drive no cars, use almost no drugs, don't participate in the money-circulation of the banks any longer, and refrain from positivistic science etc."[17]

It is difficult to take these radical therapies seriously. It looks as if we are being asked to undo ten thousand years of human civilization and return to a precivilized tribal life or perhaps to move forward to a post-civilized tribal life. This seems beyond the imagination of most of us. But of course, it may nevertheless be true that the industrial technosphere and its expansive reproduction is fundamentally unsustainable. According to Christian Schütze, a German writer on environmental problems, "all systems in nature are geared to a low flow of energy. They are more fit for life the less energy they transform in entropy, which they have to carry off. The economic process of the industrial society with its massive "throughput" of free energy and concentrated resources, with its massive production of entropy in the form of waste

---

[14] Edward Goldsmith, "The Way: An Ecological Worldview," 183.

[15] Rudolf Bahro, *Die Logik der Rettung. Wer kann die Apokalypse aufhalten? Ein Versuch über die Grundlagen ökologischer Ethik* (Weitbrecht: Stuttgart/Wien, 1987), 230.

[16] *Ibid.*

[17] *Ibid.*, 319.

heat and waste material is the exact opposite of the natural systems which are fit for survival."[18]

But may not that which is ecologically unsustainable be technologically sustainable? Maybe it is the destiny of humankind and the telos of human history to replace primal nature through a totally artificial, self-supporting technosphere, so that the industrial system would be the culmination and consummation of the history of our race. The human potential could only be fully unfolded in a global industrial technosphere and its ongoing expansion, the end of nature being the unavoidable consequence of the full development of this human potential. But let us look closer at this industrial system.

Science, technology and capitalism are the foundation and the backbone of the industrial system. They are natural allies which need each other. Modern science is dependent on scientific progress. The development of new technologies requires huge capital investment; the accumulation of capital requires a permanent increase of productivity which is only possible through ongoing technological innovation. The German sociologist Otto Ullrich has shown that science, technology and capital not only need each other, but that a close structural affinity exists between them. All three are based on a logic of power and domination. Long before it became a productive force for the accumulation of capital, modern science was driven by a power motive. It strove for control from a distance over the complex natural processes through symbolic knowledge. The analytic-synthetic character of science produced a knowledge of domination and manipulation. As Ullrich remarks: "The scientisation of the world means that the world is taken apart and put together again in such a way that all its angles are 'straightened', that all its spontaneity, self-willedness and fortuity are eliminated, that all processes are predictable and can be planned, monitored and controlled centrally."[19]

Ullrich distinguishes two phases in the development of modern science. In its first phase it was purely mathematical and abstract, completely separated from experience. Scientists were trying to read the book of

---

[18] Christian Schütze, *Das Grundgesetz vom Niedergang. Arbeit ruiniert die Welt* (C. Hanser Verlag: München/Wien, 1989), 96.

[19] Otto Ullrich, "Counter-Movements and the Sciences: Theses supporting Counter-Movements to the 'Scientisation of the World'," in: *Counter-Movements in the Sciences, Sociology of the Sciences*, vol. III. Edited by Helga Nowotny and Hilary Ross, (D. Reidel: Dordrecht, 1979), 130.

nature which was written in mathematical language. But only the movements of the planets are pure, regular and simple enough to lend themselves straightforwardly to an explanation by mathematical principles. The natural processes on earth in their complex interdependence, their unpredictable irregularities must first be isolated from each other and purified from all disturbing influences, before they can be subsumed under and explained by simple mathematical principles and their conjunction. It is in the scientific experiment that such a pure case of a reproducible natural process is artificially produced. Modern science in its second phase becomes experimental science.

Ullrich shows that a high structural affinity exists between such a scientific experiment and an industrial process of production. "The reproducible experimentally constructed natural process is the prototype of the fully automatic industrial production."[20] Science itself in its second experimental phase is already characterized by technological rationality. Just as the scientist wants to dominate the natural process from the outside and from a distance, so the capitalist wants to dominate the process of production from the outside. And just as the scientist has to take nature apart into isolated elementary processes and has to put them together again in such a way that they become fully predictable and controllable, so too the capitalist has to take the production process apart into elementary jobs, isolate and purify them from all disturbing influences and finally has to put them together in such a way that a uniform, totally predictable and controllable production process is the outcome.

For the capitalist who wants to increase the production of surplus-value, the automatic steadily running production line which can be steered and controlled from the outside becomes the model of industrial production. Scientific technology in the form of machines is needed for the full realization of this model. "The capitalist logic finds its fulfillment only in the machinery of scientific technology and scientific technology can only develop within the framework of the capitalist production model."[21]

The modern age is characterized by a historically unprecedented dynamic, an ever increasing acceleration, a total mobilization of human and non-human resources. This explosion of growth and productivity was

---

[20] Otto Ullrich, *Technik und Herrschaft. Vom Hand-werk zur verdinglichten Blockstruktur industrieller Produktion* (Suhrkamp, Frankfurt a.M., 1979), 81.

[21] *Ibid.*, 140.

engendered by the ingenious mechanism of the free market and the capitalist money-economy. The driving motive in this economy is the transformation of natural resources and living labor into goods that can be sold on the market for a profit, and which profit is then reinvested to create more profit and so on. The use-value of the goods and the satisfaction of concrete needs is only a side-effect in this money and profit-oriented economy. Freedom of the market means freedom from arbitrary intervention in the autonomous functioning of the market-mechanism. It is the inexorable competition of the free market to which all participants are submitted that sets everything in motion. Capitalism is, in effect, an economy of war. The market is a battlefield, markets are conquered, competitors are annihilated. It is the universal, always present compulsion of competition which leads to the total mobilization of the productive forces of human and non-human nature alike.

Real existing socialism attempted to replace the anonymous compulsion of competition in a modern money-economy by arbitrary bureaucratic steering and control. The result was, in the words of Robert Kurz, a German Marxist, "a Capitalism, whose blood-circulation had been interrupted, whose circulation then had to be permanently mobilized by a heart-lung-machine."[22] The human and ecological costs of this attempt of a planned money-economy have been horrific.

Modernization has been and still is in many parts of the world a very painful process for the subjects who are to be transformed into modern money-subjects. Traditional value-systems, age old forms of life and traditional community-structures are destroyed, the emancipated subject finds itself handed over and at the mercy of anonymous economic and bureaucratic forces. The natural environment is degraded and uglified by industrialization. Modernization is virtually equivalent with alienation. Alienation from the traditional environment, human and natural, from others and society, from oneself. Alienation brings with it the other typical afflictions of modernization, e.g., a rising crime rate, and other forms of deviant social behaviour, psychic disturbances and depressions, addictions etc. But in spite of all the pain, frustrations and massive unhappiness, the process of modernization seems irresistible. The secular

---

[22] Robert Kurz, *Der Kollaps der Modernisierung. Vom Zusammenbruch des Kasernensozialismus zur Krise der Weltökonomie* (Eichborn Verlag: Frankfurt a.M., 1991), 119.

faith in the Holy Trinity of scientific progress, technological innovation and economic growth have become the new world religion.

All cultural formations rest ultimately on a religious foundation, on a conception of the holy and of the ultimate ends. If a cultural formation loses its religious legitimations, if it can no longer mobilize the religious energies of its members, it is doomed. The religion of modernization is radically human-centered and worldly. Self-deification and worldly self-salvation are its two fundamental articles of faith. God is almighty; he knows all and because he knows all, he can control, manipulate, change all and fabricate everything imaginable. The progress of science and technology is the becoming of the almighty human God or the godly Human. But as almighty human God humankind can take its salvation into its own hands. We do not have to wait for some act of grace from above, we can deliver ourselves from hunger and misery, from scarcity and disease, from war and violence and eventually, who knows, from death. With the help of modern science, modern technology and the industrial forces of production we are on a steady path to earthly paradise where there is plenty of everything for everyone, from an electric tooth-brush to a whirlpool. It should be stressed, however, that material welfare and abundance was indeed a central element of the utopian vision of the modern age, but its vision included as well liberal democracy, human rights, rational and peaceful solution of conflicts, efficient, equitable and accountable government etc.

A remarkable contradiction, however, is to be noted in this religion of the modern age between its optimistic belief in progress, its belief in progressive human self-deification and self-salvation on the one hand and its pessimistic anthropology on the other. Human history had to be submitted to an anonymous and autonomous economic mechanism that has created a permanent war-like competition, a struggle for survival between the economic subjects. Without this anonymous whip of competition, without the constant appeal to and the stimulation of greed, jealousy and aggression, humankind would quickly fall back into their pre-modern sloth. The liberation from first nature, the liberation from tradition, modern freedom and autonomy, were only to be had at the price of total submission to the iron laws of the second nature, those of the industrial capitalist system.

The spectacular achievements of the modern age regarding science, technology and economic growth cannot be denied. It can even be argued that the rule of reason, law and decency has won some terrain from the forces of violence, despotism and brute repression. The priests of modernization try to keep the faith alive by contrasting these successes with a horrifying picture of pre-modern times. The fear of regression is deeply rooted in the modern mind. For this mind, "modern" is equivalent with urban, civilized, enlightened, wealthy and above all with the freedom of the individual from oppressive traditions and closed collectivities. The modern age is seen as the only recently reached highest stage of human history after a long march through dark ages of large-scale misery, oppression, obscurantism and general primitivism. But in spite of certain incontestable achievements and the strong attachment to the modern way of life, once the process of modernization has struck roots, the suspicion grows as to whether this whole process of modernization may not have lead the human race into a dead-end or, worse, that the hubris which the secular faith of modernization involves will not inevitably be punished by death.

The dynamic of modernization, driven by science, technology and a capitalist world market has a more and more destructive character. Paradoxically enough, it is its success in certain respects which is the cause of large-scale destruction and of growing human misery and alienation. The fusion of science, technology and capitalism, their gradual liberation from all culturally determined constraints, the total submission of the life-world to their autonomous logic, in particular to their most fundamental imperative—the imperative of efficiency and productivity—have unleashed a delirious dynamic of growth and change. The productivity of the capitalist forces of production has reached such levels that the global unemployment rate is above 50%. The markets are shrinking because purchasing power is being destroyed. Whole countries and regions are being de-industrialized because they are falling behind the ever increasing global standard of productivity. The battle for the shrinking markets becomes ever harder. Over-production, suffocation, is a constant threat for the capitalist production. Marx had already recognized that the capitalist production-machine cannot wait for the demand, but has to create and stimulate it. One of the biggest industries today is the industry which produces the demand for the goods supplied by this production-machine. The second answer to the threat of over-production is product-change. Technological innovation makes possible a rapid product-change which can be sold as product-improvement. The flow of goods is constantly accelerating, the moral waste increases.

This whole over-productive, over-active, aggressively growing system of total mobilization and acceleration which, according to the German physicist Peter Kafka, seems to follow the cancer-principle of the "ever more of the same ever quicker"[23] leads to chaotic instability. The anonymous mechanisms of this system are pitiless against the losers in and the victims of the global monopoly-game. There is something totally irrational about this seemingly highly efficient and productive enterprise of the increase of scientific knowledge, technological power and economic productivity. It is the irrationality of the means that are developed at the price of their ends. It is the irrationality of the tools which become their own ends independent of the needs and aims of the owner of the tools, even worse the owner has to serve the reproduction of the tools.

These metaphors, however, can seriously mislead us. The global industrial capitalist system is not something like a tool which can be put to good or bad use by a sovereign subject. It is not a machine which unfortunately has gotten out of control and over which the collective human intellect and will have simply to reaffirm their mastery. The abdication of this mastery belongs rather to the essence of this modern form of social reproduction. The historical process has been handed over to the mutually reinforcing logic of what used to be culturally constrained subsystems of society.

So far we have concentrated on the modern industrial capitalist system and its exterministic logic. But, of course, the question can be raised whether human history has just recently taken such a destructive turn. According to Lewis Mumford in his important work *The Myth of the Machine*," civilization as such has been the way of the machine, of centralized power and steering, of uniformization and standardization, of a hierarchical and pyramidical structuring of society, of military might and police force which together with a bureaucratic machinery implement the will of the king through all the layers of society down to the smallest communities. The way of civilization, of large-scale, centralized organizations leads to an enormous enlargement of the collective possibilities and of the collective power of humankind. But a horrific price was to be paid for this increase of centralized power which made possible the stunning feats of humankind, from the pyramids to the space-flights. The history

---

[23] Peter Kafka, *Das Grundgesetz vom Aufstieg. Vielfalt, Gemächlichkeit, Selbstorganisation; Wege zum wirklichen Fortschritt* (C. Hanser Verlag: München/Wien, 1989), 65.

of civilization has been a never-ending story of barbaric atrocities, of
violence, aggression and oppression, of equally large-scale misery and
desperation. We are strongly reminded these days of this dark side of the
way of civilization when the 500th anniversary of what from a Eurocentric
point of view is called "the discovery of America" is celebrated. What,
without doubt, was proof of the organizational, technological and even
spiritual power of European civilization is at the same time one of the
darkest pages in human history. The Europeanization of the world, which
is, in effect, the modern age, begins with the annihilation of about 100
million native Americans and the enslavement of close to 20 million
Africans which were brought as slaves from Central and Western Africa
to America.

Bahro, who is strongly influenced by Mumford, has argued in his book
Die Logik der Rettung (The Logic of Deliverance) that the whole evolution
of humankind from its very beginning is characterized by an exterministic
tendency, a logic of self-destruction, as he calls it. If one begins to pay
more systematic attention to the dark side of human civilization during
its history, this thesis becomes credible. For Bahro, however, this
exterministic impulse reaches back into the emergence of the specifically
human life-form, when the arrow of cultural evolution began to diverge
from the arrow of biological evolution. According to Bahro, cultural
evolution from its germinating beginnings was built on a strategy of
arming, conquering, exploiting and safeguarding. The peaceful living
together of the pre-civilized primitives with each other and with nature
is a myth for Bahro.

The German philosopher Odo Marquard has shown that the notion
of compensation is a key concept in modern philosophical anthropology.[24]
The human being is homo compensator: a *"Mängelwesen"* (a being
defined by its defects), a *"Defektflüchter"* (one who runs away from
her/his defects), a "dilettante of life," the "retarded living being" that has
to find surrogate solutions to compensate for her/his vital defects. This
anthropological concept of compensation plays a vital role in Bahro's
reconstruction of the logic of self-destruction. Tormented by dread from
a chaotic inner life of uncontrollable impulses and hallucinatory visions
as well as from an overpowering and threatening outside world, the

---

[24] See Odo Marquard, "Homo compensator," in *Philosophische Anthropologie.
Arbeitsbücher 7. Diskurs: Mensch*, Willi Oelmüller, Ruth Dolle-Oelmüller, Carl
Friedrich Geyer (eds.) (Ferdinand Schöningh: Paderborn, 1985), 317-330.

emerging human beings developed their intellectual abilities into compensating instruments of power and control. Suppression of anxiety and compensation: the ultimate ground of all this never-ending violence and aggression are, according to Bahro, unresolved original problems of the human psyche. It has never come to terms with its traumatic anxieties of its own inner nature, of outward nature and of the other human being. If the ultimate roots of the present crisis reach back into the origin of our species then it follows that only a kind of anthropological therapy will be able to deliver us from evil. Bahro talks of an anthropological revolution, an evolutionary leap, a second birth.

This anthropological revolution must bring about a new economy or ecology of the mind, a new configuration of our psychic forces and capabilities. Instrumental, calculative and manipulative reason driven by the will for power has alienated itself from the rest of our subjective forces, it has turned into a demon, an evil spirit which leads us to invest all our energy into the reproduction of the alienating and exterministic industrial machine. Bahro stresses that analytic and instrumental reason is certainly a highly valuable human capacity but that it has to be reintegrated into the ecology of the mind where it occupies a specific niche and where it is constrained and checked by other forms of awareness and knowledge.

The ecological crisis is for Bahro the final crisis of human history under the exterministic sway. This is irrevocably the last stage of this history. Either it will be followed by the silence of an ecologically devastated lifeless planet or humankind will succeed in surviving civilization and history, and will begin anew, reborn, in what Lévi-Strauss called a cold society, that is, a meditative, non-expansionist, communitarian culture with a simple reproduction rooted in a new ecology of the mind.[25] In this perspective the ecological crisis indeed has a truly apocalyptic character: it is the final chance, a final call for us human beings to come to terms with ourselves, to gain our center. As Bahro puts it succinctly in an interview: "The crisis is not in the trees, it is in us."[26]

It is not difficult to criticize Mumford and Bahro for their sweeping generalizations and speculative reconstructions of the whole of human

---

[25] Please add note.

[26] Rudolf Bahro, "Theology not Ecology," interview in *New Perspective Quarterly* 6.1 (Spring 1989), 36.

history, and Bahro's anthropological revolution does not seem to be a very practical proposition if we recognize the need for quick and decisive action to save the planet. But so far all the practical and pragmatic propositions have had only very limited success. They seem to have dealt only with the symptoms of the disease, which does not mean that they are not urgently needed. The scale of the crisis in the global human household requires a deeper inquiry into the ultimate causes and roots of the ecological crisis. A major part of this inquiry will consist in a comprehensive analysis of the industrial-capitalist system, its elements, its structure, its logic, its ideological and religious legitimations and of the modern mind, the modern subject and its motivations. But the inquiry cannot stop here. Further questions have to be raised about the course of human history as a whole, about the logic of cultural development as such, about such great transformations in human history as the agricultural and the industrial revolution, about the nature of man/woman and the possible change and transformation of this nature in the course of human history. It is only against the background of such a fundamental inquiry that we will be able to develop a more radical and more far-reaching perspective beyond a mere technocratic management of the crisis.

One of the most promising and elaborated proposals for such a far-reaching perspective we find in the school of eco-philosophical thought which was founded by the Norwegian philosopher Arne Naess and which is known under the name of Deep Ecology. Similar to Bahro, Deep Ecology tries to re-think what it means to be truly human. Its main target of critical attack is anthropocentrism. But it bases its radical critique of anthropocentrism not primarily on axiological and moral grounds but on ontological and psychological insights. Ecological moralizing criticizes our individual and collective selfishness, our ruthlessness in the exploitation and domination of nature, our materialism etc. We are exhorted to preserve nature, to restrain our greed, to reduce our material consumption etc. These moralizing arguments, however, are often hypocritical, their persuasive appeal is rather limited, they are often futile and even counter-productive. Deep Ecology claims that an ecological life-style would follow naturally from a new, ecological understanding of the self. According to Naess, "the requisite care flows naturally if the 'self' is widened and deepened so that protection of free Nature is felt and conceived as protection of ourselves."[27]

The modern subject is a social atom which, beyond the primary groups of the family, the neighborhood and a small circle of friends, is linked to other social atoms by contractual or market relations. Money is the symbol of freedom for this modern subject. The more money you have, the more independent you are. Modern ideals of emancipation and autonomy are based on an atomistic social ontology and an atomistic conception of the self which are realized by and in turn legitimize the modern capitalist money-economy. This modern self is fundamentally egoistic, it is spasmodic, never at ease, always on the alert, always anxious of contact, of the danger of losing itself—Bahro calls it an "ego-fortress." According to Deep Ecology, this is an immature form of the self, a self cut off from its full self-realization-potential, because it cuts itself off from that with which it is internally connected. In the words of Naess: "The ego-trip interpretation of the potentialities of humans presupposes a marked underestimation of the richness and broadness of our potentialities."[28] The self is not only internally related to other human beings, it is not only a social self, but it is internally related to the non-human world as well, it is an ecological self. To quote Naess again: "We may be said to be in, of and for Nature from our very beginning. Society and human relations are important, but our self is richer in its constitutive relations."[29]

The foundation of this view is a holistic and relational ontology which is strongly implicated by recent developments in different sciences but in particular, of course, by scientific ecology. This is how Warwick Fox, leading representative of Deep-Ecology-thinking, expresses this ontological position: "All subdivisions are seen as relative rather than absolute. Or in metaphorical terms, 'separate things in the world' should be thought as eddies, ripples and whirlpools in a stream ('unity of process') rather than as bricks that are totally self-contained and self-sufficient."[30] Naess himself, drawing inspiration from ecology, Ghandian metaphysics and

---

[27] Arne Naess, "Self-Realization," in *The Trumpeter* 4.5 (Summer 1987), 40.

[28] *Ibid.*, 37.

[29] *Ibid.*, 35.

[30] Warwick Fox, *Approaching Deep Ecology: A Response to Richard Sulvan's Critique of Deep Ecology*, Environmental Studies Occasional Paper 20 (University of Tasmania: Hobert, 1986), 15.

Gestalt-philosophy speaks of the "relational-field-view." Everything is what it is as a knot in a relational net and "a person is a part of nature to the extent that he or she too is a relation junction within the total field."[31]

This relational conception of the self, Deep Ecology thinkers insist, should not be mistaken as a denial of individuation, it does not mean that all is one and it does not imply an oceanic dissolution of the individual in an undifferentiated whole. But our ontological individuation is not self-referential, it is essentially relational, so that the relations cannot be separated from the individual. Maturation of the self can then be understood as the growth of the narrow egoistic self towards the social and ecological Self. The self realizes that its true self-interest reaches much further than the bounds of the narrow self, it identifies with that which it recognizes is not separable from but is an integral part of it. In defending nature in its richness and diversity against the destructive forces of the industrial technosphere, I am defending my integrity.

It is certainly a thesis worth considering that we are under-estimating our empathetic and spiritual potentialities because the capitalist money-economy is dependent upon, and therefore continuously reinforces, a psychologically immature, narrow self. But are there not limits to our ability for social and ecological identification? Can we not equally over-estimate ourselves in this regard? Besides, if I really identify as intensely and comprehensively as possible with the Other, does this not imply that I have to bear the immeasurable amount of pain and suffering in the world? But as I shall never be able to alleviate more than a tiny part of this suffering, will not total resignation and desperation be the consequence? The concept of the ecological self still needs further refinement.

The ideal of self-realization, of course, is a normative concept. It is the top-norm in Naess' and other Deep ecologists' own personal ecosophy. Naess distinguishes between eco-philosophy and ecosophy. Whereas the first is a theoretical academic discipline, the second is a personal philosophy, a total world view on the basis of which a person decides and acts. Such a total view is, if articulated, a hierarchical system of derivations from ultimate normative and factual premises. According

---

[31] Arne Naess, *Ecology, Community and Lifestyle*, translated and edited by David Rothenberg (Cambridge University Press: 1989), 56.

to Naess, we always decide and act on the basis of such a total view: "All we do somehow implies the existence of such systems, however elusive they may be to concrete descriptions."[32] Naess maintains that it is of great importance in the present crisis-situation that people try to articulate their total views as clearly and as systematically as possible even if a complete articulation is in principle impossible.[33] We should try to articulate what our ultimate values and norms are and how we derive other lower norms from them with the help of certain factual hypotheses. This means in fact that we should try to articulate in the form of a hierarchical normative system what we really and ultimately want and what we really believe in. We should "announce our value-priorities forcefully,"[34] but do without any dogmatism. "To accept a particular norm as a fundamental, or basic norm, does not imply an assertion of infallibility nor does it claim that the acceptance of a norm is independent of its concrete consequences in practical situations. It is not an attempt to dominate or manipulate. As with descriptive statements, we should retain a principle of revisability. The cult of obstinacy in the realm of norms renders calm debate practically impossible."[35] On the basis of a systematic articulation of a total view, rational and meaningful debate about value-priorities becomes possible. There can be and there should be quite different total views as the ground of different lifestyles and different cultures. There is no end to meaningful debate and interaction, to clarification and modification regarding our total views. The present system of industrial production and consumption, and this seems to be Naess' conviction, would turn out not to be supported by any coherent total view. Therefore it is particularly important that we urge the defenders and representatives of this system to articulate their total views and involve them in a debate about their and our total views.

A more detailed presentation and discussion of Naess' sophisticated normative-system-technique as well as his and others' concrete elaboration of the total view of a Deep Ecology, down to rules for a Deep-Ecology-lifestyle, for Deep-Ecology-politics and -economics, is

---

[32] *Ibid.*, 68.

[33] Regarding the impossibility of a complete articulation of a total view see, Arne Naess, "Reflections about Total Views," in *Phenomenology and Phenomenological Research* (Sept. 1964).

[34] *Ibid.*

[35] *Ibid.*, 69.

beyond the scope of the present essay. In this essay I did not present a tight logical argument for a certain position. As do most of us, I feel overwhelmed by the complexity of the problem of what, in short, is called the "ecological crisis," and consequently I am extremely anxious not to fall prey to simplifications. However, it seems obvious to me that philosophy and philosophers can not remain indifferent to a crisis of such magnitude inside the human household and its relation with non-human nature. According to the young German-Italian philosopher Vittorio Hösle, "it is one of the most urgent tasks of present-day philosophy—which it thus far has hardly taken up—to understand the meaning of this crisis."[36] But what exactly does a philosophical understanding of the meaning of this crisis consist in? What can and should be the role of philosophy in the unfolding drama of accelerating nihilistic destruction and desperate rescue-efforts?

Philosophy itself, of course, is in complete disarray. It does not speak with one voice. It is itself in a permanent and deep crisis regarding its purpose, its tasks and its methods. Some would even say that philosophy has come to an end and that what lives on in university departments under this name is only a ghost of the deceased. However, the ecological crisis seems to offer philosophers a chance at new public relevance. It is widely recognized in the various diagnoses of the crisis of our age that polluting industries, inefficient technologies, out-dated political structures and economic mismanagement are only superficial causes of this crisis. Most often something like a crisis of purpose and values is diagnosed as the fundamental cause and the common denominator of the various symptoms of the disease. The latest report of the Club of Rome diagnoses a deep moral crisis, a dissolution of traditional value systems—"only materialism remains today as a strong, all-penetrating negative value"[37]—a break-down of ideologies and a lack of global vision. According to the report "If we want to survive, human wisdom must quickly be called up."[38] Presently, "we are rich in [scientific and technological] knowledge but poor in wisdom."[39] But to be poor in

---

[36] Vittorio Hösle, "Über die Unmöglichkeit einer naturalistischen Begründung der Ethik," in *Wiener Jahrbuch für Philosophie*, Bd. XXI (1989), 15.

[37] Alexander King, Bertrand Schneider, *Die Globale Revolution*, 65.

[38] *Ibid.*, 64.

[39] *Ibid.*, 129.

wisdom seems to be the same as to be poor in philosophy if philosophy is what its name says: love, search and care for wisdom.

This wisdom, I think, has to be something rather close to Husserl's phenomenological reason but purified from its Eurocentric and absolutist connotations. It will be something between arbitrary and idiosyncratic world-views and a rigorous science and so come close to Naess' conception of a total view. It will incorporate firstly an understanding as comprehensive, as radical and as deep as possible of the character and of the roots and causes of our present predicament, secondly a metaphysics and anthropology, and thirdly, grounded in the foregoing, a normative vision about the future course of human history. These integrated disciplines of wisdom have to be developed in an open-minded, non-dogmatic, self-critical and communicative inquiry beyond the narrow bounds of positivistic science. This inquiry will be phenomenological in two ways:

1. It will not be calculative and constructive, quantifying and modelling but it will rather be intuitive, meditative and hermeneutical. It will ground its knowledge-claims in intuitive evidence and reasoning.

2. It will not be objectivistic. Instead it will be fundamentally subject-oriented, engaged in a continuous process of self-examination, self-interpretation and self-knowledge.

To raise the question of wisdom is equivalent to raising the question of the subject. And if there is a point towards which the fragmentary considerations in this article converge, it is the crucial importance of this question. It may be, as Bahro remarks, that we need something like an act of grace to help us. But grace is something like a field of spiritual energy which needs to be charged from our side as well. "A society of depressive junkies will not be met by grace."[40] The avalanche has to be stopped—miraculously—from the inside. As Ivan Illich said so succinctly: "To face the future freely, one must give up both optimism and pessimism and place all hope in human beings, not tools."[41]

---

[40] Rudolf Bahro, *Die Logik der Rettung*, 307.

[41] Ivan Illich, "The Shadow our Future Throws," interview in *New Perspectives Quarterly*, 6 (Spring 1989), 23.

Chapter 8

Phenomenology and Ecofeminism

Don Marietta
Florida Atlantic University

Abstract: *Ecological feminism is an important aspect of environ-*
mentalism and is making important contributions to environmental
ethics. It is also of particular interest to phenomenologist for several
reasons. Ecofeminists use methods very similar to those of
phenomenological analysis. Also the emphasis on contextual and
pluralistic ethics calls for attention to the voices of many different
types of people who had been ignored. This gives phenomenologists
opportunity to look at many types of lived worlds; this can add
intersubjective richness to our "returning to the matters themselves."

Ecological feminism is an important aspect of environmentalism. It is important to the environmental movement and to the development of environmental ethics. A growing body of literature attests to the significance of ecological feminism. The active work of feminists in the cause of preserving the natural world, as seen in the World Women's Congress for a Healthy Planet in 1991 and in the activity of women at the World Summit in Rio, shows the important role of women in environmental concerns. Phenomenologists who are not already doing so should take note of the important work done by ecogical feminists. We should, as human beings who are dependent upon a natural environment and share some responsibility for the health of the planet, be knowledgeable about important aspects of environmentalism, and ecological feminism is important. As phenomenologists we will find that we have some significant affinities with ecofeminism, and will find that we are able to understand this approach to environmental philosophy and ethics at a depth not open to those who have not acquired a phenomenological approach.

M. Daniel and L. Embree (eds.), Phenomenology of the Cultural Disciplines, 193–210.
© 1994 *Kluwer Academic Publishers. Printed in the Netherlands.*

This has been my experience. My work in the discipline of ethics, with a focus on environmental philosophy for over a decade, has been greatly enriched by basic phenomenological methods, especially in dealing with the relationship between knowledge and moral obligation. Some years ago I was made aware of the close connection between the attitudes and beliefs which lead to the destruction of the natural environment and the oppression of women. Later, especially in the work of Karen J. Warren, I found an approach to ethical theory which seemed similar to mine. Now I realize that phenomenology offers a good perspective for grasping the significance of feminist ethical theory and feminist environmentalism. Perhaps we will be able to contribute something of importance to feminist ethics using the tools of phenomenological analysis.

First let us see why feminist thought is important to environmentalism. Then we will look carefully at some reasons why phenomenologists, as phenomenologists, can relate significantly to ecological feminism.

I

Françcoise d'Eaubonne coined the term *ecofeminisme* in 1974, giving a focus to a movement of growing importance. A central aspect of this movement was indicated by Rosemary Ruether[1] when she pointed out the historical connection between the domination of nature and the domination of women. In the 1970's, Carolyn Merchant and a few other writers explored the significance of ecological feminism. The literature became more abundant in the 1980's and 1990's. *Environmental Ethics* has published significant work in this area, including two papers by Warren.[2] Warren also edited a special issue of *Hypatia*[3] on feminism and environmentalism with articles by many of the leading voices in ecofeminism. The American Philosophical Association Newsletter on Feminism and Philosophy[4] has several articles on ecological feminism and

---

[1] Rosemary Ruether, *New Woman/New Earth: Sexist Idiologies and Human Liberation* (New York: Seabury Press, 1975)

[2] Karen J. Warren, "Feminism and Philosophy, Making Connections," *Environmental Ethics* 9:1 (1987):3-20 and "The Power and the Promise of Ecological Feminism," *Environmental Ethics* 12.2 (1990):125-146

[3] *Hypatia* 6.1 (Spring 1991).

[4] 91.1 (Spring 1992).

outlines of courses in environmental ethics taught from a feminist perspective. Articles on this aspect of environmentalism will probably continue in the philosophical and feminist journals. We may expect anthologies of papers on ecological feminism to be forthcoming.

Environmentalists will find feminists at the forefront of the movement to save the earth. Environmentalists are discovering that a crucial feature of environmental action is economic. This concern for economic realities has been called the "fourth stage of environmentalism," following the first stage of resource conservation led by Theodore Roosevelt and Gifford Pinchot, the second stage of concern over pollution which followed Rachel Carson's *Silent Spring*, and the third stage of recognition of moral obligation to the natural environment itself. The fourth stage does not displace the concerns of the first three stages, but it recognizes that environmental preservation is not unrelated to the economic realities of the world's nations. It involves the economies of the developed nations which use most of the worlds irreplaceable resources and cause most of the pollution. It also involves the economies of the poorer so-called third world nations, whose people destroy the forests in a desperate search for fuel and space for agriculture and livestock. The concerns expressed by women at the Rio Summit and at the World Women's Congress show that they have been pioneers in the fourth stage of environmentalism as they joined efforts on behalf of the natural environment to their social agenda.

It is imperative that humans achieve a sustainable society on this planet. If we fail to do this, we will destroy the most highly developed life forms, if not all life. Achieving a world society which can last into the foreseeable future will not be easy. The greatest difficulty is not technological. It is the problem of human motivation. Life in a sustainable economy will require more frugality, more simplicity, than people in the developed countries are accustomed to. For people in the less developed countries, it will mean forgoing the dream of living like affluent Americans.

It seems obvious that the "macho" ways of thinking and acting which lead to plundering and polluting the planet cannot motivate acceptance of the ways of life which will be necessary in a sustainable society.

A deeper look at the "macho" attitudes will be insightful. The attitudes and approaches which led to wasteful use of natural resources and despoilation of the natural environment include the excessive use of force in undertaking very large projects and completing them in a short period of time. The technologies employed have involved large machines which use vast amounts of fossil fuel energy. These technologies are highly entropic because large amounts of fuel are used in order to do things quickly. This has been contrasted with "human-scale" technologies and labor-intensive technologies. Not only is the employment of heavy machines to get the job done quickly quite entrophic, it leads to environmental damage because vast changes are made in the natural environment so quickly that there is no chance to notice and respond to "feedback," the indications that undesired effects (so-called "side effects") will follow the changes which are made. Unexpected damage to the natural environment can occur before project designers can take warning.

A clear example of this is the Aswan Dam, a major project designed to improve the economic life of Egypt. The changes made in the Nile River had some undesirable consequences. The impounded waters lead to a major health problem (the parasitic disease schistosomiasis) and to changes in the salinity of the Mediterranean Sea, with its effect of fisheries. There were, of course, effects on village life which were unexpected, discounted, or ignored.

In southeast Florida, where I live, the hydrology was altered in a fairly short period of time by the U. S. Corps of Engineers. Vast drainage projects made more land suitable for agriculture and provided land for development. Now we are discovering the "side-effects," threatened eutrofication of Lake Okechobee, threats to the Everglades, and periodic water shortages, not to mention the many problems which come with very rapid population growth.

These examples of changes made too quickly in the natural environment show some of the qualities associated with a "macho" approach: employment of force very aggressively, impatience, lack of concern for effects upon human life and society, along with lack of feeling for natural systems.

What lies behind the "macho" approach? What sort of thinking allows such recklessness. There was a combination of atomistic thinking, hierarchical thinking, and what Warren has called the "logic of domination."[5]

Atomistic thinking is the opposite of the holistic thinking of much contemporary environmentalism. With this atomistic thinking everything

---

[5] Karen J. Warren, "Feminism and Philosophy," 6; "The Power and the Promise," 128f.

is seen in terms of individuals, disconnected atoms seen as having only an accidental connection to other parts of the systems of which they are a part, if the system is considered at all. This has led to a separation of people from nature, a separation of people from each other, and a division within the person with mind having its significance apart from body. Mental life was itself even seen as divided, with intellect separated from emotion and will. When things are thought of as separate units, there is a failure to see the richly complex unity, composed of many webs of interdependent organisms and functions, which is the system of nature.

With atomistic thinking the land is not thought of as the foundation on which all life rests. The land can be seen as merely real estate, lots and acreas to buy and sell, to use without regard for the life which is dependent upon the land. The dredging of a canal is not seen in relation to the swamp which may be destroyed. The swamp is not seen as a fertile source of biological diversity. The hunters who killed passenger pigeons treated each bird as an individual thing, and unwittingly destroyed a species.

An important aspect of atomistic thinking is the defining of things in terms of separation from other things, rather than in terms of connection and interdependency. This type of definition defines a species on the basis of what is unique to it, neglecting the qualities which it shares with other species. This exaggerates the importance of the point of difference as it neglects the shared attributes which can tell us most about the nature of the species. So it was that the human was thought of as the thinking animal, even though thinking is not all that we are. The effect of this is to make light of the other qualities of humanness, the feelings, hungers, lusts, vulnerabilities, and enjoyments which also make us what we are.

Atomistic thinking made it easier for us to focus on narrowly defined goals and to overlook "side effects." We could see the dry land for farming which our drainage canal could provide, but we did not see or could easily ignore the undesirable consequences of making the canal. Atomistic thinking gives us focus, but it makes us lose our bearings. We can be clear about our steps while we are blind in regard to our journey. We could see the dry land for agriculture and development in southeast Florida, without seeing that we would need to undo and correct our canals in a few years. We humans did not see ourselves adequately, so it is little wonder that we had such a faulty view of our natural environment.

Atomistic thinking had a significant effect upon ethical thought. We thought in terms of individual rights and duties, and we did not manage collective responsibilities very well. When we were not able to assign a duty to a specific person, we did not know how to handle the moral responsibility. In the face of environmental disasters, we could not find relief from the court system without finding a specific claimant who would suffer definite financial loss. Terrible things were done to the natural environment when our individualistic approaches did not match the realities of what was happening.

Closely related to atomistic thinking is the tendency to think in terms of gradations or value hierarchies. When we had to think in terms of relationship between things, we did not think in terms of systems, but in terms of graded relationships, with one thing above another and thereby more important than the lower. Some of these over/under relationships were very basic. Mind stood above body, while culture was over nature. The male ranked above the female, as the strong were over the weak, the rich over the poor, and the white over the black. Since the time of Aristotle, at least, we have placed the animal over the vegetable, with the thinking animal above all. It is not too difficult to see how this gave an exaggerated importance to some things and denied the full significance of others. The things of nature took second place to things of culture. Since women were associated with nature, as Nietzsche finally expressed the commonly held bias, women were associated with things thought to be of less importance than those which were in the purview of men. It is not surprising that economic achievement ranked above preservation of natural things. The impact of this can be seen in classical economic theory, which attributed value to human work and considered natural resources to be "free goods." John Locke actually held that nature contributes a very small amount to the value of things which humans have raised or manufactured. Even now, the economic strength of nations, measured as gross domestic product, looks only at marketable goods and services, but does not take account of diminishing natural resources. Destruction of forests does not show up in the assessment of economic health. Loss of fisheries does not get reckoned until a smaller sale of fish affects the statistics.

An interesting sidelight on the discussion of economics is the etymology of the word 'economics.' It comes from a Greek term for the management of a household. 'Home economics' is really redundant. Is it not an anomaly that economics has lost sight of that part of itself to which women have traditionally contributed more than their share? Now economics thinks of the household in terms of the grocery store shelves. It has not yet accepted its role in regard to the household which is the biosphere.

Hierarchical thinking led to the "logic of domination." It was taken for granted that the higher had a right, or even a duty, to dominate the lower. So males, who were assumed to be more rational and were engaged in important matters of culture, dominated females, who were engaged in matters of nature such a childbirth and nurturing activities, activities which required little use of reason. This "logic" was based on faulty assessments of value and on a questionable ethical principle, but this did not prevent it from having powerful effects upon human relationships and upon the way nonhuman animals and natural things were treated. Until very recently very few people thought that humans had any direct moral responsibility to nonhuman lives or to the system of nature. Only human interests determined the right uses of nonhuman entities, and atomistic thinking greatly limited understanding of human interest. Only a few people of remarkable insight realized that destruction of natural things diminished human life. Few were the people who realized that the domination of women diminished human life.

Before we leave this matter let me make an important point. The objection to hierarchical thinking on the part of ecological feminists is not the insupportable rejection of all judgment of relative worth which some superficial thinkers might support. Warren makes it very clear that she is not rejecting all judgment of worth. It is within the context of the "logic of domination" that judgment becomes vicious. What happened in specific cases of injustice was that biased judgment was based on inadequate knowledge. In general the fault in hierarchical thinking is limiting thought to the relationships of superior to inferior, which obscures other important relationships. For example, in the context of the working of natural systems, the decomposers, which are the small and usually not very attractive insects, worms, and bacteria which break down dead organisms into their constituent materials, are at least as necessary to the system of nature as the larger members of the system. Even though we place more value on our fellow humans, there is a sense in which a decomposer is no less important than a human. The significant point is that a judgment of relative value is useless in this context.

Judgments of relative value have a limited application. Thinking in terms of systems is necessary if we are to have a sound understanding of things.

It is easy for us to see how "macho" thinking developed and seemed to its practitioners to be right and proper. The aggressive pursuit of one of the higher values seemed to justify use of extreme means and lack of concern for peripheral matters. Certainly it would be right to seek the economic welfare of a community, by whatever means could accomplish the goal most quickly. To see that this would not be right and would most likely be counterproductive requires a holistic way of thinking in terms of complex systems. To understand the evils of sexism, racism, and jingoism requires a way of thinking which most people seem still not to have.

Our survival as humans and the survival of life on the planet, certainly the more complex forms of life, require the widespread adoption of holistic thinking. The development of a sustainable human society requires growing beyond the "macho" ways of thinking, feeling, and acting. Feminism can help develop the kinds of character and ways of thinking and acting which fit into sustainable modes of human life. It can help channel aggression into constructive paths. We are not advocating a general "wimpishness." Aggression is a problem when it is extreme and destructive, especially when it is directed against persons. Aggressive pursuit of solutions to problems is not problematical if atomistic approaches are avoided. Aggressive pursuit of a career is not harmful if responsibility to other people is not denied. Aggressiveness can be beneficial to the society if joined to awareness of other people and a concern for their welfare. The aggression which grows out of anger and frustration must be corrected by removing as much as possible the causes of the anger. A sustainable society must pay careful attention to the welfare of its children. Angry adults are sick adults, who may have been mishandled children. A caring society can address either problem. Violence is destructive, even when it is the use of violent means to bring about a desired state of affairs. Holistic thinking will show us better ways to achieve the good without destructive "side effects" in nature or in society.

Let me describe two specific contributions which feminism can make to a sustainable society, which must be a society which gives humans a satisfying and rewarding life. One may seem trivial at first, but it will be seen to be very important. One contribution is providing for recreation which will not harm the natural environment. The other is providing ways for the society to handle its affairs and solve its problems.

Too many people now feel a need to use motorized vehicles in their recreation. Their sense of personal significance seems often to be tied to certain accomplishments in the use of vehicles, or perhaps just in the possession of such expensive toys. It will be a mistake simply to force people to give up that which gives them satisfaction and a sense of their own importance. Somehow people must be taught the joy and satisfaction to be found in activities which do not require an inefficient petroleum fueled motor. This simply must be done, and a feminist approach to life provides the sorts of values which can give satisfaction and significance to a life without offroad vehicles, water skis, and cigarette boats. Of course we will fail with a large number of people; they have been made unfit for an environmentally sound life. They will lay aside their "macho" toys only when the fuel is unavailable or too expensive. The society may need to find cheap ways to entertain them until they die.

The majority of people will be less intransigent, and the various modes of public education can teach new ways to find pleasure and satisfaction. To realize what some of these are we need turn back no further than the period before the second world war, when few children played with motorized toys or realistic looking space-age weapons. Many things developed since that time are not energy intensive and do not serve primarily to express hostility. Many people are now enjoying making things, growing things, cooking things, and learning things. Others are writing, painting, singing, and acting. We do not need a total revolution in ways of living, just a shift away from those activities which are at odds with a sound natural environment and a sustainable society.

A critical aspect of finding a sustainable society is learning ways of governing ourselves which are just, peaceful, and conducive to personal growth. Feminism is now identified with ways of leadership and group activity which can contribute to living in a sustainable society. These are really much older than feminism; I was taught some of them years ago in group dynamics and leadership workshops. Feminists have adopted these democratic ways of group activity, however, so I have no qualms about referring to them as feminist.

The "macho" approach insisted that every group needs a forceful leader: "every ship needs a captain." Studies of leadership methods can demonstrate that a forceful leader can seldom get from a group the creative contributions which can come from a democratic group, even one without an official leader. Democratic leadership does not require a vote on all issues and rule by the majority. Seeking common consent does lead to resolution of issues, and it usually does it quickly. This may be a paradox, that patience saves time. When consensus is patiently waited for, those who eventually change their position or go along with the rest do not feel beaten or overcome. This form of group life promotes peace and creativity. One of the reasons for the creativity is that not having a heavy-gaveled chair allows people to come forward when their area of competence is called for; this allows a kind of temporary leadership of the group which uses each person's skills and knowledge most efficiently. Again we find a paradox, that making efficiency the first priority is seldom truly efficient. More values can be realized when a group is person and process oriented rather than goal oriented.

I have been a part of democratic, process-oriented groups for many years. I know they are creative and conducive to harmony and good will as well as effective. A sustainable society must be a just society and a society of people who are satisfied with their lives. So-called feminist group methods can make a tremendous contribution to this. The patriarchal ways of the past have left us thinking that wars are inevitable and that any society must be built on the expectation that a few must win and many must be losers. The feminist way can help us give peace a chance. This is worth the careful attention of all responsible people. Feminism is too promising to leave to a few women. It is time for all of us to receive its benefits and contribute to its development. Ecological feminism is too significant a part of the effort to save the life on this planet for any thoughtful person to ignore it.

It is important to see that ecological feminism is not a radical, fringe movement in environmentalism. It does not advocate a way of life which is basically different from that which holistic environmentalists have been advocating for some time. Arne Naess's concept of deep ecology incorporated many features of an ecological feminist treatment of the earth. He combined respect for the earth with social concerns. He advocated equality between all living things, with the right of all to exist respected. He was also concerned with equality between persons and groups of persons, with a respect for differences. He was a champion of biological diversity and of cultural diversity. He saw a close connection between world peace and a holistic view of life on the planet. Caring, respect, and other feminist values were clearly incorporated in Naess's ethical view. Other environmental philosophers have advocated living in the world as a member of a life community. What these environmentalists see as a proper way to inhabit the earth in terms of simplicity, avoidance of a grasping attitude, willingness to live and let live, and a careful avoidance of doing damage to the planet is substantially the kind of life an ecofeminist would advocate.

How then is ecological feminism different? One difference is the realization of the historical connection between the suppression of women and the abuse of the natural environment. Deep ecologists seem to be insufficiently aware of this significant factor, one which continues to have an effect upon our social life. It is important that our pursuit of a sustainable society not be too reductive and seen only in biological terms. Ecological feminists will make certain that issues of peace and justice are not neglected.

Ecological feminists are doing what all holists have known in theory. Humans are a part of nature, as much so as any other creature. It has been all too easy to lose sight of this, to lose sight of the social aspects of the preservation of the earth. Ecological feminists keep the environmental movement reminded of the human factors.

In large part the difference between ecological feminists and those environmentalists who do not feel an affinity for feminism seems to be a matter of style, and I do not mean this in a derogatory way. Ecological feminists speak more of the underrepresented peoples of the world and the need to listen to them. There is more concern about communication and understanding other people than is found in the writing of many environmentalists. Ecological feminists seem to be interested in understanding why people think and act as they do, and this leads to openness and toleration and a desire to help people work things out together. It is a typically feminist thing to care more about correcting what is wrong through cooperative interchange than about placing blame.

It is probably a bit early to see all of the differences between ecological feminists and other environmentalists. I cannot clearly identify in myself those differences which are ecofeminist, in part, as mentioned, because I had been influenced by leadership practices now identified with feminism long before I knew how important feminism or environmentalism would be to me. It is possible that the ecological feminist style will show up in political and economic actions. What would we look for? A clear indication of feminist influence would be more effort to work out differences cooperatively and less willingness to run roughshod over the opposition on the basis of a majority vote. Whether this will happen in a large way is yet to be seen. Cooperation requires some effort on both sides, and those who profit from the abuse of the natural environment have not always demonstrated a willingness to cooperate. In some situations, however, developers have been willing to negotiate, and when they are willing, feminist social skills might be very valuable. Then again, will ecological feminists be inclined to engage in boycotts? The actions of feminists in defense of reproductive freedom may give some indication of how environmental struggles would be approached.

II

There are special reasons why we phenomenologists, as phenomenologists, should take an interest in ecological feminism.

One affinity between phenomenology and ecological feminism is in the stress on context in feminist ethics. Warren has described the ethical approach of ecological feminism as contextual. Rather than base ethics solely on the implications of abstract principles, contextualists look at the actual contexts within one acts, as they are experienced by the people involved. These actual experienced contexts help determine the appropriate behaviors. My interest in phenomenological analysis, more than anything else, made my ethical approach contextual. Contextual ethics, including the contextualism of ecological feminist ethics, provides rich examples of attending to actual experiences, which I see as examples of going "back to the things" to test thought against lived reality.

In my work in ethical theory I have stressed the important role of individual world views in the development of moral beliefs and in the assessment of moral opinion. Warren seems to refer to something similar, if not the same thing, when she talks of "conceptual frameworks. Phenomenology can greatly clarify our understanding of how the individual's lived world works in the structuring of that person's view of the world. From an understanding of how a lived world is constituted, we can explain more clearly the formulation of moral beliefs. Why does the racist not feel the same horror which many of us feel when police beat a person of color unmercifully? In some way the racist's beliefs about the inferiority of nonwhite people affect a constitution of the situation in which what is done is fitting. We need to understand this aspect of the formation of values and moral beliefs as clearly as possible. The same sort of effect is at work when women and men constitute situations differently. Why do some men find something funny when most women find it offensive? Phenomenology can go a long way in explaining this. Abstract principles can explain far less than can be understood through a clear perception of the context in which something occurs.

Careful attention to the factors which shape a person's constitution of a lived world can help us evaluate the adequacy of that person's world view. We need not be stymied in our search for understanding when we find disagreement over basic moral beliefs. Holding an opinion does not need to rest as a given. We can explore the formation of individual world views and discover certain obvious inadequacies in the way they were arrived at. Since world views are not often, if ever, the product of logical deduction, we need not be afraid of falling into an *ad hominem*. We need to make a distinction between premises which prove logically that a proposition is true or a logical inadequacy which shows the proposition to be false and observations which give us strong grounds for accepting or rejecting a proposition when there can be no proof or disproof. When we find that superstition, scientific error, or reliance on a discredited authority played a significant role in shaping a world view, we have good reason to be suspicious of the view. The bare logical possibility that the view could be correct need not disturb us. When we find dishonesty, avoidance of unwelcome information, or self-serving selection of data we have other reasons to reject a world view. We are not put into the position of having to accept uncritically every constitution of a lived world which claims to be a true picture of the world. Specifically, we are not left with mysterious differences between men an women which we will never be able to fathom. We can understand very well why women respond to certain situation in ways which are quite different from those of many men. Furthermore, we can make a sound judgment about which responses are appropriate and which are not, which responses foster civil society and which do not.

In the context of the natural environment, just as we can be confident that debilitating fear of garter snakes is not reasonable, we can spot inadequate beliefs about the environment. If someone advocated destruction of a forest because such wilderness areas are frequented by Satan, a view firmly held because a psychic had said it is so, surely we would not need to accept that view as on all fours with the views of ecologists. If we found that the advocate is in the lumber business, we would have even further grounds for denying any credence to beliefs about unhousing his satanic majesty. The importance of context has a number of dimensions. One of the strengths of feminist ethics is its recognition of the significance of context, including the kinds of thinking which give rise to various moral claims.

Ecological feminism shows the influence of Carol Gilligan's report that women do not think about moral matters in the same way men do.[6] Ecological feminists support the feminist demand that different voices, those of women and oppressed people everywhere, be heard. This entails the rejection of a monistic system of ethics based on the following of rules which can be derived from one rationally argued foundation. In one of her recent papers on ecological feminism, Warren says that feminist ethics must approach ethics from a context of dialogue and practice which includes the voices of people who have different moral experiences from living in different circumstances.[7] Jim Cheney opposes monistic rationalistic ethical schemes, which he calls "totalizing theories." He favors a feminist environmental ethics which is grounded in "bioregional narrative," an "ethical vernacular," through which the "multiple voices of this Earth" are heard.[8] Warren does not deny the importance of theory in ethical thought, but in ecological feminism we see an ethical approach which is experience-driven, rather than driven by abstract theory.

The pluralism of ecological feminism, seen in its listening to many voices, attending to many narratives, gives to an understanding of the human condition reports from many lived worlds which have not been considered much before. Surely the lived worlds of about half the human race are worth the attention of philosophers. Indeed, without them, any understanding of what it is to be human is bound to risk incompleteness and one-sidedness.

The feminist interest in the concept of narrative gives phenomenologists and feminists an important area of shared interest. Phenomenology can benefit from a broader awareness of how large numbers of people perceive the world, feel about the world, intend to act in the world, and communicate to other people the doxic, pathic, and praxic features of their lived worlds. For its part, phenomenology can contribute to feminism through it's experience and technical skill in attending to

---

[6] Carol Gilligan, *In a Different Voice*, (Cambridge: Harvard University Press, 1982); *Mapping the Moral Domain* (Cambridge: Harvard University Press, 1988).

[7] Karen J. Warren, "The Power and the Promise," 139, 142f.

[8] Jim Cheney, "Postmodern Environmental Ethics: Ethics as Bioregional Narrative," *Environmental Ethics* 11.2 (Summer 1989): 117-134; "The Neo-Stoicism of Radical Environmentalism," *Environmental Ethics* 11.4 (Winter 1989): 323f.

the specific features of experience. Phenomenologists can help clarify many aspects of the functioning of narrative. Each narrative grows from a distinct constitution; phenomenologists have the tools for exploring and better understanding the ways in which the world is constituted and experienced. People who see the world as unjust, unfriendly, unrewarding, threatening, or evil had to come somehow to see the world the way they do. Coming to understand why these people constitute the world as they do is the beginning of an appropriate response to them. The factors which influenced their view of the world might be things which can be corrected. If the view of the world is the result of misunderstanding, confusion, or unreasonable expectations, this also indicates how the matter should be approached.

Feminists have sought to understand the development of a feminist consciousness. Important work on this has been done by Sandra Lee Bartky.[9] Bartky holds that to be a feminist, one must become a feminist, which is a transforming experience, including an altered consciousness. She describes a feminist consciousness as an anguished consciousness which recognizes the possibility of the transformation of an intolerable condition. Feminist consciousness is that of undeserved and offensive victimization. It is a divided consciousness which is aware of weakness and of strength and experiences confusion and guilt because the victim is also better off than most people in the world. The feminist consciousness suffers psychological oppression; "harmless" things can become sinister and social reality becomes deceptive. This consciousness suffers category confusion, is unsure how to categorize things, including one's own behavior. The feminist becomes vigilant, suspicious, and wary. She realizes the deceptive character of social situations, which might indicate opportunities to struggle against an unjust system, and which makes many social occasions into tests.

This sort of consciousness has been largely neglected in phenomological studies. It warrants study because it is the way life is constituted by a large number of people. We see aspects of this throughout our society. Some people are offended, even threatened, by what others consider amusing or harmless. Women report feeling unsafe in situations in which most men are aware of no danger. All of these aspects of lived worlds must not be ignored or minimized as the results of women being "too

---

[9] Sandra Lee Bartky, "Toward a Phenomenology of Feminist Consciousness." *Social Theory and Practice* 3 (Fall 1975) 425-439

sensitive" or "overreacting" or "taking things too seriously." These lived worlds of half the human race warrant attention and study. If we phenomenologists want to really do phenomenology, here is rich material for study. More important, however, is that this is an opportunity to use the insights and skills of phenomenology for human betterment.

There is an important contribution which ecological feminism can make to philosophy and phenomenology through the contextualism and attention to practical aspects of feminism. Some of us are becoming aware that philosophers, including phenomenologists, have had a tendency toward intellectualization of all questions and issues. Even though phenomenological analysis has shown a close connection between thought, feeling, and volition, most attention has been given to issues of thought and cognition. Now feeling, volition, and action are getting a bigger share of attention. Ecological feminism has opposed the reduction of all concerns to intellectual discussion. It has avoided separating thought from the other dimensions of consciousness and active life. Perhaps phenomenologists can learn from ecological feminists. At the very least we have in feminist thought examples of holistic, integrated approaches to the several facets of human life.

This integration of thought, feeling, and action has bothered some philosophers who wanted a priority to be given to thought. It may have seemed to some thinkers that validation of feeling and attention to context more than general principles would result in irresponsible emotionalism and a failure of sound governance of life. That some people have favored such an approach to life is clear, and distrust of such a non-rational and anti-intellectual way of living is justified. Is ecological feminism such an irresponsible philosophy? I do not think it is, and any tendency to move in such directions can be corrected. As an ethical theorist I have defended contextualism and pluralism. The more threatening approach for many philosophers is pluralism, which holds that moral principles which cannot be reduced to a common principle or be shown to be logically derivable from a common foundation may legitimately be employed in different sorts of situations.

To make this claim responsibly one must explain why this resort to unrelated principles is justified. This can be done, I believe, by showing exactly how various situations are different and how several types of moral judgment are different. It is not too difficult to show that such activities as deciding what one ought to do, evaluating the past behavior of oneself or another person, assessing the moral character of a person, and deciding upon principles of legislation are very different sorts of activity. The demand that one moral principle serve all these functions is not obviously a sound demand. It can also be shown that contexts can differ so significantly that one principle might not be the most appropriate one to use in each context. Deciding how to handle issues which arise within a family, where each person is known individually and where promises can significantly be made and where there is great opportunity to affect the life of an individual person, is a different matter from deciding what one should do about starvation in a remote country.

Different again is the making of laws to govern just distribution between a large number of people, most of whom are strangers to each other. To demand that the principles which provide ethical guidance in dealings with business associates should guide decisions affecting wild animals and future generations seems doctrinaire. To acknowledge that a number of moral principles can be employed, whether as an expedient until we know a great deal more about the world or as a permanent necessity in a world which we will never fully understand, is a morally responsible stance. It is not the same as claiming that I have a right to do whatever I wish. It is not the same as saying that two different actions in the same context can be morally right, one right for me, the other right for you. This kind of subjectivism is not justifiable, even if one can make sense of such use of the concept of right, but the alternative to subjectivism need not be an equally extreme monistic moral absolutism.

What ecofeminists are saying, it seems to me, is that we must hear the many voices of the world's people, to heed their personal narratives. This says to me that we must be far broader and more open in attending to the lived worlds of other people who share this planet. This does not mean that we must eventually judge them all to be of equal value. It does mean that we must not be too hasty in rejecting the voices of those who cry out to us in their suffering, in their anxiety.

The complaint that philosophy has placed too great a weight on intellect and has had too little respect for the emotional aspects of conscious life can be misunderstood. What some people seem to hear in this is a call to replace thought with emotionalism. Indeed some feminists may have become far too uncritical, far too willing to trust emotion blindly. This is not the case with all feminists, however, and it is not the only alternative to the excessively abstract rationalism which feminism opposes. Blind emotionalism is simply an excessive reaction to another excess. As a feminist I say my curse upon both houses, and I am not taking a maverick position in doing this. For me one reason for seeking to understand the role of emotion instead of rejecting it out of hand or considering it an incubus to be overcome is the phenomenological insight that thinking, feeling, and intention to act are intertwined in most activities. The idea of learning through the emotions is not absurd, even though we have not mastered the doing of it very completely. Probably we can all recall experiences in which we did not realize the full value of something until we realized the emotions which it aroused in us. In anger or in tears we learned something about what we valued which we had not realized in calm intellectualizing.

To reject critical thinking or to reject feeling would be to reject part of what we are. Historically, we philosophers have favored intellect. In reaction to that some of us might say things about emotion which sound extreme when interpretted literally. It will be a good practice, I think, for us to avoid recoiling in horror at what appear to be extreme statements until we find out whether the person means them to be taken in the extreme sense. If we do not practice this precaution, we may fail to hear something very useful and valuable.

I believe that a cooperative endeavor, taking advantage of the intuitive and adventuresome work of feminists and the technical skill and exactitude of phenomenologists, will reward the overcoming of any initial trepidation of either side in relating to scholars whose styles are different. Both sides will benefit from overcoming any annoyance or frustration with the style of the other. These two approaches to philosophy are too important and have too much to give the other for us to avoid an encounter which might go a long way toward improving the way we practice philosophy. This cooperation might also contribute heavily to our efforts to save human life, all life, on the planet.

Chapter 9

Ethnic Studies as Multi-Discipline and Phenomenology

Stanford M. Lyman and Lester Embree
Florida Atlantic University

Abstract: *Ethnic Studies is a type of academic program that includes a multiplicity of cultural disciplines (education, ethnology, history, literature, philosophy, political science, psychology, sociology, etc.). The history of the integrationist (or assimilationist) and the pluralist (or multi-cultural) conceptions of race and ethnic relations chiefly in the United States is examined first and then the attempt is made to show how research within and philosophical reflection upon this multi-discipline can be phenomenological.*

Introduction

Ethnic Studies has been established only recently in major American universities. There seems to be a growing need for something of the sort elsewhere in the world. Since Phenomenology appears to offer an approach that could be taken within the disciplines that make up this multi-discipline, and since phenomenological philosophy is likely to benefit from reflecting on Ethnic Studies, an entry on it from a phenomenological perspective was desirable for *The Encyclopedia of Phenomenology* (to appear from Kluwer Academic Publishers). Nothing in an explicitly phenomenological way has been published yet with respect to Ethnic Studies as such, but I had a world authority on it among my colleagues, and he agreed to try to write such an entry with me. After discussion about how to proceed, we sat down with an outline and a tape recorder. The transcription of what we had to say proved to be several times the length desired for an encyclopedia entry, but it seemed to have merit in its own right and for that reason has been revised into this double

M. Daniel and L. Embree (eds.), Phenomenology of the Cultural Disciplines, 211–249.
© 1994 *Kluwer Academic Publishers. Printed in the Netherlands.*

essay. (The encyclopaedia entry has been distilled from this longer and more conversational version.) Stanford M. Lyman leads off in the first part of the following exposition and I follow, my prompts and reactions being expressed in italics. This pattern, including the significance of italics, is reversed in Part II. Dr. Daniel, who also has deep interests in Ethnic Studies, agreed to our including this unusual text in the present volume. *Lester Embree*

I. Ethnic Studies as a Multi-Discipline

Stanford, how might we generally characterize Ethnic Studies? I would say that Ethnic Studies arose in response to the American racial situation and reflects changes and shifts in orientation within the consciousness of, as well as the attitudes toward, certain racial and ethnic groups in the United States. *Are these the so-called 'minority groups'?* Yes. Essentially, Ethnic Studies arose as a counter to what had come to be seen as the marginal utility of assimilation theory. Whereas, especially in the discipline of sociology, virtually all previous theoretical work had been narrowly devoted to studying the processes associated with assimilation, and to assuming that assimilation was both the likely and the desirable outcome of racial contact in America, alternative projections of the character as well as the solution to the "American Dilemma"[1] had been relegated to subterranean arenas of thought and scholarship. Recently, however, the assimilation thesis has come to be recognized as less than valid, a desideratum rooted more in hope than in a positivistic science's requirement of predictability. Ethnic Studies arose less from inside the academy than from responses to academic orientations by Black, Asian American, Hispanic, and Amerindian students and interest groups primarily concerned about American history and culture and their presentation in the university. *This sounds to me like the Civil Rights Movement. Did it follow the integration of the schools in 1954?* Yes, a few multi-cultural academic programs did begin after the early successes of

---

[1] Gunnar Myrdal, with the assistance of Richard Sterner and Arnold Rose, *An American Dilemma: The Negro Problem and Modern Democracy* (New York: Harper and Brothers, 1944). See also David W. Southern, *Gunnar Myrdal and Black-White Relations: The Use and Abuse of* An American Dilemma *1944-1969* (Baton Rouge: Louisiana State University Press, 1987); and Walter A. Jackson, *Gunnar Myrdal and America's Conscience: Social Engineering and Racial Liberalism, 1938-1987* (Chapel Hill: University of North Carolina Press, 1990).

the Civil Rights Movement had taken hold, but the programs also arose out of the recognition in the late 1950s that the conventional textbooks in history, especially, as well as those in sociology, anthropology, and literature did not represent the range or the reality of population groups in this society.

Ethnic Studies put itself forward in two directions. These partly contradicted one another, but each sought a remedy to the situation. One trajectory aimed at integrating the history and literature texts, i.e., providing a documented presence of Blacks, Asians, Hispanics, and American Indians in and with respect to the general run of American history and letters. The other direction asserted the need to legitimate the prideful rectitude and unique character of each of the hitherto neglected ethnoracial groups through academically respectable special studies programs. Both approaches went on simultaneously, although the latter one, an assertion of an ethnoracial *thymos*, seems to have made greater headway. Conventional history courses have made attempts to bring the ethnoracial "other" in; however, these attempts are uneven in their quality. One result has been a decline of consensus history or consensus historiography, and a similar decline in monocanonical literature in American studies. Still another result has been a rise in claims by other human groupings—to take two examples, those that represent women and those seeking legitimation for different sexual orientations—each making its demand for representation and dignity on the educational curricula in the humanities and the social sciences. *I suppose we might call these 'insurgent academic movements', Ethnic Studies, often taking the form of Black Studies, Asian American Studies, Chicano Studies, Native American Studies, etc., being the first, and then the others, beginning with Women's Studies, taking the former as a model as they faced similar problems.* Exactly right. Ethnic Studies arose first, and the others modeled themselves on it.

What distinguishes ethnic differences and relations in the United States from differences among and relations among the several 'nationalities' in Europe? I don't think these phenomena are the same. Indeed. This is a central issue. The ethnic-studies issue might be conceived as a special variant of an older idea, viz. "American exceptionalism." That is, the conceit that what is said about America historically and sociologically is exceptional with respect to the way one talks about seemingly similar matters in Europe. In Europe, beginning in 1919, the breakdown of the Ottoman and the Austro-Hungarian Empires was followed by the enunciation of Woodrow Wilson's idea that institutionalizing national self-determination would produce an ethnically homogeneous nation state for each ethno-national group. This project was, as it turned out, unrealizable, so that multi-national states appeared, especially in Eastern Europe, each claiming a national identity. Such had already existed in places like Belgium.

And less overtly in England and elsewhere? Yes. In Great Britain there had long coexisted—uneasily—the four cultures of Albion.[2] A European resolution of this arose in the early days of the League of Nations with the adoption of "minority treaties." Minority treaties were signed by various states, each promising to protect the minorities in their territorial domain. A few attempts were made to secure minority rights in constitutions, and special legislation. And from that resolution of the problem the word "minority" began to come into social scientific prominence. It had not been a prominent term before that era. In Europe, "minority" referred to groups that had not yet achieved and were not likely to achieve national statehood and political independence.[3] Soon, the term was transferred to America to refer to racial and ethnic groups, but, with occasional exceptions to be noted, the conceptualization in America was that the racial and ethnic groups would find a place within the society and under its already established political jurisdiction. Out of this development there arose two sociological theories about ethnoracial groups in America, one dominant, the other subordinate.

The dominant approach, which had been presented as early as 1913 by Robert E. Park (1864-1944), and owed its origins to an even earlier formulation by Sarah Simons,[4] put forward what was in effect a promissory note that assimilation would be the eventual outcome of the contact of peoples and races in America. Moreover, Park believed that assimilation would be the eventual outcome of race contacts throughout the world. That view became the prevalent perspective. It fit in with Woodrow Wilson's assertion, for example, that in America there were to

---

[2] David Hackett Fischer, *Albion's Seed: Four British Folkways in America* (New York: Oxford University Press, 1989).

[3] J. A. Laponce, *The Protection of Minorities*, University of California Publications in Political Science, 9 (Berkeley: University of California Press, 1960).

[4] Sarah Simons, "Social Assimilation," *American Journal of Sociology* 6.2 (May 1901), 808-815; 7.1 (July 1901), 53-79; 7.2 (September 1901), 234-248; 7.3 (November 1901), 386-404; 7.4 (January 1092), 539-556.

be no permanent minorities, that everyone was to become an American. In the same year, Horace Kallen (1882-1974) published a short essay, which he would later expand into a book, *Culture and Democracy* (1924), in which he put forward a countervailing thesis that would be sub-dominant for the next forty years. That was his assertion that a democratic society would prove its own worth if it preserved and protected the cultural identification of all its peoples. Kallen's was an attack on America as a "melting pot" and was seen as such. However, for Kallen, as for Park, the United States was conceived as a single polity. Kallen did not believe in national self-determination for each ethnoracial group. Rather, he was concerned with cultural preservation within a democratic state society.[5]

An exception, of course, might be the American Indians with their quasi-autonomous but socially and economically impoverished reservations, but perhaps that is a small exception in terms of the numbers involved, although one that is quite striking symbolically.

The exceptions here are actually two: that of the Indians and that of the Blacks. In the case of the Indians, in the beginnings of their contact with Europeans invaders and settlers of the Americas, they regarded themselves and were regarded by the Europeans as distinct nations. Hence, their early relations with the people and government of the United States were bound by diplomatic treaties. This practice continued for almost a hundred years after the founding of the United States of America, and some are still in effect. This meant, in effect, that the Indians, unlike every other people in the United States, were not subject to the general operations of the federal system, that is, the several states did not have the same kinds of control over them that they had, say over Blacks, Asians, or Europeans. This jurisdictional difference was clearly stated in a Supreme Court case in 1831, *Cherokee Nation v. Georgia*. In the opinion delivered in that case by Chief Justice Marshall a unique status for America's Indians was established, a status under which they have lived ever since. Marshall declared the Indians to be "a domestic dependent nation." And so, since 1831, every issue touching the juridical relations of Indians to Whites has turned on the hermeneutics entailed in those three fateful words, "domestic dependent nation."

---

[5] Horace Kallen, *Culture and Democracy in the United States: Studies in the Group Psychology of the American People* (New York: Boni and Liveright, 1924; reprint, New York: Arno Press and the New York Times, 1970).

But these 'nations' occupied a different status than that of European nationalities? Exactly, and that was because the key term "domestic" modified the word "nation." Whereas the Indians had wanted to be treated in the same diplomatic manner as a European state, that was the one thing that was not going to be allowed. They were, the Supreme Court opinion went on to say, "in a ward-like relationship" to the Federal Government, and, because of their unique status, in "pupilage." Those words have resulted, by the way, in a separate body of law, i.e., forty volumes or so, of American Indian law, in which only some lawyers are specialists.[6] The original compiler of that jurisprudential tradition is the late Felix Cohen (1907-1953), son of the philosopher Morris Cohen (1880-1947).

The American Blacks are in a still different position, but it also modifies the pluralist-assimilationist debate. Periodically, from the 18th Century on, various forms of secessionist movements arose among Blacks in America. Some took the form of back-to-Africa movements, e.g., one in 1714, another in 1817, another in 1859, another from 1896 to 1915, and, in the 1920s, Marcus Garvey's Universal Negro Improvement Association and its movement. In other cases, secessionists looked not to Africa, but rather to some place outside the United States where Blacks might live freely—e.g., various islands, Canada, and such like. A third case occurred only once to my knowledge, in the late 1800s. Some Blacks proposed that the New Mexico and Arizona territories be set aside as locally autonomous Negro states inside the United States.

Was that on the model of Oklahoma as Indian Territory? Not quite. The declaration of certain territories as "Indian Territories" was made by the United States independent of the wishes of the Indians. The Oklahoma territory resulted from the "trail of tears" followed by the Cherokees driven out of Georgia in 1829-1831. There were also proposals to deport Blacks to Africa, the Caribbean, and Latin America. The American Colonization Society, started by President James Monroe, founded Liberia and peopled it with manumitted slaves. President Lincoln also proposed more than one deportation plan.

Looked at over the long haul, such secessionist movements seem to arise when Blacks despair of ever becoming full-fledged citizens, come

---

[6] For a summary, see Stephan L. Pevar, *The Rights of Indians and Tribes: The Basic ACLU Guide to Indian and Tribal Rights* (Carbondale: Southern Illinois University Press, 1992).

to doubt the efficacy of integration, and deplore their lack of civil rights inside the United States. The year 1896 provides a classic instance. In reaction to the Supreme Court's decision in *Plessy v. Ferguson*, which established the "separate but equal" doctrine, Bishop Henry McNeal Turner (1834-1915), Bishop of the African Methodist Episcopal church, and a staunch integrationist, a leader who was, in effect, the Martin Luther King, Jr. of his day, began to be attracted to "Ethiopianism," took up its motto, "Africa for the Africans," and became active in a back-to-Africa movement.[7] With respect to absorption in American society and cultures, Blacks constitute a swing group. The main thrust of Black programs, life and letters has been toward assimilation, here meaning the acquisition of what W. E. B. DuBois (1868-1963) called every "jot and tittle" of civil rights; nevertheless, their experience in the United States and throughout the Occident has raised serious geo-political, cultural, and indentificational questions in their hearts and minds.

What is the standard view of what race and ethnicity are? Race seems fairly clear at first but can become obscure. The phrase you habitually use is 'Race and Ethnicity.' What does the whole phrase mean?

Here again there is both an historical and a conceptual answer. The historical answer is that up through the early years of the 20th Century, and all through the 19th Century, the word "race" was used rather loosely to encompass what today we would call racial, national and ethnic groups. The word "race," conceived as a term in physical anthropology, refers to phenotypical groups that possess at least one hereditary trait, or a cluster of traits, that is used as a characteristic sign of a race. The term has lost much of its value to science, but survives in social, personal, and political life. A trait is—and here arises the social aspect of it—a visible one—e.g., skin color, hair texture, eye shape. There have been occasions when students of human biology and endocrinology were able to show that peoples differed in a patterned and hereditary way in gland secretions, but these distinctions were . . . *Socially invisible?* Yes, they were socially invisible. Essentially, and for all practical purposes, racial traits have been conceived to exist in skin color, hair texture, and eye shape and in the other features of human anatomy that the ordinary person can readily observe. "Races" are social constructions used to distinguish human types.

---

[7] Stephan Ward Angell, *Bishop Henry McNeal Turner and African-American Religion in the South* (Knoxville: University of Tennessee Press, 1992), 198-252.

"Ethnic" of course is an old word that comes out of the Greek, *ethnos*, a people, and has arisen, especially strongly in America, to refer to the social and cultural aspects of life for various national minorities from Europe and, for national and racial minorities from Africa, Asia, and among the aboriginal Americans as they find expression in everyday life. At present, members of what physical anthropologists of earlier days would have designated as the "Negroid" race would consider themselves ethnically as Blacks or African-Americans. These terms are the socio-cultural expressions, respectively, of a racial or national peoplehood. *Among such expressions language and religion come first to mind for me.* Yes, language, religion, and other relatively visible and audible aspects of culture, e.g., food, music, dance, clothing. *Which are all cultural, i.e., historical, learned, and able to change.*

Yes. This issue in America sorts itself out along a line that is interesting for public policy and for such disciplines as political science, that is, along the line that divides the public from the private. Assimilation theory, when modified by a pluralistic accommodation, holds, in terms of public policy and the national economy, that people will ultimately participate in the public arena as persons, as individuals. In terms of the content of private life, i.e., such activities as family get-togethers, social gatherings, recreational occasions, and religious observances, non-acculturated elements will be the order of the day. This division makes a kind of practical sense, but critics point to the ways in which socialization into the public economy and polity insinuates itself into the ways of the "old World" cultures, erodes their values, undermines their quality, and diminishes their meaning to those who participate in them. Moreover, the prerequisites for maintaining a separate culture are not available to every culture group. To these critics, society is said to have an obligation to insure the availability of a heritage and culture for each people. In part, Ethnic Studies arose as a way to meet this need.

I have the impression that Ethnic Studies programs began in California. Is this correct? Yes, they did. I was a part of that movement and very close to it. It started in history at the University of California at Berkeley when the State of California, struck by its recognition of how racially and ethnically unrepresentative the public school textbooks were, commissioned the revisionist historian, Kenneth M. Stampp, to form a scholarly committee and examine the textbooks used in the American history courses in the public schools in the state. This occurred in 1956-57. The

"Stampp Report" that resulted sparked a movement that took off. *California at that time already had substantial Chinese and Japanese communities, a large African American population, an ever growing proportion of Hispanics, "Chicanos," and Latinos, as well as many people of Euro-American background, e.g., Irish-Americans, Italian-Americans, German-Americans, Jewish-Americans, etc. So the Ethnic Studies programs developed in what was already a "multi-cultural" situation, a situation of peoples living "beyond the melting pot."[8] Then it spread across the country, in various programmatic forms established in colleges and universities.* This is correct.

In terms of broad points of view within the field of multi-disciplinary Ethnic Studies, my thought is that there are probably three major perspectives. One is Marxist; another derives from Positivistic social science and is heavily quantified; finally, there is a qualitative-interpretive point of view that in fact is your own approach. I am most sympathetic with the latter perspective, one that relies on ethnography, i.e., where one goes into the community and talks with people, sympathizes and empathizes, and produces discursive representations of how the people researched view and relate to things, people in the same and in different groups included.

This is right. Let us start with the Marxists. It's interesting how their approach challenges everyone, both the assimilationists and the pluralists. In a Marxist approach emphasis is placed on the significance of class over all other forms of social solidarity. For Marxists, then, the rise in racial and ethnic consciousness is considered to be a false consciousness and one that must somehow be either understood contextually, explained away, or seen as a stage in the development of interethnic consciousness and, hence, in its own ultimate disappearance. Let us make a further distinction between Academic Marxists and the practicing Communists of an earlier era. The latter had a terrible problem in attempting to recruit the various national, ethnic, and racial minorities into its international organization. In late nineteenth century America, followers of Marx (1818-1883) and Engels (1820-1896) organized themselves in ethno-national cells representing the different language groups from Europe. Friedrich Sorge

---

[8] Nathan Glazer and Daniel Patrick Moynihan, *Beyond the Melting Pot: The Negroes, Puerto Ricans, Jews, Italians, and Irish of New York City*, 2nd. Edition. (Cambridge: MIT Press, 1970). Fore the pre-national period, see Joyce D. Goodfriend, *Before the Melting Pot: Society and Culture in Colonial New York, 1664-1730* (Princeton: Princeton University Press, 1992).

(1828-1906), Marx's correspondent in America, organized an all-German group. He succeeded in gaining Marx's support for expelling from the Internationale the cell headed by Victoria Woodhull (1838-1927), who had run for the President of the United States on the Equal Rights Party ticket with Frederick Douglass (1818-1895), the ex-slave, as her vice presidential running mate. Ironically, the Communist Party ended up having its own Black cell, which, on the one hand, seemed to reinforce racial segregation, and, on the other, responded to the linguistic and national divisions of 1920s America. As far as I can tell, the Communists were never fully able to resolve the contradictions contained in this issue. The thrust of their argument was that Blacks were to conceive of themselves as members of a multi-ethnic proletarian class and that they ought to deemphasize race issues as they drew together with Whites to engage in a common class struggle.

Then there was an ambivalence about assimilation? Yes. Ironically, Robert E. Park, who was not a Marxist and was opposed to Marxism, asserted in 1939 that the onset of complete assimilation, which would occur as the culmination of the irreversible and progressive cycle of race relations that he had postulated, would herald the beginning of the class struggle. That is, he thought of assimilation as solving a problem that had originally been recognized by Engels in his famous letter to Sorge in 1893.[9] Sorge had written Engels, asking "Why is there no socialist movement in America?" Engels replied by calling attention to the fact that work, in America, is organized on the basis of ethnicity and race, into work groups composed of Irish, Germans, Czechs, Poles, Italians, Scandinavians, etc. Then he adds, "And the Negroes." He continues, arguing that so long as work is organized in this way there will never be a class-based proletariat. Park's race relations cycle, with its promissory note that assimilation would be its final stage, solves Engels's problem by saying, in effect, that separate ethnic, social, and economic organization is temporary. Ultimately, assimilation would result in a society composed of ethnoracially integrated classes "and the class struggle will begin"!

My thought, in the light of recent events, is that Engels was not very insightful about European societies and failed to recognize that in England, for example, some businesses were Welsh, some were Scottish, and so on, so that he had a somewhat over-idealized notion of a homogeneous

---

[9] Reprinted in *Marx and Engels: Basic Writings on Politics and Philosophy*, edited by Lewis S. Feuer (Garden City, N.Y.: Doubleday Anchor, 1959), 458.

European bourgeoisie and homogeneous proletariat that I doubt existed there any more than they did in America. I think you are right about that. It is interesting how in Europe collective identity is almost always framed in terms of nationality or in relation to the nationality question and in the efforts to resolve it in a territorial-political way. Nathan Glazer and Daniel Patrick Moynihan edited a book on the subject in 1975.[10] In that work, writers from Europe and Asia, as well as America, describe the ethnic divisions in their own societies and conclude that the whole world is organized ethnically. Moreover, the number of self-proclaimed ethnonational peoples is many times greater than the number of existing nation-states.

I think that with the formation of the nation state, this modern geopolitical entity, this institutionalization of the Enlightenment, we have the legitimation of the legal person, i.e., the person with all the ethnicity, gender, and class characteristics stripped away. To put that into effect would have required the suppression of ethnic differences. It would have been much easier for those in power to consider themselves as having "normal" characteristics, to consider the other groups "deviant," and to urge that the ethnic differences of the others be replaced with "normal" ones. This is how I understand the acculturation-assimilation-integration model. In this way, some if not all Western European nation states turn out to be 'ethnic empires' in which one ethnic group or an oligarchy of ethnic groups dominates the others. Yes, and it was long this way in the United States, with the W.A.S.P.s. *Until very recently.* Yes.

A prominent Austro-Hungarian sociologist of the late 19th Century, Gustav Ratzenhofer (1842-1904), predicted, much to the chagrin of the assimilation oriented American sociologists, that there would come a time in the United States of America when the population would become dense and the struggle for existence difficult. At that moment, he claimed there would be a reawakening of ethnic and national consciousness. He said this in 1893, the very same year that Engels wrote to Sorge about the permanence of ethnic separateness in America. Albion Small (1854-1926), the founder of the Department of Sociology at the University of Chicago, adapted Ratzenhofer's sociology of the American situation, but he balked at the idea that ethnic consciousness could be revived. That, Small insisted, would never happen.

---

[10] Nathan Glazer and Daniel P. Moynihan, eds., *Ethnicity: Theory and Experience* (Cambridge: Harvard University Press, 1975).

There are, hence, two acculturative challenges—viz., an assimilation challenge that, as a political policy and a social and cultural process, will foster national unity within a Capitalist-industrial society, and the Marxist variant of assimilation that is going to facilitate the onset of a conflict-ridden class structured society. Park defined assimilation as a "co-ordination of sentiments." I have interpreted this to mean, among other things, the coming into existence of a people claiming a common historical heritage, a process described in the motto found on the American dollar bill, "E pluribus unum." In plain language, this process of assimilation is indicated when, for example, the "founding fathers" recognized by, say, a Chinese American, are George Washington, Thomas Jefferson, and Benjamin Franklin, just as they are for a W. A. S. P. They share founding fathers; that shows both are assimilated.

I think that genetic explanations are extremely popular in common-sense as well as scientific understanding. You explain what something is by telling why and how it grows, where it came from or sprouted, etc. I don't know if all societies do it. It may come from agricultural life. Nevertheless, in our common-sense we often proceed this way. And what strikes me as related to this is how when we have, at least in the academic context, a group of insurgents, the first thing they want to do is discover or recover their history, e.g., Women's History, Black History. History is part of making and building group identity, establishing that we are who we are because we came from some particular source. This source provides the root of the group's heritage. Sometimes, this is done through confecting an origin myth.

Perhaps this is a good place to introduce a conceptual distinction that "the Durkheim School" introduced. Emile Durkheim (1858-1917) and Marcel Mauss (1872-1950), his nephew and colleague, differentiated between "civilization" and "culture." Civilizations are cultural expressions that transcend geo-political boundaries. Christianity, for example, is a civilizational phenomenon, as is Buddhism. I would argue, and I suspect that Durkheim would agree, that the cultures that grew up from the European seaborne empires, the Spanish, Dutch, French, English, etc., are civilizational, institutional ways of thinking, believing, and behaving that moved beyond their original "national" boundaries. For Durkheim and Mauss, "culture" was a term that referred to those ways of life and thought that were bounded geo-politically.

From this point of view, the dilemma of culture and civilization provides the basis for one problem of ethnicity in America. Thus, for example, in America, Mexican-Americans, Puerto Rican-Americans,

Cuban-Americans, Salvadorian-Americans, all except the aboriginal Central and South Americans, can lay claim to the Hispanic civilizational complex. People often use the word "culture" here, but if they were using words as Durkheim and Mauss proposed, the proper word would be "civilization." At the same time these same peoples will differentiate among themselves: Mexican-Americans claim heritages very different from those of Puerto Rican-Americans, etc. The culture/civilization confusion is built into the very structure of ethnic expressiveness.

I would be inclined to use "culture" where they use "civilization" and then recognize specifications. However the terms are used, the distinction is very useful to point to some of the dilemmas and contradictions that exist in the various geo-political units. Thus, if one takes the seaborne empires as one of the beginnings for the issues that we are discussing, there is the fact that these empires established various forms of polity and hegemony over many parts of the globe. In the second half of the Twentieth Century, the colonies established by these empires asserted themselves, claiming not only nation statehood, but also varying degrees of cultural self-determination. Inside the United States there are representatives of virtually every one of the seaborne empires, in their subcultural divisions as well as in civilizational totalities. Since the late 1950s, there has been a variety of assertions of cultural self-determination, some with political overtones, but none is seriously directed at nation statehood or secessionism. *Right, each wants its share of the socio-cultural pie, social, cultural, and economic justice.*

Let me go back to something. Besides music and food and clothes and religion and language, doesn't locality or place have something to do with ethnicity in many cases, who you are being related to your 'turf'? Very good point. Here, there is a very real phenomenon as well as the disintegration of it. There are often two places of identification: The place of origin of the people, some country or some village in Europe or Asia, for example, or some ethno-cultural region in Africa. The latter can be very problematic for Blacks. The second type of place, long of interest to American sociologists, is the territorial enclave established by a people inside the regions and cities of the United States. *'New England,' for example.* Yes, but even more significantly, the various Little Italies, Polonias, Germantowns, Little Tokyos, Little Havanas, Chinatowns, etc. These become loci of cultural expression, personal well being, and, often, of a limited kind of culturally specific economic activity. They also developed varying degrees of community autonomy vis-a-vis the state.

Probably the highest degree of this autonomy, a virtual community self-government, did arise in the old Chinatowns of the late nineteenth and early twentieth centuries.[11] The lowest degree occurred, probably, for German and Dutch areas of settlement in the United States. Of course, the Amish provide another example of such autonomy.

Would you say that the degree of autonomy correlates to the number of different ethnic characteristics, the most when the language, religion, etc., are all different from those of the adjacent majority and the least when there are the fewest differences, including, say, linguistic ones that are hardly more than dialectical, such as among, I think, uneducated English and Dutch in the 18th Century, which practically disappear in perhaps only one or two generations? They are all Protestant, all Northern Europeans ethnically, speak Germanic languages, and on and on, and are thus well on the way to merging into the WASPS. A Chinatown would be different in almost every way, especially if there was continual immigration.

Of course, change comes via the inter-generational process and its sociological effects. By the time one reaches the third, fourth, and fifth generations, there has usually occurred an economic mobility that also becomes geographic, taking these later generations out of the ghettos or enclaves of their forbears and into the middle class neighborhoods of the larger city and the suburbs. Such movements break down ethnic group solidarity. At this point in socio-cultural expression, other types of symbols remain. *My guess is that at that point you get a leap in the frequency of inter-ethnic marriage.* Yes. There is good evidence for inter-ethnic marriages in a study carried out of three generations of New England inhabitants in the 1940s. Marriage licenses from that period show that in the third generation intermarriage among the various European peoples was very high, but inter-religious marriage was very low, so that "Catholic," "Protestant," and "Jew" become the "holding companies" of ethnicity.[12]

---

[11] Stanford M. Lyman, *Chinatown and Little Tokyo: Power, Conflict, and Community among Chinese and Japanese Immigrants in America* (Millwood, N.Y.: Associated Faculty Press, 1986), 109-224).

[12] Ruby Jo Reeves Kennedy, "Single or Triple Melting Pot? Intermarriage Trends in New Haven, 1870-1940," *American Journal of Sociology*, 49:4 (January, 1944), 331-339; and "Single or Triple Melting Pot? Intermarriage in New Haven, 1870-1950," *American Journal of Sociology* 58.1 (July 1952), 56-59.

My next guess is that analysis of such data would also show significant Northern European vs. Southern European and probably also Western vs. Eastern European differences. Kennedy did not report on that. But there is some data to suggest a more recent breakdown of the religious walls; for example, the number of intermarriages among Jews and Gentiles has risen so much that Egon Mayer, a contemporary Jewish sociologist, has written a cautionary book on the subject. *Probably even more between Catholics and Protestants, especially in cities.* Maybe. I do not have data on that. What has been most startling is the high intermarriage rate between Whites and third and fourth generation Asians in America. It reaches the 70% level. *Imagine the agony of the older generations to see this.* One can occasionally see backlash in the newspapers, but not all of the older generation are opposed. And, although amalgamation was once seen as both the sign and effect of complete assimilation, it is noticeable that the very generations that are intermarrying are also asserting ethnic revival.

To turn to another topic, and again please correct me, but my impression is that American positivistic social science, going back to the 1930s, with its heavy reliance on statistical analysis, arose among those with an aversion to facing up to the stresses involved and patience needed for ethnographic or qualitative-interpretive investigations of ethnic issues. In place of the latter, one could turn it all into numbers, counting and measurement, and thus avoid empathizing, e.g., with someone who wished to marry someone of another racial or ethnic group while her grandmother is agonizing over such a prospect. Am I speaking to an issue?

I think it originated differently. The quantitative approach in American sociology was championed by an anti-miscegenationist, Franklin Henry Giddings (1855-1931), one of the founding fathers of sociology in the United States. He developed a concept, "consciousness of kind." Although he tried various ways to hide the fact, for him the phrase meant ethnic kind. He thought he could quantify and measure the degree of consciousness of kind through surveys and he hoped, also, to promote it. Not all of his disciples took that angle, but he made it a centerpiece of his teachings.

There is another aspect of the quantitative approach in sociology that deserves attention. Quantification counts units. Now the unit that is counted by a sociologist is the *socius*, i.e., the socialized individual. The relationship of his or her attitudes, status, and class are the problems investigated by this kind of sociology. Thus, in one sense, quantification celebrates the Enlightenment ideal of the individual by making it the unit to be investigated. However, sociologists have rarely treated the socius as a "person" in the American Constitutional sense, i.e., one who is entitled to due process and the equal protection of the laws, regardless of status, race, color, or religion.[13] Examination of the socius sought to show the relationship of that unit to other units in an aggregate. Thus, a particular socius shares a status or an attitude with another socius by reason of his or her race or generation or other measurable characteristics. Thus quantification could be seen as advancing the project of the Enlightenment—except, that the emergence of the individual as an entity independent of all ascriptive associations is something that has never been found. Nor has that status been investigated thoroughly.

But there was an essay written in 1955 by Leonard Broom and John Kitsuse, two sociologists, called "The Validation of Acculturation,"[14] Their essay asked the question, How would one recognize a fully acculturated person? Their answer, which I believe resonates with American as-similationist approaches, was this: An acculturated person would be one who could neither credit nor blame his or her status in society on anything inherited, on his or her race, ethnicity, or religion. Such a person would have to boast about individual merit or lament the lack of same. *It's all a matter of individual merit and accomplishment.* Exactly. *So it's neither transmitted culturally nor biologically hereditary?* Yes, the acculturated person could not say, "I'm Black and that's why I didn't get a good job," or "I'm Jewish and that's why . . . ." *Or I'm the son of a rich man and for that reason have had many advantages.* The full validation was the emergence of the meritocratic individual, who would have to stand or fall by claiming "If I did not get the job, I guess I was not good enough." Of course their theory assumed that the society would have eliminated all biases based on birth, heredity, race, color, and heritage.

If one looks at Parsonian sociology, i.e., the perspective associated with the late Talcott Parsons (1902-1979), America's ideal is summed up in the

---

[13] See Stanford M. Lyman, "Race Relations as Social Process: Sociology's Resistance to a Civil Right's Orientation," in *Race in America: The Struggle for Equality*, edited by Herbert Hill and James E. Jones, Jr., (Madison: University of Wisconsin Press, 1993), 370-401.

[14] Leonard Broom and John Kitsuse, "The Validation of Acculturation: A Condition of Ethnic Assimilation," *American Anthropologist* 57 (February 1955), 44-48.

phrase 'A universalistic achievement society.' Universal values prevail and achievement is meritocratic. Parsons's contrast is with China, which he calls 'a particularistic achievement society.' Whether he is right about China is not important for us at this point. Its the contrast that is so interesting, because, as Parsons is grudgingly forced to admit, particularist elements are also present in American society, and Americans do not wish to dispense with them. There are times when Americans want to fall back on particularism; there are times when Americans do not want to be meritocratic. In effect, the classic hypocrisy is to credit all one's successes to one's individuality and to blame all one's failures on others—a conflation of meritocracy and particularism.[15]

Plainly, we always grow up with some sense of ethnicity, our own and someone else's, difficult as it may be to recognize its exact nature. Do you see this classic hypocrisy in present-day America? Oh, yes, I do see it here. I do too, and I think that one of the reasons that there is so much anger coming from the middle classes of this country is that people claim individual credit for their own advances, but, when things go badly, they do not want to take personal blame for the fall; instead, they blame the system, blame others, blame the government, blame anybody but themselves. Of course, there are both individual and social factors for both success and failure.

In 1945, Talcott Parsons wrote an essay trying to explain anti-Semitism and anti-Negro feeling.[16] He explained them both in accordance with his social system analysis. America, as a society built around the values of universalism and individual achievement, would evoke what Parsons calls "strains" within its own system. The effect of these strains would be felt in the hearts and minds of quite ordinary middle-class White individuals. Not all of those who seek to live up to its values and norms would get ahead in the society. Some of these would become frustrated and attribute their failures to Blacks or Jews. Both groups fulfill the need for scapegoats; for Jews, so the argument runs, seem to get ahead effortlessly, while Blacks seem to enjoy—and are subsidized for—living below the level of Occidental civilization. But the White working and middle classes

---

[15] Talcott Parsons, *The Social System* (Glencoe, Ill. The Free Press, 1951), 109-112.

[16] Talcott Parsons, "Primary Sources and Patterns of Aggression in the Social Structure of the Western World," *Essays in Sociological Theory*, rev. ed., (New York: Free Press, 1964), 298-322.

can neither get ahead, despite their efforts, nor can they afford the luxury of being lazy and promiscuous. This is how Parsons accounts for anti-Semitism and race prejudice having become features of Western civilization and its discontents. His approach entails a sociological Freudianism: It is rooted in the idea that Occidental meritocratic civilization is inherently frustrating and that its frustration will lead to racist and anti-semitic aggression.

I don't have to ask you about the qualitative-interpretive perspective because your whole approach, indicated here, illustrates it. But, I know you have of late been reading in the works of the Post-Modernists, trying to figure that perspective out, trying to respond to it, trying to extract what good you can find in it. Is Postmodernism a force in Ethnic Studies today?

Postmodernism has taken on an American form. It has suited itself to the problems we have been discussing, more specifically, the dilemma of how, on the one hand, there could arise a justified ethno-cultural renaissance, and how, on the other hand, the several renascent ethno-racial minorities will fit themselves into American society, with all the rights and privileges accorded thereunto. What the ethnoracially oriented post-modernists—and, I should add, there are not many of them but the few make up in brilliance what they lack in numbers—do is borrow the theme of dis-privileging texts from Jacques Derrida, the father of Deconstructionism. But, they have not taken up the universal dis-privileging that is inherent in Derrida's fundamental position. Instead, they will borrow the deconstructive approach as a critique in order to dis-privilege the hegemonic texts of the WASP cultures and in order to make a place for the neo-privileged texts of their own ethnic canon-to-be. *In parallel with raising their own ethnic group to be on the same level as other ethnic groups.* Exactly.

This fits in with my thought that the modern nation state is typically an ethnic "empire." They are in effect fighting the ethnic hegemony of one ethnic group and acquiring a new lexicon in order to be able to say that the United States of America is a multi-ethnic society. It is an ethnic empire, but instead of being ruled by an ethnic monarchy, its power base is rooted in an elite ethnic oligarchy that holds sway within an ostensibly democratic republic. *Right. An ethnic oligarchy of groups. I often say that "White" is the most racist word in the American language because there is an oligarchy of Euro-American ethnic groups jointly lording it over the little more than 20% of the whole population in the so-called minorities.[17]*

The core of this ethnic oligarchy are the W.A.S.P.s, who stem from the older English, Dutch, and German settlers of the North East going back to the 17th Century and seem to have developed some racial and cultural homogeneity that sets them off both from the European groups they came from and from the Southern European, Slav, and Mediterranean as well as of course the Native American and African American groups that have been present for a comparable duration, not to speak of more recent immigrations. Probably the Low Land Scottish fit easily into the WASPs, but not the Highlanders and Irish who are not only Celtic but Catholic. Now, however, in discussions about America, "White" almost always signifies "non-Black, non-Hispanic, non-Indian, non-Asian." Perhaps now the wealthy Hispanics—of New Mexico, for example—are as much into the hegemony as the Italians were in the 1950s. Nevertheless, if we look more closely America has hundreds of ethnic groups. There would appear to be this grand coalition against the others, the latter regarded as the more different ones. The "Whites" suppress their internal differences, while they intensify the differences distinguishing them from the "others." Thus they evoke an us-them structure.

The foremost scholar among the ethno-racially concerned post-modernists in America is Henry Louis Gates, Jr., at Harvard.[18] He has posed two problems. One I just mentioned, is concerned with how to dis-privilege what is called the White canon and to neo-privilege or "canonize" the as yet unrecognized complement of Black texts. However, the nature of what he wishes to include in the Black canon requires that he adopt an even more complex post-modernist orientation. One the one hand, he wishes to claim that what appears to Whites and to sophisticated Blacks as vulgar Black speech, ghetto patois, and the argot of the inner city slums, is in fact a literary expression that on proper examination ought to be regarded as canon-worthy. On the other hand, he wishes to subject

---

[17] See the entries on the hundreds of ethnic groups that reside in the United Sates in Stephan Thernstrom, ed., *Harvard Encyclopedia of American Ethnic Groups* (Cambridge: The Belkap Press of Harvard University Press, 1980).

[18] See three works by Henry Louis Gates, Jr.: *Figures in Black: Words, Signs, and the "Racial" Self* (New York: Oxford University Press, 1987); *The Signifying Monkey: A Theory of Afro-American Criticism*, (New York: Oxford University Press, 1988); and *Loose Canons: Notes on the Culture Wars* (New York: Oxford University Press, 1992).

Black texts that already have been recognized as such to a special mode of deconstructive analysis that exorcises from them the characteristics of the era of White hegemony in which they were produced. These texts are not truly Black, because they have been racistically contaminated. To extract the Blackness that is hidden below the White, hegemonist, facade of those Black texts requires a structuralist archaeology of the works in question. Gates is a cultural "archaeologist" who is "digging into" older Black literary works.

Of course, if one wants to form or strengthen a group around its own self-proclaimed ethnicity, these techniques can be effective. Take, for example, Jewish literature in the United States and peel off its "American-ness" to get at its Jewish core; or, take Latin American literature in the Southwest, written in English, and remove the Anglo element to uncover the pure "Chicano." This helps build group consciousness and solidarity, it gives it a literature, it gives it a history . . ." What Giddings called 'a consciousness of kind.' *Right.*

One more thing, an hypothesis of mine, relating to philosophical-political theory: If one examines the two idealized images of American society, the assimilationist and the pluralist images, it seems each has its own internal contradiction or dilemma. We know enough about the assimilationist process to recognize that, in its pristine promissory vision, it rarely occurs. In its place are the several ethnoracial senses of identity and community with varying degrees of legitimation, thymotics, and power. If these many consciousnesses of ethnoracial kind, with their appropriate senses of pride, were to develop as each group would like its own to do, what would constitute the basis of the general Social Contract in this country? Here is my tentative answer. I think it is a philosophical answer, but one that requires both a sociological understanding and a political-juridical institutionalization of some sort. Hidden beneath this struggle over ethnicity is an unwritten covenant that defines the limits of ethnoracial expression and action. In the absence of such a covenant, which may, of course, be implicit, one risks the current situation of the former Yugoslavia. The presence of it, so far, is revealed in the praxes of ethnoracial conflict and accommodation in the United States. In the latter, though, the terms of the covenant are being modified grudgingly and without a clear understanding of the terms that limit either expression or action. I see a social contract problem in all of this.

There seems an assumption here, namely, that making such a contract explicit would be good. Maybe leaving it implicit, in such tacit rules as

'Don't make trouble for them so they won't make trouble for us', perhaps that is how we are able to function as we do. By 'making it explicit' I only meant explicit for and to intellectuals. *Yes, we have to figure it out, but for the society to try to overcome the contradictions in how the several parts of the contract are to be defined, and how they are to be rationalized, given how emotional the issues are, explicitneess might harm more than help and might tax the people's abilities beyond their capacities.*

The final thing I think we need from your side, is a general statement of the issues in Ethnic Studies. By the way, I move from the adjective to a noun, "multi-discipline." I treat Ethnic Studies as a multi-discipline. It is not, strictly speaking, a discipline; it is in fact a coalition of disciplines. Correct, and, incidentally, I like that expression. *Environmental Studies is another multi-discipline and so is Women's Studies, as are Cognitive Science, Science Studies, Technology Studies, and, perhaps, Religious Studies. These are all multi-disciplines.*

The very word "multi-discipline" describes the central issues: The issues are the same in life as they are intellectually when one is working in a multi-discipline, i.e., how do you bring together orientations, points of view, and particular questions that have arisen out of various sources—viz., what might have arisen out of older debates wherein the issues were couched in terms of Political Science vs. Sociology vs. Anthropology vs. History, etc. In non-academic life the same kind of issues arise out of the various contacts and varying confrontations of different racial and ethic groups and of the individuals who are willy-nilly designated by ethnic or social signifiers. Happily, Ronald Takaki, a scholar in the forefront of ethnic studies as a multi-discipline, has informed our discussion with his latest book.[19]

You are saying that we have acquired perspectives as trained intellectuals that pertain to our disciplines, but, when we try to work together, its the same thing as growing up Hispanic or growing up Polish and then trying to work together on the City Council and . . . The problem of the multi-disciplines is the problem of the plural society; it's a microcosmic mirror image of the problem of the plural society. But, just as working together on the City Council is possible, so also is multi-disciplinary work.

It might be that this contrasts with the idea of what might be called an "inter-discipline," which translates the assimilationist view and believes

---

[19] Ronald Takaki, *A Different Mirror: A History of Multi-Cultural America* (Boston: Little Brown and Co., 1993).

that the disciplines can be easily merged and lose their distinctive characters.
This also has not worked, often because one discipline proceeded to try to
assimilate the rest, which makes me think that integration is again the
ideological mask for disciplinary hegemony on the part of, say, history in
past attempts to build Women's Studies programs or biology in attempts to
build Environmental Studies programs. In contrast, a multi-discipline would
respect the uniqueness and distinctness of the constituent disciplines, strive
for equality of all disciplines within it, and not gloss over the difficulties of
maintaining mutual respect and cooperation.

Whether called a multi-discipline or an inter-discipline, Ethnic Studies
will project itself according to the problems posed. Many Ethnic Studies
departments are staffed by historians, sociologists, anthropologists,
psychologists, and experts on literature. Within the framework of the
curriculum, each approaches the topic from his or her disciplinary
outlook. Ethnic Studies is, at present, organized along lines similar to
the old "area studies" programs, e.g., "Japan Studies," "Middle-Eastern
Studies," etc. No intellectually distinctive Ethnic Studies *perspective* as such
has yet emerged. I discount those chauvinistic claims that are empty of
everything but rhetoric about a "Black," "Hispanic," "Asian American,"
or other outlooks.

While we have focussed on a development of the last four decades in
the leading universities in the United States and the centuries of interna-
tional (especially European) cultural events and over a century of cultural
scientific debate (especially in sociology) that are behind them, could you
sketch what you know about similar intellectual responses elsewhere in the
world to what I guess we might call diversities of ethnic groups? I suspect
that Great Britain, for example, has been moving in the direction, especially
since World War II, adding substantial new racial and ethnic groups from
outside to the internal set it has had for centuries. Are there academic-
institutional as well as theoretical developments that relate to this? I cannot
speak about multidisciplinary developments in most of the world.
However, it is worth noting that in countries that were once part of the
English seaborne empire, and are now part of the British Commonwealth,
the term "British subject" modifies jural and civic identities of ethnona-
tionals. In Africa, a few Anglophone and Francophone scholars are
struggling to define a continent-wide civilizational culture that preceded
imperialism and that has survived both colonialism and separate na-
tion-statehood.

The institutional and theoretical developments in such a situation can be expected to be different from those in America we have focused on, but may also learn from this aspect of the American experience, and, in any case, comparisons and contrasts with what has been going on here should be useful to all.[20]

---

[20] The model that emerges for me (Embree) from this discussion also fits some recent events in American academic-professional philosophy and pertains to what might be called schools of thought or "philosophical orientations." I am not sure how many colleagues would welcome comparison of what might be called an "orientational rights movement" with the Civil Rights Movement or the outcome of what might be called "multi-orientationism" with multi-culturalism, but the similarities are striking enough to me to wonder whether the structure fits struggles among schools of thought within other academic disciplines if not among all cultural groups and in any case this seems a good occasion to show briefly how the model fits.

When Phenomenology had become established by about 1960 as an imported or migrated tendency finally adapted to a flourishing American academic situation and was large and growing enough no longer to be ignored (Dorion Cairns had called it an "exotic" in 1950), it was still a minority orientation. It was the eldest in a recent continuing series of orientations diffusing if not migrating after the war from Europe. Deconstructionism is the latest in this series. Phenomenology had been transplanted to North America before the war. Neo-Scholasticism and also Logical Empiricism (*Positivism*? Yes.) had preceded it and flourished, but with the war there was discontinuity until Europe was back on its feet and students could again go to France, Germany, etc. to study. This diffusion was accelerated when the demand for college teachers outstripped the supply from the departments of the dominant orientation from the late 1950s through the early 1970s.

Lyman: For sociology, World War II was also a watershed. Before the war, American sociology had a "trade deficit" with Germany and France and had relied on the ideas of the *Methodenstreit* School in Germany and the Durkheim School in France. For the first three decades after 1945, the "trade deficit" was reversed as Europrean sociology recoverd by adapting to Parsonian structural functionalism and Lundbergian positivism. Ironically both had originated in pre-war Europe but been elaborated in America. Only in the 1970s and thereafter did Europe once more export sociological ideas to America.

Embree: There was little conflict among the various and still lowly populated post-war European tendencies in American philosophy up into the 1980s. Indeed, Don Ihde at SUNY Stony Brook and I independently and simultaneously invented in 1978 the application of the title "Continental Philosophy" to the tacit alliance of these small imports. I doubt that this "Rainbow Coalition" was 20% of the whole of American philosophy then, although it might be today. The dominant orientation was and is called "Analytic Philosophy," but is now increasingly called, in more historical terms and from outside, "Anglo-American Philosophy," i.e., it is now suffering the loss of status of being named by others and having its major recent foreign source area identified (earlier there was input from Austria as well). At the same time, what calls itself Continental Philosophy is often still called "Phenomenology and Existentialism" by the Anglo-Americans. Many non-Analysts suspected that

Analysis considered philosophy to be just what it did and, while some of us non-Analysts always thought that what Analysis did was philosophy, we refused—to put it in terms of logic—to convert an "A proposition."

The opposition to the academic-professional hegemony of Analytic Philosophy, which included not only domination of the American Philosophical Association but also various accrediting agencies, and the grant committees at the National Endowment of the Humanities, took the form of a "Committee on Pluralism," which included Catholics, Historians of Philosophy, and Pragmatists as well as Continentals (Feminist philosophy was still emerging and benefited considerably from this struggle without participating in it as a distinct group). The pluralist movement showed that Analytic Philosophy was actually not a majority but only a plurality among all American philosophers and was even, eventually, seen to be itself an unstable oligarchy of tendencies that put on a homogeneous face toward minority tendencies, much as the Euro-Americans do in the wider American society.

This effort took an interesting amount of agitation, articles in the *New York Times*, amusing-to-remember stereotyping and even some bad manners on all sides, probably some equivalent to race baiting, directed at non-Analytic Philosophical orientations that preserved or enhanced solidarity within the White-like oligarchic hegemony that was challenged. But, there also emerged an increasing manifestation of tolerance on all sides and a questioning of any assumption that a single philosophical orientation had established once and for all that it had developed the superior if not the sole valid approach (which seems a manifestation of the "professional contract," as one might call it), etc. After all, that philosophy smacked of dogmatism and ideology, which it is most unphilosophical to find oneself smacking of.

After almost ten years, the playing field had been leveled enough through formal changes in the election rules and committee structure as well as through a series of explicitly pluralistic presidential elections in the Eastern Division of the American Philosophical Association that it began to seem to me at least that the continuing differences among factions were more understandable in terms of the stronger universities, sheer numbers and the constant proportion of talent, as well as the age of members, i.e., the issues dividing philosophers were not chiefly orientational differences, although there were differences affected by an orientationally structured past, especially resentments about how Analysis continued to dominate the big schools in the North East. Whether the widespread tokenism that emerged recently will expand to something like proportional representation is for the moment an open question.

It has also seemed to me curious that, once some ground had been gained and more members recruited, the more or less tacit alliance within so-called Continental Philosophy began to break up. Especially among the devotees of the later Heidegger, Deconstructionism and its allies seem of late to believe themselves coterminous with Continental Philosophy and, with support from similar thought in such non-philosophical disciplines as literature, they have attempted to exercise hegemony over other Continental tendencies. A few years ago, one attempt was beaten back within the Society for Phenomenology and Existential Philosophy and, at the time of this writing, there seems an uneasy truce among Continental tendencies each of which would have difficulty arguing it is either dominant or

II. Phenomenology in and of Ethnic Studies

Lester, how do you see Phenomenology as relating to Ethnic Studies? The way in which the special sciences, which include the cultural sciences, are related to in Phenomenology can be extended to such a multi-discipline. We can speak of Phenomenology "in and of" a discipline or, now, multi-discipline, including how phenomenological philosophy can benefit from reflecting on Ethnic Studies. There is some work in this vein prior to the rise of Ethnic Studies program. Firstly, there are the immediately relevant ideas in Hannah Arendt (1906-1975),[21] Aron Gurwitsch's (1901-1973), *Human Encounters in the Social World*,[22] Alfred Schutz's (1899-1959), "Equality and the Meaning Structure of the Social World,"[23] and Jean-

---

subordinated (but, of course, I know of a certainty that Phenomenology is the oldest and best, and that the others should be understood as merely branching from its stem!). Such a dynamic of orientational tendencies is what one might predict in the histories of other disciplines, something which is of no small interest for an historically oriented philosophy of the natural or formal sciences as well as of the cultural disciplines. Does this work for sociology?

Lyman: Sociology for some years has been in a "crisis," i.e., it suffers from the unmistakable decline of the dominent paradigms—structural functionalism and quantitative positivism—and from the enormous retrenchment in govenment funding. A number of other subordinated approaches—Marxism, feminism, ethnomethodology, symbolic interactionism, deconstruction, postmoderism, etc.—are vying for recognition and desciplinary domination. None seems likely to win out over the others. Sociology is likely to remain pluralistic.

[21] Hannah Arendt, *The Human Condition* (Chicago: Chicago University Press, 1958) and *The Jew as Pariah: Jewish Identity and Politics in the Modern Age*, edited by Ron H. Feldman (New York: Grove, 1978).

[22] Aron Gurwitsch, *Human Encounters in the Social World*, (Pittsburgh: Duquesne University Press, 1979), originally written in Berlin during 1928-31..

[23] Alfred Schutz, "Equality and the Meaning Structure of the Social World," in Lyman Bryson, Clarence H. Faust, Louis Finkelstein, and R. M. MacIver, eds., *Aspects of Human Equality* (New York: Harper & Brothers, 1957) and reprinted in Alfred Schutz, *Collected Papers*, Vol. II, ed. Arvid Brodersen, (The Hague: Martinus Nijhoff, 1964). Probably the first phenomenological move towards the issues of Ethnic Studies as a multi-discipline is a related manuscript, "A Search of the Middle Ground" (1955), Chapter 18 in Alfred Schutz, *Collected Papers, Vol. IV*, edited by Helmut Wagner and Fred Kertsen, (Dordrecht: Kluwer Academic Publishers, forthcoming).

Paul Sartre's (1905-1980) *Anti-Semite and Jew*.[24] Then there is much in the way of approaches and concepts in the other basic literature of Phenomenology. With the rise of the multi-discipline, new reflections are called for.

I think I should start out with some remarks about what my orientation is in general, not because it is still a novelty but rather because many people currently seem to consider Phenomenology merely a body of literature or a set of texts by Husserl, Scheler, Heidegger, Gurwitsch, Schutz, Berger, Sartre, Merleau-Ponty, de Beauvoir, etc., in which one can just read around to get ideas. We all do this, of course, in work outside our own discipline or orientation. But it is a mistake to consider any account we develop from such sources *ipso facto* phenomenological. Would one's position become Marxist if one adopted a few Marxist ideas?

For work to be considered phenomenological, an approach needs to be taken not only in the examination of what classical authors of the movement advocate but also in attempting to explore new areas. There are, to be sure, methodological squabbles and several towers of Babel within Phenomenology, for it is a century-old school of thought represented by hundreds of active participants in a variety of disciplines who publish in English, French, German, Italian, Japanese, Spanish, and other languages. *This sounds to me like a "civilizational" phenomenon.* I hope so! Also, since we believe we are on the right track, we are not surprised to find work before and, so to speak, "beside" self-conscious Phenomenology that we can consider phenomenological. *Are you saying that there is work which is said to be phenomenological but is not and work which is not said, at least by its author, to be phenomenological, but still is?* Exactly. In principle—I cannot think of a case, but it is possible—an author could even explicitly deny that her work is phenomenological, and yet it might be.

---

[24] Jean-Paul Sartre, *Anti-Semite and Jew*, translated by George J. Becker (New York: Schocken Books, 1948) [originally published as *Réflexions sur la question juive*, Copyright 1946 by Paul Morihien, Paris]. See also Maurice Merleau-Ponty, "From Mauss to Levi-Strauss" and "The Philosopher and Sociology," *Signs*, translated by Richard C. McCleary (Evanston: Northwestern University Press, 1964), 98-113; and Laurie Spurling, *Phenomenology and the Social World: The Philosophy of Merleau-Ponty and its Relation to the Social Sciences* (London: Routledge and Kegan Paul, 1977), 76-109, 191-182.

It all depends on the actual approach by which results were produced and are then to be comprehended, examined, corrected, and extended. Interestingly, "Phenomenology" can actually be a misleading name. Husserl sought to establish a primal science in which other sciences and their applications would find their ultimate transcendental grounding and chose a name for this essentially philosophical project like the name for a science. But if we are thinking about an approach that can be used in a variety of disciplines, including practical ones, such as nursing, as well as empirical-theoretical ones, such as sociology, then we need a name more like "Positivism," which I can gloss as the attempt to account for social and cultural matters as if they are physical things. This name is a characterization in terms of a research strategy or approach. To re-name it analogously, Phenomenology seems better called "Reflective Descriptivism."

Phenomenology is emphatically descriptive before it is argumentative or explanatory (It is a conversation stopper among philosophers of other schools of thought to mention that Phenomenology basically offers descriptions, not arguments!). Being descriptive signifies that it attempts to answer questions first of all about what the matters at issue are. All disciplines have this descriptive component, but it is often implicit, taken for granted, and yet the explanatory parts of the effort presuppose and are affected by the descriptive part. *Can you give an example of this?* A concrete one at this point would be difficult, but let me say that if one has not distinguished between positive willing for and negative willing against alternatives and between objects willed for their own sakes, i.e., as ends, and objects willed for the sakes of other objects, i.e., as means, then teleological explanations are very difficult to construct. Then again, it is difficult to account for cases of life if we do not recognize a difference between feeling or valuing, on the one hand, and willing or action, on the other hand. Such considerations may be on a more general level than those of empirical science, but philosophy of course tends toward such a level.

Descriptions in Phenomenology are typically not factual descriptions but rather universal claims that we call "eidetic descriptions." The results include definitions, which are not just semantic conventions postulated for the sake of an argument but rather descriptions of the kind of matter referred to. One can err in what one takes to be the defining determination and discover and correct an error in the same perspective. Such definitions are presupposed by factual statements of what happened where, when, and why. Phenomenologists can and do make factual claims, e.g., about whether the world exists, but the emphasis is on establishing eidetic claims, e.g., an answer to the question "What is a multi-discipline?"

Fundamentally, Phenomenology not only emphasizes description but also reflective observation or reflection. This means that it alternatively considers how objects present themselves to subjects and how subjects relate to (or are "intentive to") objects. I know this is an abstract formulation. It may become somewhat concrete if we begin to ask how Protestants relate to Catholics and Catholics relate to Protestants in Northern Ireland. Religion and other matters, such as family size and structure, attitudes toward abortion and divorce, taste in food, clothing, literature and music, and perhaps even typical racial features obvious to both (but difficult to discern by members of third groups), are all valued such that each group considers its own better than those of the other. There is then to be recognized in-group self-valuation and out-group other-valuation. And in this case, it is not a matter of "better" like "apples are better than oranges" but rather of right and wrong with all the depth of love and hate that is possible in religion. Outsiders to both groups may find it easier to recognize the different values that the two groups have for themselves and for the other groups and also the different valuing processes involved than insiders of the groups do.

This example also shows that subjects can be collective as well as individual and that one can reflect upon other's as well as upon one's own individual or collective life, on 'you all' as well as 'thou' and on 'us' as well as 'me', to paraphrase Schutz, who accepts the in-group/out-group distinction from Sumner[25] as a powerful distinction. Such reflecting can occur on the common-sense level within and between ethnic groups in everyday life and it can also occur on a scientific level when done by sociologists, historians, political scientists, economists, etc. either "uni-disciplinarily," so to speak, or multi-disciplinarily. In the latter case, the scientific one, the reflecting is for theoretical purposes, while common sense is for practical purposes.

I am beginning to see how you can consider research that relates values to the group doing the valuing as well as to the group that is valued

---

[25] William Graham Sumner, *Folkways: A Study of the Sociological Importance of Usages, Manners, Customs, Mores, and Morals* (Boston: Ginn and Co., 1940 [1906]), 12-15, 493-508.

phenomenological whether or not the researchers consider themselves phenomenologists. This means that a great deal of social science has always been phenomenological, doesn't it? Yes, and, while much of what a phenomenologist may assert will be familiar because already explicit or implicit in, say, sociology, we think it useful to try to pursue not only an explicit but also a comprehensive and systematic presentation. This may help get some distortions and exclusions recognized and corrected. For example, negatively speaking, there has been, for generations, a great deal of interpretationism and naturalism in cultural science. Interpretation and language are played up by these tendencies, even though the bulk of our lives and interactions, even for intellectuals, is non-linguistic and, unless "interpretation" is broadened beyond recognition, non-interpretive. Stand back and look at the big picture and one can see, positively, that the valuing and willing by virtue of which we have cultural objects is neither linguistic nor interpretive. It is often heard that cultural objects are meaningful, but is this "meaning" as in the meaning of a word? Is encountering a chair as something on which to sit always and only merely the interpreting of a physical thing?

In worrying about the big picture, one needs to worry also about the crucial distinctions within it. Thus, how objects (which are not just focal objects but also situations and indeed cultural worlds) present themselves and, correlatively, how subjects are intentive to such objects can both be analyzed and discussed in terms of *awareness* (perception, recollection, expectation, eideation, pictorial, indicational, and linguistic representation, etc.) and what Husserl calls *positionality* (positive, negative, and neutral believing, valuing, willing, and objects as intrinsically and extrinsically believed in, valued, and willed). Although my stress is on the distinction between perceiving or willing, on the one hand, and the object as perceived or as willed (its sometimes called the "ing/ed" distinction), on the other, might be considered unusual, I am using these expressions in close to the ordinary significations.

Members of ethnic groups in conflict might believe that those of the other groups are unclean and immoral and hate them. *Or, they might believe the others are supra-moral and meta-hygienic and envy them but avoid them out of shame.* I hadn't thought of that, but it fits. And this is not a matter of something that, in one signification, might be called "objective," such as emumerations of how many baths are taken per capita per week or how many bars of soap one used, but rather what "we" believe (and value, and will, and are aware of) about "them." Or, members of one group might be bothered by the gestures, e.g., the flamboyant hand movements, of members of another group, while the others are bothered by the posture, e.g., the stiff necks, of the first group, and these cultural encounters, as we might call them, involve types of awareness as well as types of positionality and are analyzable in terms of these types.

(Figure 1)

Noema   Noesis

I. Awareness
  A. Presentational Awareness
    1. Perceptual
    2. Memorial
    3. Expectational
    4. Etc.
  B. Representational Awareness
    1. Indicational
    2. Pictorial
    3. Linguistic
II. Positionality
    1. Belief
    2. Evaluation
    3. Willing

To elaborate on this, let me comment on the attempt at a reflective-descriptive systematic classification or big picture expressed in Figure 1,[26] which might possibly be of use in cultural scientific approaches to race and ethnicity. Across the top, "Noesis," which can also be called "Intentiveness," generically includes mental states, processes, and attitudes as they relate or are intentive to objects and "Noema," also called "the object as intended to" or "the object as it presents itself," includes everything that reflection on objects discloses. "Objects" here include

[26] See Lester Embree, "Some Noetico-Noematic Analysis of Action and Practical Life," in John Drummond and Lester Embree, eds., *The Phenomenology of the Noema* (Dordrecht: Kluwer Academic Publishers, 1992).

other collective and individual subjects, so we can value your willing, you can will our valuing, etc., etc. Figure 1 also reflects the awareness/positionality distinction. The sub-specific types with Arabic numbers should help to make the specific types clearer. But let me add that "representational awareness" is awareness in which, on the basis of awareness of an indication, depiction, or linguistic expression, one becomes aware of something else.

I should add that each of the three types of positionality have positive, negative, and neutral modalities, so that believing, disbelieving, and suspending judgment are parallel to liking, disliking, and being apathetic, and also to willing for, willing against, and remaining neutral in one's action. Then again, "willing" here has a very broad signification, whereby it emphasizes deeply habitual reactions and thus far more than merely the deliberate making and executing of decisions. As for awareness, one might begin with how some people typically remember how others sang and the sounds of their voices while other people might typically remember how others wrote and the cogency of their arguments. Hence different people have, as a matter of deep cultural learning, different aspects of objects that they routinely attend to recollectively.

Let me also add that "reflection" connotes in part the analyzing behind the descriptions but more importantly that it connotes, in a broad way, observation, which is to say that one reflectively looks and ponders and avoids empty or blind as well as unreflective thinking. The point, of course, is only to make distinctions where there are observable differences. In this respect, Phenomenology grows out of the classical English philosophical tradition called Empiricism that includes Locke, Berkeley, Hume, the Mills, and even William James. And thus, concerning the phenomenological slogan of "Back to the matters themselves!" we can ask what we are to go back *from* and answer that it is a preoccupation with propositions and their forms apart from the matters they are about. What it goes back *to* is how objects are related to and how they correlatively present themselves.

This Reflective Descriptivism does not preclude hypothesis formation and testing; what it does is emphasize the observing, seeing, or evidencing of matters that is fundamental to testing as well as concept and theory formation. It also does not preclude explanation, which can take teleological as well as aetiological and other forms, but, again, the emphasis is on the description in general terms, the eidetic description, that explanation presupposes. *I suppose that this does make clearer what phenomenology in general is and, if terminology does not get in the way, how much social scientific work that does not know it is phenomenological actually is. Also, I can see how clarifying these basic terms can be useful for analyzing phenomena. Is this part of the "in and of" Ethnic Studies you referred to?*

It is difficult not to start getting into that. But to sort things out, phenomenological approaches are taken *in* cultural sciences when there is reflection on the positing and the posited and on, so to speak, the "awaring" and the object as "awared" in cultural encounters, but they can also be taken when cultural scientists reflect on their own disciplines, for example when historians discuss what history is about. (And, getting ahead of myself again, let me mention that participants in a multi-discipline would seem especially prone to reflecting on similarities and differences among their disciplinary perspectives, which includes philosophy and thus questions about how it is different from and possibly hindering as well as helping cultural scientific research.) Thus there can be Phenomenology in a cultural science with respect to how it approaches its subject matter and also with respect to how it is concerned with itself. "Phenomenology of," by contrast, pertains to phenomenological philosophy and divides into how philosophy can relate to a non-philosophical discipline or, now, multi-discipline, and into how it can benefit from doing so.

My description illustrated by the Protestants and Catholics in Northern Ireland, vague as it is, might be a case where inquiry can be considered phenomenological, whether or not the inquirer knew it, if there was distinguishing and describing of the positive and negative values that the groups have for themselves and for one another in relation to the valuings of various types and modalities. I easily recognize that there are individuals, who are abstractions, and groups, which are made up of concretely interrelated individuals. Fascinating methodological questions arise, but let us assume that one can observe and describe groups and how they relate to their worlds and, correlatively, their worlds as they present themselves to groups.

It is also easy for me to accept that there are socially visible determinations of various sorts that support cultural characteristics. What one is originally aware of I would call "naturalistic determinations." These include psychic as well as somatic determinations and does not preclude many of both sorts from being artifactual in the signification of being affected by human efforts. There is mind building as well as body building. "Determination," incidentally, includes relations as well as properties. Among naturalistic properties relevant for Ethnic Studies, which we can perceive, remember, and expect, there is indeed skin and hair color and texture, eye shape, and so on, which are biologically inheritable. There are also biological relations between mates and between parents and offspring. How biological rather than artifactual social relations are otherwise is not clear to me, e.g., whether it is genetically determined for men to protect women and children. Human behavior is far more plastic than that of lower animals.

Many naturalistic determinations are artifactual, not biologically inherited, but shaped during the lives of individuals and above all through social interaction. These include language, religion, food and dress habits, but it also includes learned ways of walking, standing, sitting, making love (How did "the missionary position" get its name?), gesturing, facial expressions, etc., which are not purely somatic, as they can appear in a physicalistic attitude, but rather what I like to call somato-psychic. I have no scientific typology or terminology to offer at this time, but I have seen more than enough German glares, French shrugs, and ingratiating head movements pertaining to various ethnicities, not to speak of ways of waving hands, to believe such gestures are cultural and can be classified. Attachments to place, to technological and ecological systems (railroad men and fishermen earn those titles in ways akin to those by which the ethnicities they specify are constituted, it seems to me), and also ease and its lack in types of social relationships, e.g., with mothers-in-law, I would also consider artifactually naturalistic psychic determinations.

There seems a tendency to refer in speech to the naturalistic determinations that cultural characteristics are founded upon. But of course we learn to be aware of somatic characteristics as relevant for racial classifications. Puerto Ricans whose so-called Negroid characteristics are noticed by Whites in New York might well have learned back home to pay more attention to postural and linguistic subtleties indicative of ethnic or class membership. One might believe that naturalistic determinations are equally noticed by both insiders and outsiders of a group, but I wonder about that.

Only a little reflection discloses that naturalistic determinations are believed in, valued, and, in a broad signification, willed as well as "awared." This may be clearer concerning psychic determinations. To say that most Faubokians are lazy is literalistically to assert affirmatively that members of that tribe typically have that character trait and one is able to do this because one has come, with or without a foundation, to believe in that trait as belonging to them. But we readily comprehend such a seemingly cognitive assertion as additionally if not originally a value judgment that could be explicated with "and are thus despicable, i.e., negatively valued" (or, if the predicate was instead "prefer a leisurely life," it could be explicated with "and are thus admirable, i.e., positively valued"). Similarly, "Italians are great lovers" might be not merely a cognitive or even a value judgment but rather a practical recommendation, even though the statement has the form of a factual assertion. If it is a recommendation, it relates to willed purposes and thus the use of members of one ethnic group for a specified purpose, use, including end-use and means-use, correlating intentively with willing and being thus different from belief characteristics and values in objects as they present themselves.

Probably the most interesting issue for our concern is how some determinations are selected and used for the recognition of people as belonging to ethnic and racial groups. I could imagine most researchers approach this in terms of language and concepts or interpretation, but what I would pursue is the pre-predicative and non-verbal structures of awareness and believing and how these are motivationally connected with valuing and willing. One can change the language and the rules rather easily, but it is in these deeper areas of life that change is difficult and yet necessary. If there were no problems at this level of cultural encountering, language and thinking would follow easily.

The point is that somatic and psychic artifactual and non-artifactual naturalistic determinations of various sorts, whether clearly seen or not, are believed in, valued, and willed in different deeply learned (or conditioned) ways in different ethnic groups. In other words, they are cultural. That they are learned signifies that they are deeply habitual, so deep that it can be doubted by those who have these habits that they are learned at all and instead are believed to be natural in the biological way, which seems easiest believed in within traditional agricultural societies or at least parts of societies concerned with such things as breeding. In that case people will often fight to the death over what I am trying in general to describe.

In any case, perhaps this intimates how a philosophical generalist might attempt to deepen and widen the answer to the question of what race and ethnicity are. Note that I did not get into the completely legitimate questions of the circumstances under which members of an own or other group come to have the cultural characteristics constituted in deeply habitual believing, valuing, and willing. The first direction I would go in order to develop an explanation would, however, be with respect to the structures of the value system and the use system for the group, the higher and lower values and purposes it has, which has to do with teleological accounts. But, plainly, one does not have to be engaged in philosophy to comprehend and examine whether what I have outlined is at all insightful. Is this analysis in terms of visible naturalistic determinations and their cultural characteristics as constituted in deeply habitual believing, valuing, and willing by insiders and outsiders at all relevant?

I would say that is a necessary feature of any sound approach to the analyses of race and ethnicity. Indeed, without such analysis, I do not think the work could be pursued.

Thanks. Another type of "phenomenology in" pertains to disciplinary self- consciousness. I have previously reflected on particular disciplines, say psychology,[27] but now I am fascinated by the cultural multi-discipline of Ethnic Studies. I think a multi-discipline is different from an inter-discipline by virtue of the continuous explicit recognition of differences between the component disciplines with respect to how they approach race and ethnicity. Thus, economics, education, sociology, linguistics, nursing, political science, literary theory, psychology, anthropology, etc. can be said to focus on different, but perhaps overlapping and hopefully complementary, systems and processes of individuals and groups as ethnic. Then again, psychology might differ from sociology and political science in emphasizing the individual rather than the group.

Hence, the reflective-descriptive approach sketched that I have can be taken to the central problem of Ethnic Studies you identified above as like the problems of multi-cultural or multi-ethnic life. The cultural scientist participating in a multi-discipline would have a self-understanding not so much in terms of just how her own uni-discipline differed from others but also how there can be interaction and, it would be hoped, complementarity with other disciplines represented in a multi-discipline. This would be more emphatic when hegemony by one or a few disciplines was not permitted to subordinate and peripheralize the others. How members of disciplines are aware of, believe in, value, and will themselves and others individually and collectively and, now, uni-disciplinarily and multi-disciplinarily in their awareness, believing, valuing, and willing lives in the habitual way characteristic of professionals trained in disciplines can be reflectively observed and described in an arguably phenomenological way in cultural science.

It sounds to me like phenomenology of *that kind of science.* Well, actually, it is phenomenology "of" it as done "in" it! What is basically signified by the in/of distinction is the difference between the level of

---

[27] Cf. Lester Embree, "Aron Gurwitsch als phänomelologischer Wissenschaftstheoretiker," translated by Alexandre Métraux, *Zeitschrift für allgemeine Wissenschaftstheorie* 5 (1974), 1-8; "Gestalt Law in Phenomenological Perspective," *Journal of Phenomenological Psychology* 10 (1978), 112-27; "The History and Phenomenology of Science is Possible," *Phenomenology and the Understanding of Human Destiny*, edited by Stephen Skousgaard (Washington, D.C.: Center for Advanced Research in Phenomenology and University Press of America, 1981), 215-228; "The Natural-Scientific Constitutive Phenomenological Psychology of Humans in the Earliest Sartre," *Research in Phenomenology* 11 (1981), 41-60; "Teleology in Human-Scientific and Natural-Scientific Psychology and Psychotechnics," in *Current Issues in Teleology*, edited by Nicholas Rescher (University of Pittsburgh Center for Philosophy of Science and University Press of America, 1986), 120-128.

special disciplines (and multi-disciplines) and the level of philosophy. Something has already been intimated in this respect where the generality of philosophical concepts and the scope of the big picture we try to outline are concerned. I see two components to the philosophy of Ethnic Studies. Such a philosophy does not have to be phenomenological in orientation, but for reasons that this is not the place to get into, Phenomenology is, I believe, the most fruitful approach here as elsewhere. One component of phenomenological philosophy has to do with how another discipline is best approached philosophically and the other component has to do with what phenomenological philosophy gets from doing so.

Of course, philosophical phenomenology is also basically an effort at reflecting and describing. "Basically" does not signify, however, that this is all that it does, for it is also concerned with a type of explanation that consists in attempting to justify, and for some phenomenologists it culminates in action, so that philosophy is a practical discipline. We are here concerned with Ethnic Studies. This is a multi-discipline and philosophy, it seems to me, ought to be included in multi-disciplines if only because of how, being more different, so to speak, than the other participant disciplines, it can help keep recognition of differences and limitations in view, although this might not always foster harmony. Philosophy is different, not only with respect to the generality and scope of its thinking but also because, on my view, it is not concerned with exploring race and ethnicity from a different point of view among specialized points of view, but rather by being oblique and pondering how such a matter can be approached from many interrelated points of view. Thus it is reflectively critical as well as general or comprehensive in its purpose.[28]

For phenomenological philosophy, that is a philosophical task that can be performed in Ethnic Studies. *But you are now concerned with phenomenology of Ethnic Studies.* That is right. But I want to make a contrast with what I have just said: The phenomenological philosophy of Ethnic Studies proceeds from an outside point of view. It can attempt to observe and describe what it is to be a multi-discipline in contrast to a uni-discipline and to an inter-discipline, compare Ethnic Studies with Women's Studies, Environmental Studies, and other multi-disciplines, etc., whether or not these reflected upon matters include philosophy. They should, I would say, but probably often do not, and it can at least be wondered whether that is a good or bad thing. And just as a third point

---

[28] See Lester Embree, "Methodology is Where Philosophers and Human Scientists can Meet," *Human Studies* 3 (1980), 367-373.

of view can understand the valuing and values, the willing and purposes, etc., of Catholics and Protestants in Northern Ireland, an outside perspective may be able more clearly to observe and describe how Ethnic Studies works.

I think the central issues for a phenomenological philosophy of Ethnic Studies would be, on the one hand, how the perspectives of the various disciplines in the multi-discipline interrelate, have a common purpose, develop and overcome conflicts among themselves, etc., and, on the other hand, how the one huge subject matter, which can be called race and ethnicity, can present itself in different but, hopefully, complementary ways in the different perspectives. Thus, how does a relationship of ethnic groups, again say the Catholics and Protestants in Northern Ireland, present itself to an economist, to an historian, to a sociologist, to a political scientist, and so on? Researchers are trained in disciplines like these to observe and theorize in different ways. Like the three blind men in India touching the elephant, who found a tree trunk, a snake, and a rope, do they find different things or is it one thing with different parts or aspects? My guess is also that members of each discipline are like members of ethnic groups, i.e., conditioned each to consider her outlook as the fundamentally correct one. This might be called "disciplinary chauvinism." If so, why are they participating in a multi-discipline? Ideological and careerist motivations aside, could it be that they have gone on to suspect that they can learn from other disciplinary points of view because their own is not as ultimately adequate as it pretends to be after all?

These are central questions a philosopher of a multi-discipline would ask. Another sort of question, which also pertains to Women's Studies and Environmental Studies, has to do with the place of biological factors. Could there be anything to Sociobiology? Do endocrine secretions affect gendered behavior and relationships? In the last-mentioned multi-discipline, the problem seems one of getting naturalistic scientists and engineers even to consider the cultural elements that, at least for the cultural scientists who are not overwhelmed by naturalism, consider central. Whether or not participants in multi-disciplines are concerned with such questions, they proceed on the basis of answers to them, and philosophers can raise such questions and probably do so best from outside. *One answer to the sociobiological issue has been given recently by a historical sociologist, Kenneth Bock, who claims that sociobiology imposes a "naturalistic" imprimatur on questions better answered by a discerning history.*[29] Such an inter-orientational discussion is very like the discussions between members of different disciplines or different ethnicities.

Finally, let me ask myself what philosophy as a uni-discipline might itself get out of reflecting on Ethnic Studies. I have already addressed the question of what phenomenological philosophy can get from using the cultural disciplines in an oblique way in order to learn about and act within cultural worlds,[30] so this amounts to a specification. Here we can proceed from an outside point of view established when the functions that philosophy might have within the multi-discipline are suspended. The questions of whether philosophy is affected by unexamined assumptions relating to class, gender, and modern Western "Enlightenment" civilization are already alive in my field. What needs further to be asked is to what extent philosophy has also had an unexamined ethnic aspect to its life as lived thus far. After all, we easily speak of American, English, French, German, and other national traditions in the history of philosophy. In particular, to what extent is the fact that Husserl is a Jew converted to Protestantism in the late Austro-Hungarian Empire, the fact that Heidegger is a Southern Catholic German, the fact that Sartre is a French intellectual of Protestant background, etc. reflected in their positions? Here emerges the possibility, always lurking in the background, that philosophy might be reduced to ideology, which is say relativism. In contrast to that is the question of how philosophy, like other disciplines, might arrive at positions that are culturally non-relative. This question cannot be answered by ignoring it. Ethnic Studies can help the phenomenological philosopher face it.

Finally, to summarize all this a bit more explicitly, I see Ethnic Studies as able to be related to phenomenologically in four ways. In disciplines of the sort that would make up this multi-discipline, (1) there can be a reflective-descriptive component that can be recognized and perhaps enhanced through explication, systematization, and refinement of expression and, (2) when such disciplines combine into a multi-discipline, phenomenological philosophy might be included for reflective and critical questioning of the benefits derived therefrom. As philosophy of Ethnic Studies, Phenomenology can (3) attempt from an outside perspective to understand both how disciplines in a multi-discipline might work together and how, correlatively, one subject matter, here race and ethnicity, might present itself differently in different disciplinary perspectives and (4) it

---

[29] See Kenneth Bock, *Human Nature and History: A Response to Sociobiology* (New York: Columbia University Press, 1980), 61-88.

[30] Lester Embree, "Reflection on the Cultural Disciplines," this volume.

can use Ethnic Studies in order to act more rationally in cultural worlds, including its own participation in multi-disciplines.

Chapter 10

Technology and Cultural Revenge

Don Ihde
SUNY, Stony brook

Abstract: *Of all the cultural products of Euro-American recent history,* technoscience *is probably the most powerful. No nation has ever successfully colonized the world, nor gotten its language to be the* lingua globale, *nor even gotten its arts universalized. But technoscience is a* world phenomenon *which has* englobed *the earth. In this essay I focus upon one particularly powerful strand of contemporary technoscience—its "image" and communications technologies—which play an especially important cultural role in a technologically englobed earth. We are all familiar with and use them: television, cinema, telephonic and computer networks, even the newer "fax" technologies. Correlated with these image and communications technologies, I focus upon the emergence of what I call* pluriculture, *a unique form of late modern crossculturality. And, within pluriculture, I shall examine a set of theses around "technology and cultural revenge."*

I. Technoscience

In the philosophical literature concerning technology there is a deep irony. On the one side technoscience—as it has begun to be called—is extolled as a *unique, distinctly Euro-American phenomenon of modernity.* According to this intepretation, it originated out of the specifically European philosophical penchant for theory-turned-science, which was welded to a technological embodiment in experiment and instrumentation, and became a cultural form which spread throughout the world itself. In one sense, one could claim that this development is the high point of the Western philosophical, scientific and technological phenomena.

M. Daniel and L. Embree (eds.), Phenomenology of the Cultural Disciplines, 251–263.

On the other hand, and particularly with the European early philosophical critics of technology—sometimes heralded as the forefathers of "philosophy of technology"—there also emerged in the second generation, a kind of dystopian tendency to see in technoscience the signs of the very "end" of the Greco-European traditions of metaphysics. The first generation of *Technikphilosophie*, occured with the neo-Hegelians. Ernst Kapp first used the term (1877), but Kapp was overshadowed by Karl Marx, who related technologies to modes of production and introduced the notion that technologies change social structure. But in the period between the World Wars, technology again re-emerged as a theme with Friedrich Dessauer and Martin Heidegger in the late twenties. Then, again, immediately after WW II, in a "second generation," the dystopian themes began in earnest. Here I refer to such thinkers as Ortega y Gassett, Nicolas Berdyaev, pre-eminently Martin Heidegger, and, later, Herbert Marcuse and Jacques Ellul.

Their critiques, not unlike the metaphysical traditions which they analyse, are *variations on a theme*. They saw, in the rise of technological culture, *a threat to some deep aspect of European culture in its highest values*. Ortega saw the emergence of a "mass man," Berdyaev a loss of "spirituality and community," Heidegger saw "calculative thinking" and all as "standing reserve," Marcuse saw "one-dimensionality," and Ellul the replacement of "nature" with "technique." To phrase it succinctly: modern technology was seen to be *acidic to traditional or "deep" culture*.

There were two other generic features to these critiques which I should like to note: (a) all were "high altitude" critiques, perhaps best exemplified by Heidegger's interpretation of "Technology" as itself an "applied" metaphysics, which would be also the "end" of metaphysics, and (b) all were *internalist* critiques. Whatever the fate, or negativity, which clung to "Technology," it was seen as internal to the history of European thought. Other cultures played no role in any of these critiques—with the exception of an often muted or indirect critique of "America" as an exemplification of both high technology and "popular culture."

Were one to take a socio-anthropological view of this tradition, it might be possible to characterize it as itself highly "Eurocentric." Its perspective was that of the highest altitude "theory"—metaphysical. Its proponents were intellectual mandarins, cast in the mold of the magisterial professoriat. And—with Heidegger sometimes adding workshop and peasant romanticism—they were elitists regarding "high culture."

This is not to dismiss the insightfulness of the second generation critiques. These pioneers who forefronted technology (a) saw that technologies are *historically-culturally embedded*. And, while there are elements of Eurocentric chauvinism implied, in the main the development of technoscience as a peculiarly Euro-American phenomenon remains correct. (b) They also saw, again correctly in my view, that technologies are *non-neutral* and with changes in technologies there are changes in cultures, whichever way the causal patterns drive. And (c) they saw, although sometimes only implicitly, that late modern technologies are again *acidic to traditional cultures*. I agree with and affirm each of these findings which occur among the forefathers of philosophy of technology. But, in retrospect, it now seems to me that their aim was off, both with respect to altitude and target.

If "technological" culture was acidic to traditional European culture, according to these critics, one might also expect any cultural *"revenge"* to take place at the level of a different metaphysics. That has not occured—does this then, leave us, like Heidegger, "awaiting a god"? But this is to treat "revenge" in its older, explicit form. To get revenge is to get back at someone or group which has injured you or your group, to enact an equal or greater injury upon the perpetrator(s).

If the "internalists" are correct, the injury here has been self-inflicted. Technological culture is our own invention, so the metaphysical tradition holds. And, from this perspective, the injury can only be a variant upon a deterioration theme—high values are replaced by low ones. Nietzsche sounded these, followed by our second generation dystopians. In this version high culture is replaced by "popular culture" and the enemies are symbolized by "jeans," "MacDonalds," "rock," "Madonna," and "MTV," (often not too subtly associated with the "American.") Thus the irony is one of having both high technology and "low" or popular culture.

In what follows, I shall argue that this tradition misreads both the nature of the challenge to Eurocentrism and the histories which lead to the present situation. I shall propose a different reading which finds in crossculturality a much deeper history of "cultural revenge" and the emergence of a "pluriculturality" which is a distinctive contemporary form of the cross cultural which has too often been occluded even in our critical perspectives.

The macrothesis is one which sees in technoscience a dominant movement outward from European, then Euro-American roots, like a tide reaching all the shores of the world. But, like all tides which reach steeply sloping beaches, there is also always an undercurrent. That undercurrent is the *other*, the other cultures, which today, in the form of pluriculture, reaches back and into the dominant movement which went out from the Euro-American. The result is a chaotic mix of dominant and recessive tidal rips which sometimes take the shape of cultural revenge.

II. Crossculturality

The pluricultural is simply a late modern form of an often underrated ancient history of crossculturality. The "others" have always been a more important element in our own histories than we give them credit for:

A. Cross cultural exchange has always been a powerful historical "motor." One cannot ignore this, particularly in Rome, where the waves of Etruscan, Hellenic, Roman, Latin, and Italian cultures shaped this part of our history. Our wars have often been cross cultural: In Europe our occluded "other" has often been Islamic. The Moors, the Ottomans, and, with Iraq, today again our attention cannot help but return to this often surpressed history. This same history even played a vital role in the rise of technoscience itself! Had it not been for the Arabic scholars who preserved, studied, and developed classical Greek science—especially Aristotle and the atomists—while our northern "barbarians," we of Germanic descent, ignored and destroyed those same culture fragments, our modern science originating in the Renaissance might not have happened at all.

Nor should we forget that the origins of technoscience *followed* by over a century, another landmark in cross cultural history—the voyages of discovery which we locate with Columbus and Vasco da Gama. 1992, was both the celebration of a more united Europe and also the quintra-centennial of the Euro-American beginning.

B. Cross cultural exchange is also material exchange. Poor Columbus died thinking he had reached the Orient. He did not know that the Conquistadors who first dominated in South America and the North American south and west, were to transform Europe itself with the material culture of the New World. Gold and silver, passionately collected, were melted down and reformed into European artifacts—but the culinary revolution which followed in the wake of potatoes, tomatoes, tobacco, corn, coffee, so totally transformed the European diet that today we almost take this result as indigineous. Nor should we forget that the cross cultural exchange went the other way as well—if alcohol typically decimated much Native American culture, feral ponies from Coronado's heartland explorations were the major factor in changing the Sioux from a more sedintary, forest dwellers into the famed horseback hunters of the migrating buffalo.

C. Technoscience itself, however, has its roots within Euro-American histories. These postdate the discovery of the New World, but the first moves were within Europe and the rise of Renaissance science which also presupposed the already sophisticated technologically textured Medieval developments. The accelerations which occured in the seventeenth and eighteenth centuries—Modern and Enlightment Periods—were even more technologically enclothed with the nineteenth—the Industrial Revolutions.

If the first moves were Continentally originated, there was a very early and little noticed movement of the same technoscience into the North American context, now wedded to a new entrepeneural and competetive spirit. Technologies soon began to be more refined and radical than their European ancestors—the clippers and then the "America," a pragmatic racer faster than all of England's racing yachts, won the cup of the same name and held it (with one temporary loss) for the nearly century and a half of naval design development.

And, if technoscience as the unique weldment of science and technology, is today regarded as needing an educational, theoretical, and productive basis to be autonomous, its production has not yet spread far beyond the Euro-American confines even today. Most of the European countries, North America, Anglo-European Australasia, Israel, and South Africa, to which we may now add only a couple of Northern Asian countries—Japan and Korea—with all the other, often largest populace countries, still struggling to attain autonomy, are the only originating producers of technoscience. The remainder of the englobed world is the recipient.

The recent Gulf War is our most dramatic, contemporary example: The USSR, many European countries, and the USA, were the suppliers of Iraq with the war technologies needed to undertake a peculiarly vicious form of cultural revenge. If Iraq's was the world's fourth largest army, its weaponry was non-indigineous. Nor was the attempt to forge the War into an Islamic/West conflict successful. Insofar as the conflict was cultural and confrontational, the dangers were unlikely to have been successful as revenge. It is not here that I shall focus the inquiry into technology and cultural revenge.

III. Cultural Conquests

If histories teach, they do not do so in any simple or straightforward way. The same applies to cross cultural histories. Insofar as cross cultural exchanges occur, there is a muliplicity of patterns. In what follows I shall draw from naval design histories and the often deeply engrained building traditions which they exemplify—but the same patterns could be found in many other practices:

A prominent pattern is simple intermixture. The Arabs invented the lateen sail; Mediterranean and Europeans used simple square rigs. As the sailing cultures met, the Europeans mixed the two—square rig forward, lateen sternwards (to help direction). This was to continue for centures of ship building.

The Vikings invented a new hull design, flat and broad and shoal, which was very fast, also clinker built, but from an easy to row craft, they adapted the simple square sail from southern Europe and never evolved beyond this. New hull/old sail.

There are also simple failures of meshing, or cultural rejections. Early Europeans sailing the Pacific for the first time encountered multihulls: catamarans and trimarans of the Pacific Islanders. And even though they recognized the superior speed and manuverability of these craft, they never—until recent high technology racing—adapted any multihull design. One possible explanation lies with the difficulty in transforming an entire practice of ship building which would have been required.

In technological histories there are also many examples of dominance leading to extinction or near extinction of alternative types. In naval history the most widely known example of this might be the virtually total displacement of commercial sail by powered ships. A less known example comes closer to the type of cultural transformation I will soon focus upon—the Eskimo kayak, a more recently adapted white water design, has virtually eliminated the older open design boats used for the same purposes.

And the last example also illustrates a subtle form of minor cultural revenge, better exemplified in the revival and adaptation of Pacific multihull design to high tech versions which today have exceeded virtually all sail racing records previously set in traditional monohull design. And, recently in Sydney, even the hydrofoils were replaced by high tech versions of faster catamaran ferries. Here the "conquered" returns in the repressed as the "conqueror."

What I am suggesting here is not only are there multiple patterns of the outcomes of cultural changes, which range from those of a highly successful conqueror dominating a supressed culture, even, perhaps extinguishing it, to many modes of mixtures, but also to the conqueror itself becoming the conquered, but that such encounters also motivate much historical and cultural change.

In the history of technology this is well documented—Lynn White Jr. has shown how much medieval technology was an adaptation of borrowings from the Far East, in particular, to European power needs. If windmills were originally prayer wheels; they became windmills and windpumps, etc. What may be less noted, however, is that periods of high technological development may also be associated with high trade and intercultural moments as well.

The Hellenic and Roman Periods were considerably more productive of engineering and technological innovations than the Period of Classical Greece. The post-Orient Medieval Period already cited in White was also noted. In both cases, the history of technological innovation could be associated with equally high periods of intercultural exchange. And, to anticipate a later observation, such periods in pre-technoscience times often *did not correspond to the periods of high theoretical development, such as that of the rise of philosophy in ancient Greece or the development of high logics among the late Medieval theologians.* Eclectic, cosmopolitan, and high trade periods seemed to favor multiplying technologies with or without high theory!

IV. Pluriculture

Many of the above examples, however, are pre-technoscience and more general with respect to cross cultural phenomena. "Pluriculture," I contend, is a distinctly late modern form of cross culturality. It is the type of cross culturality which takes shape and embodiment in the enviornment of contemporary "image technologies."

Pluriculture is dominantly shaped by *technological mediation, that of the image and communication technologies.* First, we must note several features of this mediation:

Image technologies mediate "representability" in a kind of quasi-realism different from the traditions of "handwork" art imagery. I shall begin with photography as an example: early representations of Native Americans by artists travelling with the Conquistadors, look like

Europeans with or without feathers. They take shape in pseudo-Renaissance style, and while one may recognize the accoutrements of dress, it is hard to determine what is distinctive about the "native." The exact parallel naiveté occurs in the earliest Japanese reproductions of Westerers—prints of Perry's ships and sailors in Japan, look Japanese. Indeed, the more realistic paintings of Catlin and Remington in North America, *follow* the invention of photography. And the ultimate painterly reproduction of photographic (quasi-)realism is itself, "photo-realism."

This quasi-realism of respresentation is also a kind of mediated *immediacy* which is enhanced by the literal immediacy of image availability. The photo, the fax, the television tape, can be reproduced, disseminated, and broadcast today in quasi-"real time." We note this in a particularly poignant way in times of crisis. Watching the Gulf War, very little was coming through. What there was was highly selected. And it was repeated—but overall, it always appeared as "immediate" and as if "one were there."

I have characterized the "realism" of image technologies as quasi-realism, and the immediacy as quasi-immediacy. This is because *all technologies are non-neutral and transformative* of any object referent. I shall not here develop anything like a full phenomenology of this, but several features can easily be pointed up: (a) images are "framed." Thus they are focally selective, and they *leave out or cover over* what lies outside the frame. (b) Images transform both space and time. Optical technologies *reduce* depth of field, flatten the referent. Watching a sports match on television makes the background player seem immediately behind the foreground player—the space is a kind of "Cartesian" space, but that is not accidental since Descartes took much of his geometry of space from optical technologies. (c) Image time is repeatable, reversible, transposable—as is its space. The movie "flashback" technique, or reverse time, as in *"2001, A Space Odyssey"* tracing its narrative backwards into the infancy of the voyager.

Image space-time is malleable. Its "realism" is equivalent to its "irrealism." We have seen this vividly in the technical virtuosity of contemporary science-fiction films—but the same celebration of malleability—to which is added a kind of associative fragmentation of narrative—is hyperpresent in "MTV." Here images are juxtaposed, short-lived, in a kind of associative *bricolage*. (This bricolage will become a significant factor in what follows.)

And image technologies have become *multipliers* in another sense as well. They are the everyday equivalents of cubism, surrealism, and the Picasso style of multiple perspectives within a single scan or frame. Here the multi-screen or the television control rooms are good examples. Instead of one single screen, multiple screens are situated side by side. It is a "compound vision" but not di-optic as in the multiple eyes of insects which have many lenses upon the same referent—here there is a multiplicity of referents. This, too, is a kind of bricolage which the technical editor—in the case of the TV control room—or the viewer, "composes," sorts out, selects in a usually fast and changing movement, what will be foregrounded and "revealed," and what will be left out or "concealed."

All of these technologies are familiar, part of a now ordinary lifeworld. With and through them we are now ready to note how contemporary cross-culture is technologically mediated as "pluriculture."

First, cultures (plural) are daily present. The international portion of the television news displays Middle Eastern, Asian, Irish, varieties of both Eastern and Western European culture-bits virtually every night. Old fashioned magazines and books, too, often combine these culture fragments—as particularly the *National Geographic* in Anglophone countries.

Nor is the multicultural presence restricted to imagery—it is materialized in other practices. (a) Cuisines are good examples—every international city has a range of restaurants other than local or national and these have multiplied in the last few decades. (b) Fashions, which belong to cultural semiotic systems, take their lead, not only from traditional centers such as New York or Paris or Rome, but from Tokyo, Hong Kong, with fashion borrowings from virutally every culture—African dashikis, Nehru jackets, modified Samurai warrior padding, para-military, American cowboy—all are part of the plurifashion presence. (c) Even *religions* take on this culture fragment appearance. In "Christian" Europe one finds small groups of convertees to Hari Krishna, Islam, Buddhism, even tribal or witchcraft revivals.

I have already noted in the multiple presence of cultures that they often appear as culture fragments. These are also mixed in dizzying forms of bricoleur eclectic. Nouvelle cuisine is French with an Oriental touch; Japanese "Samurai" fashion is high modern plus traditional; Bahai is Hindu with Christian forms absorbed into it. There is a kind of center of gravity which is that of culture bricolage within pluriculture.

Pluriculture also seems to favor *"surfaces."* One picks and chooses culture fragments, but one does not take with them their context or history. We become adept with chopsticks; we change our clothes with the occasion; we become religious "aesthetes" at most.

These are some of the characteristics of a technology mediated pluriculture, familiar to all of us, particularly in cosmopolitan and international centers and in the "First World."

V. Cultural Revenge

Where in all of this lies revenge? Our answer is closer to hand than thought. What we have been taking to be popular culture is only the surface manifestation of one element of pluriculture. Popular culture is a kind of bricolage, an easy picking of the cross cultural. Jeans: originally the fashion of the American gold-miner, invented by a Jewish tailor (Levi-Strauss), exemplifies the working class. Beatles: bricolage music, invented in Liverpool, England, but drawing from older jazz, blues, and swing (African American) as well as folk harmonics. And, what of "pop leftist politics"? The "Events of May" in '68 had "Maoist cells" (Chinese Marxist), later the "Che" inspired groups (Latin Marxists), or Biko (African) anti-apartheid groups. Popular culture *reflects* a kind of bricolage pluriculture. It is neither "American," nor "European," not "Asian," but, in a sense, all of these. It is "postmodern" in the sense that it is a "both/and" rather than "either/or" and constantly forming and reforming in a non-foundational mix of culture fragments.

I shall not pretend that popular culture appeals to many intellectuals. But often *literature* does. And if this is an ascent in a cultural scale, something emerges here, too, which points to pluriculture. Nobel prizes in physics have retained their virtually strict Euro-American cast—all such prizes in the last decade plus have gone to Western Europeans and Americans. But the same is *not* true of literature prizes, with nearly half going to non-Euro-Americans! Egyptians, Israelis, Africans, Latin Americans now inhabit that field.

One could—as several have done—make the case that literary production is now clearly multicultural and that the best is often non-Euro-American. But new "theories" of criticism such as deconstruction, post-structuralist, French feminism, etc., which are today popular do seem to be largely Euro-American. Does this mean that fictive and imaginative production has shifted its center of gravity outwards? and that the Euro-

American remains "theory-centered"? Or, provocatively, is the move to decrease the difference and value between literary work and critic's work in most post-strucutralist theory, a "retreat strategy"? If the artistic imaginative production now is located beyond Euro-American precincts such that we must retaliate in our own form of cultural counter-revenge?

I could multiply examples here of the new forms of cross culturality which have become powerful within a global pluricultural framework. But the leading and final question regarding cultural revenge needs to be addressed: how does pluriculture express this revenge?

The answer has many dimensions only some of which I shall point up: Reactions to technoculture have often been defensive, entailing an implicit recognition of the acidity of technoscience to traditional cultures. This is most apparent in the world revivals of the "fundamentalisms." Crude attempts to censor magazines and television, particularly in Islamic countries and often directed at "protecting" women, are likely to fail over time.

Sometimes technoscience has been enthusiastically adapted and re-molded into distinctive non-European cultural form. This is most apparent with Japan and Korea, the two Asian countries now become techno-scientific. Last year, while in Spain, I watched the televised races of mid-size sports sedans in England be led by nine out of ten *Japanese produced cars*, all outfitted with four wheel drive and often four wheel steering technologies which outdid the Europeans. What is interesting about this phenomenon is that these now superior technologies were not the result of Western production and innovation strategies, but of a consensual, infintesimal but continuous improvement strategy which only gradually achieved its result. The technology is developed upon a different cultural base.

Neither of these forms of cultural revenge, however, are as powerful as the more subtle *decentering and displacement* effect of the pluri-cultural. The very presence of a bricolage of culture fragments, from which one may and does choose, gradually decenters any privileged single choice. It makes us aware of a certain "arbitrariness" of our past choices, or of our simply having taken for granted a given tradition.

Like multiple cuisines, or musical styles, we have a larger palate, a more discerning ear available for us. Moreover, it is not the paucity of one-dimensionality which poses the problem, but the richness of the cafeteria or the music halls.

In one sense, there has already been a decentering of Eurocentrism. Our second generation critics of technology recognized that, although they did not see the source nor the level at which the decentering was occuring. And—although it may seem strange at first—our choice is not and cannot be either resistance or reaction. It could be that we can choose, to some degree, whether to read this pluriculture nostalgically, and thus negatively with respect to changes which have already occured, or celebratorily, in the recognition that a proliferation of cuisines, musics, or cultures may, in fact, enrich our own histories.

What could be called the "shape" of pluriculture, however, is distinctly postmodern. It has the richest palate of cultural possibilities in human history, but it is also something of a "floating feast" in which the older hierarchical, foundational, and core structures no longer occur. There is, or can there be, any single "best" cuisine, or music, or literature—but there can be and is a new variety out of which we may now fashion our own inventions. That is no small feat for the "end" of modernity.

VI. Conclusion

If now we look back at the itinerary I have taken, and consider the rise of pluriculture to herald a certain decentering of Eurocentrism, it will be seen that its occurance is such that it did not correspond to the internalist critiques of technological culture in the following ways:

Its content is not internalist, but cross cultural. What becomes attractive is often the exotic, the different, the other.

Its form, however, is not any coherent or single other culture, but the bricolage of culture fragments in ever new mixes, not unlike the cinematographic or televisual maleable narrative.

Its appearance was not a direct confrontation of high altitude "theory," but a growth and expansion from "below," the praxical and the popular. Its historical antecedents are more like the eclectic periods of cross cultural trade, technological innovation, exchange, than any moment of high theory.

And, as "postmodern," it is probably transitional, but its trajectory is not likely to be either the reassertion of the Eurocentric, nor the return to traditional forms.

And, if there is a danger, the danger is precisely one which may undercut the motivational sources of technoscience itself—if technoscience is motivated by distinctly "Western" cultural motivations, pluricultural ones may and often do differ.

But, in any case, pluriculture is a current manifestation of contemporary technoculture.[1]

---

[1] This article was published previously as "La technologia e la vendetta della culturea," *Paradigmi: Rivista di critica filosofia* Anno X, n. 30, 1992.

Chapter 11

The Study of Religion
in Husserl's Writings

James G. Hart
Indiana University

Abstract: *In this paper I attempt to systematize Husserl's remarks in the* Nachlass *on the study of religion. I will not be dealing primarily with his own philosophical theology which he regards as the culmination of his transcendental phenomenology, but rather with what he thinks religion is and what is studied when people study religion. I will first briefly discuss how religion is a developing cultural phenomenon which comes to have a relationship to philosophy and reason. This leads us to the consideration of a variety of senses of theology.*

I. Culture and Religion

Husserl thinks of culture properly as the intersubjective constitution of idealities in sensible materials which have an abiding validity and which shape a people.[1] He claimed also that it is the way the active life of a people objectifies itself. Objectification is not merely self-expression but also an externalization in sensibility and physical substrates of the spiritual life of the people, the meaning of which is able to be experienced by subsequent generations. Such an experiencing of culture may be an occasion for ever renewable spiritual strength or a source of distractions and burdens (Hua XXVII, 21-22).

We may note with Iso Kern that the phenomenological basis for appreciating the Hegelian theme of "objective spirit" is Husserl's concept of "indications" (*Anzeichen*). To be precise, objective spirit has to do with the formations of sensibility which indicate, a human achievement as well as occasion, by their so indicating certain types of intentional acts, e.g.,

---

[1] I wish to thank Prof. Samuel IJsseling, Director of the Husserl Archives in Louvain, for permission to quote from the *Nachlass*.

M. Daniel and L. Embree (eds.), Phenomenology of the Cultural Disciplines, 265–296.
© 1994 *Kluwer Academic Publishers. Printed in the Netherlands.*

perceiving, acting, and representing in its various forms (picturing, reading, speaking, etc.).[2]

On the occasion of writing about "cultural renewal" Husserl reserved the term "culture" for an honorific sense which he contrasted with "civilization" and "tradition." Here the ideal sense of culture refers to a communal spirituality which mediates all social relationships and which is characterized by a habituality which is always ready to awaken the received achievements to their original intuitive validity, beauty, truth, etc. Here culture is authentic culture. Civilization and tradition, however, are the inseparable milieu of culture. These are the ways achievements fall into the merely conventional and thereby become scarcely understood.[3] As such, civilization and/or tradition are not capable of reproducing the original motivations. These thus remain, for all practical purposes, dead. Thus, at least on this occasion, civilization and tradition are regarded as the realm of the inauthentic, i.e., a realm in which there is no intuitive legitimation of one's thoughts and deeds, analogous to the way our knowledge of $9 \times 9 = 81$ is automatic and uninsightful, in contrast to $1 + 1 + 1 = 3$. Culture is properly authentic or approaches a "philosophic culture" when it has the steady disposition and ability to awaken the received achievements to their original legitimating intuitive senses.

Religion for Husserl in the Kaizo essays has the specific meaning of that towards which elemental mythic culture develops. The lower level of mythic culture, proposes Husserl at least on one occasion, has to do with the way a people establishes a practical relation to, deals with, placates, etc. the cosmic powers which pervade the world. World as a totality, he noted later in "The Vienna Lecture," becomes a theme for the mythico-religious attitude. The mythico-religious objects or values are

---

[2] I deal with Husserl's theory of culture more at length in "The Entelechy and Authenticity of Objective Spirit: Reflections on Husserliana XXVII," forthcoming in *Husserl Studies*; also in "The Rationality of Culture and the Culture of Rationality," forthcoming in the *Philosophy East and West* (1992).

[3] As we shall see, Husserl's theory resembles Max Weber's view that there is a basic pattern where charisma tends to suffer a decline and to give way to powers of tradition and rational socialization. The "routinization" of bureaucracy is the well-known devolution of the charisma of the great founder. Stanford Lyman called my attention to this parallel at the CARP Conference. For a discussion of these matters, see Reinhard Bendix, *Max Weber: An Intellectual Portrait* (New York: Doubleday Anchor, 1962), 325-328.

ways in which the totality of world becomes a theme. "World" is comprised of the invisible and visible, and the visible is but a section of an apperceived invisible dimension filled with hidden malificient and benificent powers.[4] But since the mythic attitude is under the sway of invisible hidden powers, the speculative knowledge proper to myth is subordinate to the task of shielding visible life from every sort of evil fate.[5]

---

[4] E III 7, p. 2; see also my "From *Mythos* to *Logos* to utopian poetics: An Husserlian Narrative," *Journal of the Philosophy of Religion* 25 (1989), 147 ff.

[5] See *The Crisis of European Sciences and Transcendental Phenomenology*, trans. David Carr (Evanston: Northwestern University Press: 1970), 283-284; Hua VI, 330. I am grateful to John Drummond for some insights into these matters. As far as I know a phenomenology of religion as such is missing in Husserl. He greeted Otto's work on *Das Heilige* as a "first beginning" in the phenomenology of the religious dimension. He told Otto that his speculations, presumably the theory of the innate propensity for the religious, was better left out. But as a beginning it goes to the true origins of religious experience. What precisely he liked about the descriptions of the *mysterium tremendum et fascinans* we do not learn from this letter. He only volunteers that Otto does not yet offer the radical distinction between the incidental fact and the *eidos* in religious intentionality and there is still wanting a study of the essential necessities and possibilities of religious consciousness and its correlate—as well as a study of the essential necessities of its development. What he had in mind is perhaps indicated in the discussion in the body of the text.

In response to Gertha Walter's letter (in preparation of her *Phaenomenologie der Mystik* (Halle: Niemeyer, 1923/70) Husserl offers a theory how we can be touched in "the deepest depths" by noting how the strewn out position-takings and the acquired values of the heart may get reactivated by felicitous *Gestalten* so that all of one's life is gathered together in a unique synthesis. Deeper strata of the I are awakened into play and what before functioned as unrelated motives in the passive underground are awakened into a synthesis which permits infinities and powerful new perspectives to open up (A V 21, 92); cf. my "A Precis of a Husserlian Philosophical Theology," 152-154 and 166-167; (on p. 167 I gather some of Husserl's remarks on mysticism.) What perhaps Husserl has in mind, and what can bring together these considerations, is the way the religious sphere makes present the ontological-metaphysical actuality of the divine idea. We will come back to this later on in the essay.

Finally, Husserl maintains in the correspondence with Dilthey that the phenomenology of religion is, to use Dilthey's own expression, an empathic study of the inner life of religious persons and communities in terms of the various motivations and life-forms. The historical-factual serves as exemplifications of the way the pure ideal is intended. And the history of religions investigates the historical-factual but is indifferent to the essential-ideal in the same way that actual sciences of physical bodies are indifferent to the essential nature of the spatial-temporal thingliness. Thus the endless relativities of gases, particles, liquids, solids, waves, fields, etc. are all pervaded by the ideal norms of the idea of "corporeal

Religion in the specific sense is the higher level of mythic culture in which

> these transcendent beings are absolutized to deities, to the institutors of absolute norms which they have communicated and revealed to humans and in the observance of which the humans find their salvation (Hua XXVII, 60).

On this occasion Husserl offers the opinion that the consciousness of norms is what distinguishes the human animal. All animals live out their life in the context of instincts, but the human animal has not only instincts but also norms which pervade all conscious acts so that everything she experiences is right or not right, i.e., lovely or ugly, meaningful or pointless, suitable or unsuitable, etc.[6] Thus Husserl can claim that the development of the consciousness of norms and the development of religion are interwoven. In these passages he seems to suggest that the lower level of mythic culture is somewhere between norms and mere instinct. On this occasion I will not pursue an analysis of Husserl's (rather impoverished) view of "primitives" and their religions.[7]

In religion, as the higher form of mythic life, all of communal life, not merely the realm of faith and cult, but even the private sphere, is harnessed to the requirements of divine norms. In "The Vienna Lecture" he holds that religion properly excludes polytheism and that in the

---

thing." Similarly we can measure the facts of historical religions against the ideals emergent in religious experience. The truth of religion in this sense would be relative to the various historical disclosures and irrelative in so far as these would be instances of the ideal unity which is manifest in them. See "Correspondencie entre Dilthey y Husserl," Walter Biemel, ed., *Revista Filosoia de la Universidad de Costa Rica I* (1957), 101 ff.; translation in *Husserl: Shorter Works* (South Bend: Notre Dame, 1981), 203-208. See also "A Precis of an Husserlian Philosophical Theology," *Essays in Phenomenological Theology*, edited by Steven Laycock and James Hart (Albany: SUNY Press, 1986), 100 ff.

[6] Husserl's sketches of animal consciousness are worthy of a special study.

[7] It is impoverished because he does not see the richness of forms of life and collective virtue, e.g., practices of avoiding conflict, of non-violence, of child-raising and child-care, etc. in many aboriginal cultures. Everything is judged from the standpoint of whether the culture approaches the Greek discovery of *logos*. It could be that a superior kind of social-political *sophia*, if not *logos*, is in play in many of these cultures. For a sketch of Husserl's own view see Sect. 1 of "From *Mythos* to *Logos* to Utopian Poetics: An Husserlian Narrative."

concept of God "the singular is essential" (*Crisis*, 288). In both this lecture as well as in the Kaizo essays he claims that religion as the highest form of myth leads to a hierarchical cultural priestly class which administers a universal system of absolute norms encompassing all directions of life, i.e., in general, knowing and evaluation as well as the practical details of life. Husserl makes reference to the earliest forms of this religious culture in Babylon.[8]

Suffusing Husserl's analysis of culture and religion is his own pervasive progressivism. Because consciousness itself is radically teleological and facing infinite ideals, so also the "objective spirit" which it constitutes discloses such a teleology. This becomes evident when he claims that culture, in this specific sense of religion, has an entelechy in the sense proper to human development, i.e., it is not blindly functioning as a goal toward which the organism heads unconsciously in its normal growth. Rather it functions as a conscious axiological principle, a constituted ideal-goal, under the leadership of a class of priests. Each cultural form, Husserl speculates, may be envisaged as an analogous species in so far as it grows towards its specific ideal form of maturity. But because it is rooted in the conscious functioning of an axiological perspective tied to an ideal-goal, it is to be contrasted with entelechy as a way of describing unconscious organic development.

In short, religion, as an objectification of a community's consciousness in the form of an absolutization of the mythic powers in unified norms, directs all the formations of this type of culture (Hua XXVII, 63). Seemingly the culture of religion, as a unification of norms with a resultant centralism and hierarchy, would have a nisus toward the objectification of a single God in order to ward off the destablizing, de-centralizing, and centrifugal forces of polytheism.[9]

---

[8] Husserl's brief description of religion's hierarchical, centralist, and totalitarian structures parallels Lewis Mumford's monumental account of the rise of "civilization" in antiquity as the forerunner of the modern "megamachine." See *The Myth of the Machine*, Vols. I and II (New York: Harcourt Brace Janovich, 1964).

[9] That Husserl, in some sense is philosophically a monotheist and not a polytheist cannot be held against him as a form of cultural prejudice. One may disagree with this position but then one is disagreeing with his philosophical theology, and that is the terrain upon which that difference of opinion must be worked out, not on the level of claims of personal intolerance or Western chauvinism. Similarly, the view of Husserl that religious experiences have some essential features may or may not be a kind of cultural and metaphysical prejudice, e.g., "essentialism." Such a matter has to be determined by arguments which help

In so far as the teachings and commands of the priestly class are sanctified and performed in the legitimating traditional manner they are experienced by anyone raised in the religious tradition as absolutely binding. Thus the hierarchical religious culture, as exemplified in the case of the Babylonian or ancient Israelite religions, has naivety in spite of its being a matter of conscious ideals and is unfree in spite of the free pious fulfillment of the religious prescriptions. Indeed, religion in the just sketched sense is essentially characterized by compulsion, unfreedom and lack of criticism. These seeming paradoxical remarks are less jarring when we consider that in the Kaizo essays freedom and the capacity for criticism are identified (Hua XXVII, 63). In the religious culture, asking about the truth of the beliefs and the legitimation of the norms is typically regarded as manifest impiety (Hua XXVII, 64).

Note throughout Husserl's schema the hierarchy of cultures is in terms of what seems to him to involve a greater wakefulness of consciousness, thus, e.g., having norms and not mere instincts, and what approaches authenticity, i.e., the intuitive evidence which legitimates the norms. The

---

decide whether philosophy can escape some sense of allegiance to something like essences and whether the analysis of religious experiences does show, indeed, some essential features. That Husserl presumed to tell the Japanese what authentic culture was is to be expected. The theory of authenticity runs throughout his philosophy. It would be disengenuous for him to do otherwise. I take Husserl to be cautious rather than chauvinistic when it comes to the details and specific meanings of other cultures. Consider, e.g., the reserve and seeming deliberate vagueness, in spite of his clear enthusiasm, in regard to the new translation of Buddhist texts; although he claims for the Buddhist texts a genuine transcendentalism as well as a purity and depth matching anything the West has to offer, and although he is confident the encounter with Buddhism will be a determining factor from now on in Western culture, he does not make a single specific reference. See Hua XXVII, 125 ff. I, with Merleau-Ponty, do not take Husserl's Eurocentrism to be chauvinistic. He was as ignorant of other cultures as most of us are; but it seems rather clear that he was not interested in restricting *logos* to a European form of existence. Rather, he was confident that neither Europe nor America approximated the ideal of *logos*; indeed he saw clearly that the forms of rationalism that had developed under scientism and capitalism were in danger of destroying any sensibility to a genuinely philosophic culture. Cf. Merleau-Ponty's remarks in *Primacy of Perception* (Evanston: Northwestern University Press, 1968), 89. For another interpretation of Husserl's view of religion and religious studies, see R.A. Mall's "The God of Phenomenology in Comparative Contrast to That of Philosophy and Theology," *Husserl Studies* 8 (1991), 1-15. I cannot here deal with Mall's interpretations in detail; readers of his essay will see that we disagree on many points.

telos, of course, is the *telos* uncovered by transcendental phenomenology, a eutopian authentic universal community of monads.

In the context of religious culture the entelechy of human culture beckons to a higher form by way of two possible freedom movements. One is obvious for readers of Husserl, i.e., the development of free science, the liberation movement of *logos* in opposition to *mythos*, culminating in philosophy. The other is a liberation movement which religion itself may take. We will begin with this latter first.

II. Authentic Religion

Husserl claims that there are conditions for the development to the higher form of religion which is a religious liberation from religion. Husserl suggests that these conditions are cases where the priestly mediated cult and teaching fails to provide either a national or personal salvation. When much suffering is borne and severe penance performed and when nevertheless salvation still seems remote, the formerly credible explanations of a sinful departure from the divine ordinances can begin to lose their power. And on such occasions individuals may find occasion to begin rethinking their individual as well as collective relationship to God (Hua XXVII, 64).

At this juncture (*Ibid.*, 65) Husserl offers a very ambitious but concise theory of the intentionality proper to what he calls religious culture and its religious surmounting. We can get at this by noting a distinction he made (on another occasion, A V 21, 22a) between how the world or milieu is pre-given through one's own individual and intersubjective experience and how venerable tradition co-determines the disclosure of these experienced objects. He seems to suggest that tradition is a lamination of apperceptions on top of the apperceptions of our individual and intersubjective experiences. This enables people of different cultures to see in one sense "the same thing" but yet not see "it" because one does not have the "same world" as the other.[10] He seems to want furthermore to say that there is an individual and intersubjective apperception of values, upon which apperceptions also those of the tradition are laminated. This is a complicated matter because for Husserl values are founded on the objectifying, apperceiving acts through which

---

[10] See A V 21, 22a and Hua IX, 497-498; also my discussion in §1. of "From *Mythos* to *Logos* to Utopian poetics: An Husserlian Narrative."

some sense of an object is given.[11] These apperceiving acts would be an interweaving or "inter-laminating" of the individual-intersubjective and traditional apperceptions.

Basic for this theme is the central difficulty of Husserl's theory of values: the relation between the emotive and intellectual or cognitive realms.[12] It is clear that judgments aim at truth. Their home is insight and evidence "in the issues themselves." The realm of judgments is much broader than the realm of natural things because it includes ideal objects, cultural objects, and the whole range of acts, including emotive acts. Religious objects, contents, and acts of religious judgments themselves can be the theme of judgments.

When one judges about values one is not in an evaluating or appreciating attitude but rather in the attitude of one judging. Now, of course one can, in one's judging, be determined by motives of the heart (*Gemüt*). For example the motives of the heart can weaken the premises of a judgment as when one is moved not to attend to the premises or to regard them as of lesser weight. In everyday life a judgment is not an isolated formation but is interwoven with other immediate and mediate contexts of judgment which function in any particular judgment.

Husserl speaks of a disturbing and uncomfortable ambiguity with which we are confronted when we think about how judgment plays a role in perception. When we see something a judgement can be in play but often so is belief and this plays a role in determining the judgment. Even in acts of seeing involving self-validating, self-evident matters or filled intentions belief-motivations are often in play regarding the intended object or in the structuring of the empty intention and therefore in determining the judgment as referring to a filling of the empty intention. A central question for the philosophical phenomenology of religion is whether the heart (*Gemüt*) and the will, analogous to beliefs, can produce something like "grounds" or motives for judgments. Can they provide motives in the fuller sense that they move me to judge in a certain way because I "place value" on the judgment? (A V 21, 7b)

---

[11] See *Vorlesungen über Ethik und Werlehre (1908-1914)*, ed. Ullrich Melle, Hua XXVIII (Dordrecht: Kluwer, 1988). Also my "Axiology as the Form of Purity of Heart: A Reading of Husserliana XXVIII," *Philosophy Today* (1990), 34, 206-221.

[12] See Hua XXVIII and also my "Axiology as the Form of Purity of Heart." The discussion which follows is taken from A V 21, 5a ff.

There are several issues here which I wish briefly to mention. a) Husserl is fully aware that when I place value on a state of affairs, I disvalue its opposite, I wish it away, I look away from this counter-possibility and the weight it has on the matter; this is clearly harmful to the truth of the state of affairs. b) But there are other considerations, recently called attention to by William James, where belief in a certain state of affairs does not merely make the one having the belief "blessed," and where it does not merely irradiate confidence, good will, and a healing effect, e.g., to a believed-in person; but most important for our purposes is the (Augustinian) consideration that if I place value in S being p, I may be empowered to see better than one for whom the value is missing. Husserl hestitates on this occasion (A V 21, 9b) to admit this latter possibility in the case of normal beliefs about finite states of affairs and he adds that the loving seeing with "rose-colored glasses" cannot withstand the power of truth and evidence.

c) But what about when we have to do with states of affairs for which there is no conclusive evidence, therefore with judgments in which there is no conclusive verification? Here Husserl is thinking surely of the ideals of science and of an ethical-communal life. But he has especially in mind the kinds of belief that are connected with what he calls *the absolute ought* wherein the blessedness, true self, and fully satisfying life of the individual person are brought together in a unifying *Gestalt*. This absolute ought, wherein the "ought" is inseparable from an "is," has its own unique kind of evidence. Chiefly this takes the form of the "one thing necessary" for a person's life, i.e., the loss or renunciation of which means one cannot live with oneself.[13] Husserl says the demands of the heart, and, in particular, the disclosure of the absolute ought can encompass theoretical matters which have great scientific probability. Should these conflict with the absolute ought of a particular person, it is clear that this person would not then be able to believe them (A V 21, 9b).

Often enough the absolute ought will function in a hidden way and the person will not be aware of how the issues in question conflict. It can also be more explicit and one can make an identifying act in regard to a particular matter. But such an articulation of the meaning of the absolute ought might be so informed by the personal experience and

---

[13] I discuss the "absolute ought" at length in *The Person and the Common Life: Studies in a Husserlian Social Ethics* (Dordrecht: Kluwer, 1992), Chapter IV.

historical tradition that these foster a kind of selectivity and obfuscation of the evident significance of the absolute ought in the particular context. And, of course, finally, it can be that the person is in a position to make evident the meaning of the absolute ought so that it's articulation is bound to a definite content which directly conflicts with the scientific probability.

Husserl holds that the life of a society or a community involves typically an inauthentic allegiance to, as well as a genuine intuitive experience of, values and normative types. He would seem also to want to say that although the theme of the absolute ought becomes explicit in a culture under the sway of the ideal of *logos*, individuals in all cultures have intimations of the absolute ought. Further he once argued that the sense of "depth" which characterizes religious, especially mystic, experiences is to be accounted for by the kind of "gathering" experiences in which the fuller sense of our lives and our truer selves, i.e., the absolute ought, becomes explicit.[14] The experience of values and normative types which suffuse what Husserl regards as religious experience would seem to be inseparable from the irrepressible issue of the absolute ought.

In so far as the experiences of the absolute ought, values and normative types are genuinely evident experiences they have their own kind of *rationality* even though evaluation and belief guide the perceptions and the judgments. Living within a tradition means that this authentic experience of values is interwoven with or projected onto the religious realm of inauthentic beliefs and representations. Thus the religious content is inseparably "rational" or intuitive value-insight and non-rational beliefs as well.

Recall that the mythic-religious attitude thematizes the totality of world. The thematization of this totality is inseparable from core value-experiences, that is, the whole (or "world") is profiled in these value-experiences. Thus the content of the religious intentionality, religious belief, has at its center a developing core of intuitively evident values,

---

[14] This is the explanation of mystic experiences in his letter to Gerda Walther in response to her theory of mysticism. See A V 21, 84-92; cf. above, n. 5. Another relevant text is one wherein Husserl observes that if God is regarded as he proposes, i.e., as the entelechy of entelechies, "God thereby can be no object of possible experience Ias in the sense of a thing or a human). Rather God would be 'experienced' in each belief that believes originally-teleologically in the eternal value of that which lives in the direction of each absolute ought which engages itlsef for this eternal meaning" (A V 21, 128a).

e.g., the absolute ought, the saintliness of a colleague I work with, etc., but these, Husserl observes, are wrapped in an irrational facticity which comprises the particulars of the religious beliefs. For the persons who themselves are spiritually growing, the intuited value, what Husserl calls the *rational core*, which has its own evident normative necessity in the content of the act of belief, receives the chief emphasis and becomes the sustaining force of the entire act of faith. For Husserl all progress in cultural development involves this kind of distillation of the core values which he also calls a "spiritualization" (*Vergeistigung*) of religious representations (Hua XXVII, 64-65).

In summary, values and the experiences of the absolute ought, which for Husserl's theory have as their foundation objectifying acts, i.e., presentations of things in the world, themselves are inseparable from the apperceptions of the religious culture. Husserl speaks of them as projected into a realm of mythico-religious presentations. The basic perceptions of things are qualified with values, but these, i.e., the esteemed things or events, in turn are interwoven into the mythic religious representations. In the case of the maturely developing religious personality these core value experiences are given the central emphasis and sustain the entire act of faith. Indeed, all of faith here is led by the intuition of values.

But we ought not to conceive the core value-experiences apart from the fundamental "general will" which goes in advance and therefore determines the hierarchy of interests, relevance and relief in the perceptual field and which is the foundation for the articulation of the absolute ought. This is where Scheler and Husserl draw together. On the basis of the primal passive synthesis of one's life in the general life-will, the world may be said to greet us first through the values (and not the things presented through objectifying acts) and as these are mediated by the traditional apperceptions.[15]

For Husserl the case of Jesus serves as an example of where religion is called into question on the basis of the vital intuition of the core values. Such an experience poses a crisis for religion because it discloses

---

[15] Although there is much work to be done here on the issues of axiology joining and separating Scheler and Husserl, I make an initial run at a few of the issues in "Axiology as the Form of Purity of Heart." The general will is developed in conjunction with value perception in Ch. II of *The Person and the Common Life: Studies in a Husserlian Social Ethics* (Dordrecht: Kluwer, 1992).

the merely mythical-traditional framework as a residue of irrational facticity. Husserl gives an intriguing intepretation of value experiences which places what is usually regarded as "mystic experience" at the foundation of all genuine religious experience and culture:

> The unified intuition contains the character of a unity of original religious experience, therefore also the character of an original experienced relation to God in which the subject of this intuition knows himself to be addressed not by an external God who stands over against him and [in which the subject] knows himself determined to be the bearer of a communicated revelation. Rather, he knows God as intuited in himself and as originally one with him. Therefore he knows himself as an embodiment of the divine light itself and so as a mediator of the message of the divine being (*Wesens*) from out of a content of the divine nature (*Wesens*) implanted in him (Hua XXVII, 65).

Husserl is claiming that this is a description of the experience of Jesus and the experience of anyone who genuinely transcends religion religiously. (The student of Husserl has to say that in order for Husserl to make these claims either Husserl himself genuinely transcended religion religiously or that this is his interpretation of his experience of one such as Jesus.) Husserl further observes that when such an authentic religious experience happens there is a transformation of religion or, indeed, the creation of a new religious type who presents religion from out of sources which in a good sense are rational and yet are genuinely and originally religious experience (Hua XXVII, 66). And this transformation happens through the power of original intuited values and norms. These authentically experienced values and norms are evident in the world as the basis of the meaning of the world's salvation. And in the intuitive experience of these values and norms is the evidence that a world pervaded by such values would make blessed the one who would live in this world with such a belief and with such an understanding of this sense of the world. Note that "faith" here is tied to a filled intention or intuition of values as well as the empty intention of the relevance of the intuited value to all of life. This gets filled in living faith-fully. Thus he adds,

> it is this evidence which gives faith its power and which grounds faith. Faith makes blessed and it is true because it makes blessed, because it proves the meaning of the world in the living of a meaningful life (Hua XXVII, 65-66).

Faith here is basically a mode of presencing borne by a profound intuition of values. Seemingly this intuition of values itself is not achieved through faith but through a kind of experience. Further this is *not* (at least not typically?) the experience of the absolute ought. The absolute ought places one on the way to the infinite godly ideal; this experience uncovers that one is already unified with the divine.

In the case of the followers of Jesus or of the type instanced by Jesus there is no question of merely empty intentions or an external appropriation of reports, information, propositions, etc. Rather, the follower must empathically experience the experiences of Jesus or whomever posterity will call the founder of the religion. Again, it seems we must say that for the original experiences, i.e., the experience of those who come to be called the founders, there is not an experience of the absolute ought; but the experience of the founder's experiencing and his message occasion for the followers the experience of the absolute ought, the *unum necessarium*. This, as we shall see, is what Husserl calls "faith-experience." The stories of the life of Christ, his parables, his own testimonies, provide the occasion for the believer to reproduce the original power of the intuitive values and their motivations. Again, for the follower these are experienced in a mediate, not immediate relation to God.

A hermeneutical remark seems in order here. If Husserl's theory has some validity, then the student of religion also must be actively involved in acts of empathic understanding of the followers and disciples' efforts to empathically understand the experiences of the founders. Gadamer's critique of Schleiermacher's placing of empathy in the center of the textual interpretation of the Other would probably exclude what is essential for Husserl. For Husserl such acts of empathy are not merely possible but necessary for texts within one's own culture (see our discussions below); what kind of preparatory experience will make them possible in regard to other cultures would have to be spelled out.

At this juncture I wish to turn to Husserl's rather autobiographical account of the effect of reading the Gospels which is given as Beilage IV to the Kaizo lectures. This appendix would seem at least implictly to want to illustrate, by way of an example, the just-rehearsed theory of religious experience of someone who becomes a follower of Jesus—more so than the experiences of Jesus himself. What is at stake of course is how there is in both cases an original experience of the core-values. This text is of interest also because it relates religious experience to his key ontological-metaphysical concept, namely the divine as Idea, as well as to his key ethical concept, the absolute ought. These, we have claimed, are evident in the intentionality of the follower; the founder does not face infinite Ideas but rather somehow is coincident with or embodies the Idea.

Husserl notes that what moves him in reading the Gospels is not the miracles. Indeed, he himself reads the Gospels less as an historical account than as a legend. Christ confronts Husserl as a *Gestalt* of exceeding goodness. Legendary as it is, in contrast to the clear individual personality of the figure of Paul, the Christ of the Gospels awakens in him, through the various sayings and parables, a realm of perfect goodness. He says of the Gospel presentation of Christ calling us:

> I have evidence that such an action (as it is here demanded) is purely good, that to be able to be in this way would be blessed, that such [a goodness] would awaken in me love and the purest love. And Christ himself stands before me there not as someone who merely demands but himself as one who is perfectly good, as pure goodness, all-understanding, all-forgiving, as looking with pure love upon all humans as seedbeds of a possible pure goodness. And I can think of him only as an embodiment of pure human goodness: as an ideal human . . . and I empathize and become filled with infinite love for this trans-empirical Gestalt, this embodiment of a pure idea. And I am filled with blessedness knowing this infinite person lives in relation to me. And because this power streams forth from this ideal form it has already for me a reality.[16] I believe in this legendary individualised idea and it becomes a power in my life.

> And now I understand the believer who in the contemplation of this ideal figure which at the same time is given through a continuous tradition about which he may have no doubts, who believes in the historical individual Jesus and in all his miracles and all that the first tradition of the Gospels tells of his resurrection and of what he himself bore witness to regarding his relationship to God, etc . . . . The saving

---

[16] This passage, and many others in Husserl's theological writings, recalls Peirce's views on the power of ideals: whatever generates devotion and has the power to attract us irresistably cannot be non-actual and merely the outcome of development. For a discussion, see Donna M. Orange, *Peirce's Conception of God, Peirce Studies*, N. 2 (Lubboc, Texas: Institute for Studies in Pragmatism, 1984), pp. 70 ff.

power which I actually experience as emanating from the Idea lends power to the historical religion (Hua XXVII, 100).

This text illustrates several aspects of Husserl's theory. The Gospel figure of Christ, regarded in its pure essence, occasions the presencing of a core value. Indeed, in this case an ideal is presented which at the same time is burdened with more or less irrational contingencies or accidental materials. For example, Christ is tied to miracles, the one God of Judaism (who is a kind of despot for Husserl), Palestinian history, etc.

The example of reading the Christ of the Gospels does not disclose how Jesus experienced authentically the core-values but how the followers of Jesus may have authentic religious experience. And the example spills over into what for Husserl is a basic conviction regarding the ethical importance of experiencing the core value of human goodness: Ethical experiences are not made through criticisms of others but only through a concrete loving intuition of the goodness of others which announces itself in the evidence of the pure fulfillment of love-intentions as value-intentions.

> What would the human be if he could not see admirable humans, purely good people? He can only be good when he sees good humans, when he is directed to exemplary figures and through them raises himself. He can only become good through the transforming love which poetically transforms the beloved into an ideal, [through a love] which wants to see only the goodness of the beloved (Hua XXVII, 102).

In accord with Husserl's theory of emotive intentionality there is a kind of disclosive power to emotions, and foremostly, under certain circumstances, love. Of course often the lover idealizes, and to that extent therefore less authentically perceives the goodness. But idealization is also a way of seeing and thus even the creative phantasy of the artist can create the experiential basis for the basic experiences of ethics. For Husserl this dwelling on the ideal in a non-real context parallels how logic provides an experiential basis for an intuitive penetration into scientific formations.

III. *Religionswissenschaft* and **Phenomenological Theology**

And this raises the question of what provides the experiential material for "pure theology,"[17] for the true doctrine of God and study of religion. And how are these to be related to facticity? Husserl makes this proposal: If I want to judge the factual state of the world in ethical and aesthetic contexts then I need acquaintance with the pure norms in order to judge with a sober criticism the degree of approximation. The implication seems to be that any source whatsoever, e.g., even fictional literature, that is uplifting and edifying through its disclosure of core values and ideals, that source may be regarded as the material basis for "pure theology" or the essential study of religion because it enables the intuition of core values.

This parallels his theory of "idealist art." For Husserl the artist can assume the role of a seer and a prophet. All art is joy in the seen *in concreto*, in its appearing as such, but it is not thereby "kallistic," i.e., it is not thereby idealistic and normative. In such idealistic poetizing the artist does not merely look upon facts and types of the empirical world and regions of life; nor does he present ideals as typical facts or empirical types. Rather the artist is normatively engaged with the struggle of good and evil and he aspires to inflame the love of the good in our souls without moralizing or preaching. He enables the typical values of life to be transfigured through beauty and thereby raises the souls to the divinity. The artist here takes on the role of the metaphysician in that he raises the spectator to, and unites him with, the idea of the Good, the Godhead, the deepest ground of the world, through the medium of the beautiful (Hua XXIII, 541-542).

In conjunction with this extension of the experiential basis of "pure theology" to include other sources than the typical traditional sources of revelation, Husserl makes a case for an enlargement of our appreciation for the realm of intuition. And then he urges, echoing his letter to Otto, that perhaps all attempts at articulating religious phenomena fail because religious intuition presupposes the most universal intuition of *absolute* phenomena or givens which transcendental phenomenology uncovers. And

---

[17] I take "pure theology" here (Hua XXVII, 102) to be one which can grasp the essential features of religious matters through eidetic analysis and purified of the irrational contingencies of historical religions. See the discussion of the Dilthey correspondence in n. 5.

to see this we need the phenomenological reduction. He does not make explicit what he has in mind here but it would seem to be the themes recounted in *Ideas* § 51 and 58 (see our remarks below). But clearly he believes that only the transcendental phenomenological theme of eidetic intuition can do justice to the full scope of what is in play in religious intuitions—and therefore what he calls "the noble artist" is of great aid in these matters. But this leaves him unsatisfied. Although the divine and the Good (or the absolute ought) are manifest apart from transcendental phenomenology, unless the transcendental phenomenological origins are a theme, the *proper* sense of the religious themes and values, the proper sense of "God" remains forever hidden. This leads him to oberve that perhaps it is enjoined on humans to create religion in a twofold sense:

> In the one case religion as progressive *mythos*, as one-sided and genuine intuition of religious ideals surrounded by a horizon of inkling, into whose infinities one does not penetrate but before which unsearchable infinities one bows; in the other case religion as metaphysics of religion, as ultimate conclusion of the universal science, as norm for all the intuitive mythical symbolism which rules its formations and transformations of phantasy (Hua XXVII, 102-103).

The study, not practice, of the former, would be the more familiar sense of *Religionswissenschaft* and perhaps an extended sense of the phenomenology of religion. Husserl's interests or strengths did not, it would seem, move him to work in this area.

The other, second, study would be transcendental phenomenology's own rich, if never completed, philosophical theology. Again, the best formulations of Husserl's position on these matters is §§51 and 58 of *Ideas* (which I have studied elsewhere[18]). For our present purposes suffice it to say that the non-worldly, non-thingly divine transcendence to the world-pole and transcendence in immanence of the I-pole are claimed to have modes of intuitive disclosure other than the worldly or thingly—the typical mode of disclosure, presumably, of myth and religion—to which theory may adjust itself. Although one finds in Husserl's writings considerable delineation of transcendental phenomenology's basic theological themes, e.g., the divine idea, divine entelechy, and the

---

[18] See "A Précis of an Husserlian Philosophical Theology," in *Essays in Phenomenological Theology*.

unbegun and unending character of the transcendental "I," the absolute ought, etc. there is no effort I know of to indicate how the basic positions of transcendental phenomenology in effect rule over or provide the norms for the religious symbolism and imagination of historical religions. Again, this aspect of transcendental phenomenology would require an expertise which Husserl would not pretend to claim.

It is clear that for Husserl the idea of a so-called phenomenology of religion which would tabulate the strange and exotic practices of the world's cultures would hold philosophical interest only in so far as the historical-anthropological material would supplement the process of eidetic variation.

Here we can ask whether this division of religious studies into *Religionswissenschaft* and transcendental phenomenological metaphysics is a division of "pure phenomenological theology." Seemingly it is. But in any case it oversimplifies the richness of senses of theology even as it appears in the Kaizo manuscripts from which it is taken. We can best demonstrate this by looking at the various other senses in which Husserl understands "theology."

IV. Various Senses of Theology

Although "theology" refers primarily to an activity in the Western Christian tradition, the only realm of Husserl's competence, because it has parallels with what he understands "religion" to be, it would seem, on the one hand, that its various senses may serve as possible types in non-Western contexts. On the other hand, in as much as the defining senses of theology have to do with the various relations to *logos*, and in as much as Husserl thought, mistakenly it seems, that other ancient cultures, e.g., India and China, did not have anything which paralleled the ideal of *logos*, it is unlikely that he would allow the various senses of "theology" to exist outside the West in antiquity.

Husserl on several occasions discusses theology within the parameters of the conventional and traditional Western understanding of theology. In general, theology has as its foundation faith and the religious tradition as a source of revelation and faith. The religious tradition differs from ordinary tradition. The latter does not have the character of absoluteness which the former has for the religious believer. For the religious believer the religious tradition cannot become doubtful or contradicted by reason. Thus religious faith stands in opposition to all rational knowing and grounding. It claims an absolute validity which does not come from seeing and insight. Faith is indeed judgment but not mere judgment, mere *doxa*. As his colleaque Scheler might have put it: It is a matter of believing-that, but the believing-that is inseparable from a believing-in; that is, it is a matter of loyalty and trust. Therefore of it Husserl says: "The negation of faith is not merely false but also, and above everything else, sin, and fundamentally this negation is false because it is sinful not to believe in this matter" (A V 21, 4b).

In one form theology is an apologetic in that it seeks to resolve the opposition between "natural reason" and revelation and to justify the latter before the court of reason. Revelation may be referred to as knowledge stemming from a supernatural light whereas philosophy and science are knowledge originating in a merely natural light (A V 21, 3a). Natural reason has the status of theoretical truth which is prior to and existing along side of revelation. Natural reason is the center of the tradition of autonomous philosophy, which itself has the possibility of being a non-confessional, even atheistic way to God because it gives an account of the essential necessities of the world, especially in terms of its teleology. Natural reason itself, as exemplified in Aristotle, is a form of theology, a form of philosophical theology (cf. E III 10, 14b ff.; A VII 9, 20 ff.). But as apologetics natural reason and philosophy become handmaidens, organs or servants of a pre-philosophical organ of truth, faith. This pre-philosophical truth, e.g., through Christ, enters into the world at a particular time and place, e.g., in Graeco-Roman culture.

In the Kaizo essays Husserl speaks of this pre-philosophical approach to truth as "faith-experience." This would seem to refer to the discussion we reviewed earlier of experience led by an immediate intuition of values. Recall that this experience is an encounter in faith with an original tradition. This may be a matter of the tradition's representation of immediate uplifting experiences or of exemplarily good people, or it might involve the revelation occasioned by the idealizing and edifying work of a noble artist of the tradition.

It is clear in Husserl's writings that "faith-experience" is an experiencing proper to all cultures. Thus, though his examples are from the culture in which he is situated, it is not unique to the West. In the incorporation of natural reason into the faith experience, for, e.g., purposes of apologetics, the contents of faith themselves become thereby themes of theoretic judgments which follow upon faith but are not grounded in faith (Hua XXVII, 103-104). Husserl notes that faith, as in the faith-ex-

perience, thereby plays the role of *natural theoretical experience*. What
he has in mind is what we already saw in regard to himself: an
encounter with the figure of Christ in the Gospels and with the
perceptions of the early Christians as evident in the epistles, which then
goes on to inform, and be shaped by, everyday perception and life. The
figure of Christ and the perceptions of the early Christians constitute
what he calls an original tradition which awakens an intuitive empathic
faith. This, in turn, becomes the occasion for an original intuition of
values which guides experience, what he here is calling faith-experience.
This experience of value is original but not of the same kind as that of
the "founder" or the person religiously liberated from religion. The latter
is that upon which ultimately the "faith experience" of the contemporary
and subsequent follwers is based.

Faith, like experience properly understood, can be corrected, in part
through new experiences and in part mediately through thinking. The
contemporary faith-experience stands in a relationship of dialogue with
the first Christian communities and makes progress in getting to know
Christ. Here there is an enrichening and self-correcting process. But a
dogmatism and "routinization" builds up within the community which
eventually become a norm of faith and this dogmatized faith is what
characterizes tradition and is not the fruit of original faith-experience.

There thus occurs a tension between, on the one hand, the faith which
is a result of the interplay of the faith experience and rational reflection
which builds on this, and, on the other, dogma or the requirements of
the tradition. This demand of tradition acquires the status of a second
subsequent revelation. (Husserl is doubtless here thinking of types, such
as that of Roman Catholicism, where authorititative pronouncements of
the Church or pope approach the status of revelations.) Its objectional
feature is that it is an unintuitive dogmatization in which one believes,
but which lacks the original experiential character of authentic faith (Hua
XXVII, 104). Dogmatic faith in both the Christian and Jewish traditions
was inseparable from hierarchical states and a representation of an
essentially despotic God (Hua XXVII, 105).

Husserl, who on occasion referred to himself as an "undogmatic
Protestant" (Letter to Rudolf Otto), believed that the Protestantism of
the Reformation was a revolutionary breakthrough because it enabled an
original empathic faith-experience of the ancient Christian communities
and a disavowal of the tradition's dogmatism and its claims to be a
source of new revelations. We thus see, in terms of Husserl's broader

types, that the original Protestant Reformation, not the established forms of Protestantism, was a religious liberation from religion, and to that extent rational. The medieval Roman Catholic church represents rather cleanly the type "religion" as a stage of higher mythic culture.[19]

The medieval church is pervaded not only by traditionalism but also by an aspiration for a science that would be in essence theology. It no longer, however, wishes to be merely apologetic. Rather what is desired is a science of divine things and of humanity in relationship to God not only as these are experienced in faith but also as they are grasped in concepts. Now one desires a universal philosophic system, a universal theory of absolute reality and absolute norms of action presentable in an encompassing scientific theory, all of which are' related to the foundation which is revelation and its dogmas. Husserl refers to this as the ecclesial *imperium* and as the idea of *civitas dei*.

In the course of things this medieval science declared an increasing number of spheres of revealed dogmatic contents to be incomprehensible. Husserl goes on to note, perhaps with a touch of irony, that at the same time this science distinguished between the comprehensible and incomprehensible and taught how to grasp the incomprehensible in concepts and to pursue these incomprehensible matters in their deductive consequences and how they imply norms for a universal human praxis (Hua XXVII, 105). Husserl probably has in mind the Catholic theological teaching of revealed truths, e.g., the Trinity, which are essentially supernatural, i.e., essentially beyond the reach of reason and in need of the supernatural light of faith, in contrast to truths which are *de facto* revealed, e.g., the immortality of the soul, but which are not beyond the reach of the light of reason.[20]

On another occasion Husserl complicated this type of theology by calling attention to several senses in which a new sense of philosophy emerges and which changes the very concept of science. And he linked this to the names of Philo, Plotinus, Neo-Platonism, Neo-Pythagorean-

---

[19] Needless to say, Husserl overlooks the rich movements in Roman Catholicism for which the original authentic value-experiences were central, e.g., the mystical and reform movements. But these were, for the most part, indeed, reform movements which were not favorably received—at least in their own time.

[20] Husserl's meditations (in the 1920's and 1930's) on the relationship between theology and philosophy were perhaps in part occasioned by the religious turn that many of his students took or the theological interpretations of his writings by people like Eric Przywara and others involved in the Thomist renaissance.

ism, etc. Perhaps we could here add much of Eastern philosophy. Whereas the earlier Greek philosophers rested all foundation upon what is truly evident in experience, and this served as normative for all foundations, the new sense of philosophy rests either on 1) religious belief which is counterpoised to knowing and does not rest on evidence in the sense of what rises from experience or autonomous *sophia* but on revelation; or 2) it is not grounded on a positive belief in a revelation but on an intuition which enjoys an analogy with religious faith. In this case we approach not only what in this essay we have been calling faith-experience or the original intuition of values as guiding faith and reason, but also what Husserl elsewhere called "wisdom," and "Weltanschauungsphilosophie" (see the discussion below). Here the world is not submitted to a mythic hypothesis which one hopes subsequently to validate. Rather such an idea of science is no longer an ideal because it is no longer regarded as a genuine possibility; the dream (of such a science) is over. Finally there is another type 3) in which the genuine task of science is regarded as a possibility but the very possibility of this task as well as the validation of all principles of reason rests on revelation and faith. And not only do revelation and faith validate the principles of reason but they also limit it, for example, on the assumption that the divine might violate these principles, e.g., through miracles or by revealing truths which excede what can be grasped by experience and apriori principles. Clearly here Husserl has some sense of medieval science and scholastic thought in mind. What here is essential is that the rational in some way depends upon the irrational. (For these distinctions see A V 21, 21a ff.)

V. Eutopian Poetic Theology

Perhaps Husserl's most unique contribution to the study of religion is precisely the way in which he opens up a field for reflection on how the rational depends upon the irrational and how the rational requires a kind of irrationality for its sustenance. Indeed, Husserl's own attitude toward the just reviewed types of theology would not seem to be merely one of criticism. Recall the subtle strains of appreciation for world-views or "Weltanschauungsphilosophie" in the otherwise sharply critical (1911) *Logos* essay. There Husserl argued on behalf of philosophy as strict science; indeed he claimed it was the most radical necessity of life. All life was taking a position, and the position-taking stands under the norm of validation. The dilemma is that, on the one hand, the most radical need of life today is strict science which does not yield to the urgency of life's pressures and takes the time to pursue the timeless values of science, whereas on the other hand, life's pressures are such that we cannot wait on theories to legitimate our position-takings. While a good part of the essay is a sharp criticism of world-view philosophy because it confuses theoretical philosophy with world-views and neglects the strict scientific side of philosophy, Husserl must acknowledge the practical importance of world-view presentations. Further, as Boehm pointed out, philosophy as a strict science and, indeed, its fuller version of transcendental phenomenology are preceded by the awakening of ideas through world-views and ideals of such a philosophy; without these going in advance not even transcendental phenomenology would be possible.[21]

If this is true one might well wonder whether Husserl is not forced to reappraise tradition and to recognize the value of a "wise" tradition or one which embodies a "world-view" which points in the direction of *logos*, justice, and peace and fends off the deluge of violence and irrationality. He seemed to think this was the merit of the world-view of German Idealism. Of interest is that many aborigine societies, although without a passion for *logos*, have communitarian traditions which would seem to be much wiser than ours. In any case, the ideal of a "philosophic culture" is the ideal of a certain kind of disposition, therefore an ideal of a potentiality for authenticity. As such it is not active. Are active positive beliefs and practices which promote the ideals of reason, justice, community, and non-violence not needed? Husserl himself, at the end of his life, was forced to acknowledge that "the dream was over" for most of his culture regarding philosophy. Indeed, even in the early 1920's, i.e., prior to the madness of the Third Reich, he saw Europe at the mercy of *Realpolitiker*, capitalists, technocrats, and cynical academics for whom the abandonment of reason was the most obvious course (see, e.g., Hua XXVII, 117-118).

The belief in ideals, the adopting of a grand metaphor or world-view, belief in a religious revelation, etc., are all irrational at least in the sense

---

[21] Rudolf Boehm, "Husserl und der Klassische Idealismus," *Vom Gesichtspunkt der Phaenomenologie: Husserl-Studien* (the Hague: NIjhoff, 1968), 18 ff. In a later text (A V 21, 76a) Husserl is more explicit. People have to survive and act and cannot postpone decisions in the face of unresolved scientific queries. It is typical of everyday practical life that the individual must act in a context of what is scientifically unknown and unpredictable.

that the allegiance they occasion is not founded in original intuitive evidence. It is not evidence, not filled intentions, etc. which are given priority and go in advance but rather desire, will, wonder, trust, etc. It was increasingly evident to Husserl that these forms of "irrationality" were essential, perhaps dialectical, features of the meaning of rationality. The remainder of the paper aspires to synthesize his views on this matter.

In a marginal note to Heidegger's *Being and Time* Husserl noted that it was not death alone which is our ever insurmountable, fateful horizon. Rather we live in the midst and horizon of a *universum* of irrationalities and fatal events. Husserl approached the problem of the surds of life in the context of the teleological structure of consciousness in its theoretical-scientific aspirations as well as in its practical-eutopian nisus. Reflection on both reveals indeed how we live in a horizon of fate and surds. In both theory and praxis the precariousness of sustaining the "meaning of life" becomes evident. In both it becomes clear that being is not identifiable with reason and that the real is not the rational—and therefore, as Iso Kern astutely noted, ultimate philosophy cannot be first philosophy or essence-analysis of the transcendental phenomenological realm, but rather it must deal with the factual, contingent realm, of the world and, indeed, with the contingency of reason itself.[22]

Rational theory or pure science is sustained by necessity, universality, identity, sameness, predictability, non-contradiction, the fulfillment of expectations, theoretical hope of such fulfillment, etc. Indeed it is guided by the ideal of a divine knowing which is omniscient and, in so far as control of the theoretic conditions is a presupposition, omnipotent. It itself becomes a theme in opposition to the finitude and all too humanness of human knowing. Human knowing is surrounded by the universality of fate, by the universality of the endless and incaculable surds which destroy the very possibility of theoretical praxis. And, contrariwise, it is science which gives rational concepts and logical expression to the irrational and founds its logical consequences. (Cf. A V 21, 72b-77b.) In these respects rationality and irrationality are contrast concepts.

As questions are predelineations and pre-conceptualizations as well as hopes of answers, so the scientific enterprise faces the infinite ideal of the systematic fulfillment of the endless advance of ever novel questions

---

[22] Iso Kern, *Idee und Methode der Philosophie* (Berlin: de Gruyter, 1976), 342; cf. also "A Precis . . . ," 106 ff.

arising out of the endless succession of pieces of evidence within the total horizon. But between the question and answer, hypothesis and confirmation, etc. much can and does happen. Besides the recalcitrancy of the investigated matter to conceptual clarification, sickness, madness, cowardice, greed, and death, to name only the most obvious evils, can occasion that the initial motivation and legitimation of the inquiry be called into question.

Furthermore, science is capable of becoming distorted, as in modern absolutist mathematicized "scientism" and in the wedding of science to destructive technologies and economic systems. This, in turn, can breed a no less harmful reaction which can turn the only source of salvation into the appearance of a curse. That is to say, reason itself can seem to be madness, caprice, or absurdity.

The humanities as the human sciences which aspire to make sense out of individual and social life, presuppose a context of intelligibility among acting individuals. Human motivation manifests intelligibility when humans remain "normal." Yet it belongs to human life that it be governed by accidents. It is also proper that humans take them into account in the supervision of personal life, society and the world into which they are born. The context of intelligibility which the humanities studies itself is a fact which presupposes everywhere irrationalities of accident, fate, madness, etc. Husserl claims that death, whether the death of oneself, one's society, or of humanity, as the always unintelligible yet necessarily approaching certainty,[23] is that which can call into question the meaning of life itself because it undermines all our attempts to oversee life and regulate its contingency (A V 21, 79b).

A typical and perhaps most obvious formulation of the central issue of the challenge of the surds to rational life is the following text:

> In considering the *Universum* and the universality of contingencies, I can only live, I can only take upon myself a life which is riddled with sin, error, absence of actual and universal values when I believe that everything ultimately serves the good and that each radical willing of the best really serves to the good of the universe, that my free will makes a difference, etc. Only after the presupposition of this faith does my life

---

[23] Husserl here must be speaking of the proper sense of death, i.e., the cessation of personal identity; the transcendental "I," in the most basic sense, neither comes to be nor passes away. Cf. my discussion in *Time and Religion*, edited by J. N. Mohanty and A. N. Balslev (Leiden: Brill, 1992).

gain a purpose and can it maintain itself rationally and acquire zest and
a necessarily increasing value (A V 21, 25a).[24]

Indeed, Husserl adds that only with this experience of purposeful striving
is there an experience of teleology and the possibility of an argument
about the teleology of the universe. And if we think of this teleology as
the divine holding sway, it must be said that "in order to know God's
holding sway I must first believe in God" (A V 21, 25a). Now although
the very structure of intentionality, as an ongoing interplay and surfacing
of empty and filled intentions, might be said to be teleological, and
perhaps at some level this interplay and teleology are irrepressible and
not able to be called in question, still at the level of founded acts of
intentionality and at the level of the more discrete blocks of life
dysteleology and irrationality are exceedingly evident.[25]

Thus some sense of faith in reason functions at all levels of consciousness as a prior condition to reason's achievements (see, e.g., E III 4,
24b). And at the higher, founded levels of meaning-units where the surds
most dramatically announce themselves and where the agony of holding
the mind together approaches despair, Husserl advocates a kind of will
to believe and poetics of edification of faith in reason.

For this reason the theory of postulates was esteemed Kant's greatest
theory[26] and religion itself, in so far as it provides "belief in the power
of the good in the world" and "reveals in us teleology as something

---

[24] Whether Husserl ultimately believed that action was possible only if
"everything ultiamtely serves the good . . . ," etc. is not clear to me. I think his
meditation on irrationalities as well as his metaphysics point to a more Platonic
view, i.e., that the universe is rational *for the most part*, but the receptacle and/or
hyle are eternal and never become perfectly transparent to form. Furthermore, late in
his career, he raised the issue of whether there may not be values which we are
forced to sacrifice and which remain valid and are not harmonized by that for which
they are sacrificed. See A V 21, 80b ff.

[25] We may note, however, as the most basic theme of his philosophical
theology, that at the irrepressible level of proto-reason and teleology at the
foundation of consciousness, i.e., the doctrine of association in the awareness of
inner-time, Husserl is motivated to see a divine entelechy at the heart of this hyletic
facticity which accounts for how the *propter hoc* trustfully rides on the *post hoc*. See
my "A Precis . . . ." Cf. also "Entelechy in Transcendental Phenomenology: A
Sketch of the Foundations of Husserlian Metaphysics," *American Catholic Philosophical Quarterly* Vol. LXVI, No. 2 (1992).

[26] See Iso Kern, *Husserl und Kant* (The Hague: Nijhoff, 1968), 302.

steadfast which holds sway . . . through sin and error . . . ,"[27] may be esteemed as an adumbration of *Verunftglaube* or faith in reason.

If we take faith in the wide sense of naively taking something for granted (*naïve Bodenständigkeit*), faith in reason is that which establishes itself with the event of Greek philosophy and science. Religious faith, in its turn provides a world in which the believer is constantly able to affirm meaning. The surds which occur in this world are to be dealt with as something for which the believers along with their contemporaries know themselves responsible. With the emergence of faith in reason there is a new principle: No longer is the authoritative revelation of "positive religion" the source of meaning and the foundation of the teleology of life and history; rather, faith in reason itself becomes the principle and it believes itself called to give meaning to God and the world in an autonomous and responsible way (E III 4, 38b). Yet the possiblity of "faith in reason" is not always evident. *Tod und Teufel*, to use, with Husserl, the figures of Dürer's engraving (*Ritter, Tod und Teufel*—see, e.g., E III 4, 16a) are never far away. And today, as in the last decades of Husserl's life, there is a universal disenchantment with faith in reason.

There thus emerges a "poetics," a "pragmatics," a *Dichtung, Roman*, etc. which sustain and vivify the principle of "faith in reason." And thus Husserl's rather extensive meditations on the surds or irrationalities of life, both in the theoretical and practical dimensions, lead to a kind of theology which might be called "eutopian poetics" or "eutopian eidetics." Its precursor is, of course, Kant. Kant argued that, on the one hand, the project of working out a universal history of the world, which, in accordance with a plan of nature, aimed at a perfect unity of humanity, must be considered as possible if we are going to act at all. But, on the other hand, he noted that the writing of a history of how events must develop if they are to conform to rational goals of humanity could only take the form of a "novel" (*Roman*). He adds that even if evidence is lacking that nature is teleological in a way conducive to history, the fiction of a course of events *as if* this were the case is a way of constituting a horizon of hope which nurtures action and virtue.[28]

---

[27] See, e.g., Hua XIII, 508 and E III 10, 15b.

[28] See the Ninth Proposition of "The Idea of a Universal History with a Cosmopolitan Purpose," in *Kant's Political Writings*, ed. Hans Reiss (Cambridge: Cambridge University Press, 1970), 51 ff.

Husserl himself occasionally refers to *der Roman der Geschichte*. Generally the context of these references is the need to nourish the practical horizon of action and philosophy. The motivational basis for both theory and practice is the attractivenes and realizability of the "approximation" of their infinite ideas. Husserl's speculations on the development of religion out of a prior mythic stage, and the development of the liberation from religion, i.e., the story of the turn from *Mythos* to *Logos* or to original "authentic" religion, themselves are examples of the kind of poetic work (*Dichtung*) which he believes is necessary in order to create order amidst the infinite manifold of the past and to show how the *telos* is functioning, even if in a hidden way, from the start.

But this creative poetics is not limited to the securing of a teleological history of *logos* and philosophy. Rather it has its place also in the realization of action as well as in the practice of theory. The fundamental consideration for the edifying poetics is its nurturing relationship to *logos* and the infinite ideals of the philosophic life and the absolute ought. Recall that for Husserl life is teleological. As theory it heads, in the form of a research community, toward the ideal of transcendental phenomenology; as a practical community it heads toward a blessed community of monads constituting a godly person of a higher order. The infinite ideals are not among the constitutive ingredients of experiences, e.g., the essences, syncategorematicals, etc. Rather they provide the motor for the mind's wakefulness in its dealing with bodies, Others, and the self in relation to these. They provide the mind with ultimate infinite tasks and an infinity of profiles of this task. As *unendliche Aufgaben* they are not given (*gegeben*) in any action or experience but action and experience are infinitely delivered up (*augegeben*) to the direction provided by them and thereby are set in motion on an infinite journey. The sense in which the infinite idea is regulative, i.e., provides a rule, is that it invites the mind to go on indefinitely in its determination and realization of it.[29] Neither Kant nor Husserl regarded the infinite task to be the curse of Sysiphus. Why this is so cannot busy us here. But we may note that Husserl thought of a divine life as the infinite progress in the realization of values. And he even thought of the best practical good in terms of

---

[29] See Kant, KRV, B670 ff. and Hermann Cohen's synthesis in his *Kommentar zu Immanuel Kants Kritik der reinen Vernunft* (Leipzig: Duerr'schen Buchhandlung, 1907), 157 ff. Cf. Husserl, e.g., Hua III, §§ 83, 143, and 149; also, A V 22, 31-38, E III 4, 61; Hua VIII, 10-16, 33, 48-50.

the progress of the infinite realization of the infinite practical good (see Hua VIII, 350). The expectation from the pursuit of these infinite ideals is a life of blessedness. "Blessedness" is the joy of being true to oneself in pursuing the absolute ought. Such a life contrasts with happiness as the fulfillment of all one's hopes. Happiness is not to be expected in life.[30]

One's loved ones die, friends can go mad, crippling sickness can set in, death occurs, not only for oneself, but for the community whose hopes are inseparable from one's own. And science predicts the extinction of the sun. One must then ask: How can one preserve oneself? How can hope stay alive, when everywhere dire need, fate, death, etc. can and often do break in, not only into the lives of individuals but also into the life of the community and humanity at large? Is the hope in an infinite horizon of a human future possible when we believe that any work we might do will eventually vanish into nothingness. Is such a belief not madness? Is it not quite possible that humanity spurn what is true and sink to a kind of bestiality or even worse forms of degradation? Is the open infinity of the community not a capricious assumption, given that it is probable that the earth is heading toward a cataclysm? (This is a paraphrase of a paragraph of A V 21, 90b.)

Husserl's responses to such agonizing meditations take the form of proposals to hold open the horizon of real possibility for immediate agency and reflection. In the worst-case scenario, i.e., where universal doom seems certain, he offered this resolution to the task of the absolute ought:

> The practical universal goal, which up till now I regarded as the goal of a genuine life of humanity, I now recognize as illusion—and yet: I hold firmly to being genuine, I want to be true to myself, and that means: I will so live as if the goal were still a practical possibility. It can no longer be my goal in its infinity. But I am and we are; and in the actuality of our life there remains the life-horizon as legitimate, if nevertheless indeterminate, anticipation. Therein we have a stretch of vital human development and to this we dedicate in love our strength, as far as it reaches, to know as a practical possibility and in accord with this knowledge to consciously effect as a practical possibility (A V 21, 91b).

---

[30] I discuss these themes in *The Person and the Common Life: Studies in a Husserlian Social Ethics*.

To this Husserl immediately adds: "But here I have neglected all motives for the question of God; and the question arises whether compelling motives do not here arise?" And it is clear that they do in other passages (e.g., Hua VIII, 258); and in other places religion is praised as securing steadfast a belief in teleology, come what may. I will come back to this.

Another text, more sanguine and Jamesian in spirit, argues that in the absence of evidence that any attempt at ameliorization is bound to fail, and so long as a case can be made that the pursuit of what is great and beautful can be successful, a creative self-displacing into a horizon nurturing hope is in order:

> I will do best to overestimate the probabilities and to act as if I was certain that fate was not essentially hostile to humanity and as if I could be certain that through persevering I could ultimately attain something so good that I could be satisfied with my perseverance. What is theoretically reprehensible, i.e., the overestimation of probabilities or of what is only slightly likely at the expense of empirical certainty, is practically good and required in the practical situation (F I 24, 88b).

Perhaps the most sustained meditation on these matters is Beilage V of Hua VIII. Here again we find the thesis that what holds open the world within which action and theory is founded is not itself the fruit of a prior achieved evidence but both a passive-synthetic faith and trust as well as an active faith in reason. This latter is coincident with a poetic-pragmatic postulate which holds open the beauty of the Idea (of the Good and True) and sustains the will to strive for this ideal. In terms of phenomenological detail the horizon which is coincident with the hopeful will must be pervaded by real possibility, i.e., be evident as determinable. If the horizon is pre-delineated with doom, the real possibility of the futural horizon is closed; it, although lacking the determinateness of focal objects functions as determinate and one does not experience actuality surrounded by determinability and therefore one's "I-can," one's elemental sense of will and freedom, the correlate of the determinable horizon, is lamed. The life of blessedness requires progression in the creation of values and one can only so live as long as the future offers a promise. But if that promise is not forthcoming, then what?

> As long as I have an open practical horizon for which no termination is definitely predelineated, and so long as I have given to me a recognized realizable value—even if it be merely in a vague presumptive mode of givenness—which presumably can lead to new practical values in the direction of the best possible or the absolutely binding, I have the duty of acting . . . . When I believe [in the practical realizability of the *telos* of theory and practice] and make myself aware of this belief, when I freely perform this belief out of this practical source, there is given meaning to the world and my life; there is given also a joyful confidence that nothing is in vain and that all is to the good (Hua VIII, 351 and 355).

But what if this faith is not forthcoming? And whence arises this source for this free act? Faith is not a matter of caprice or will-power. I can neither force myself to believe nor can I modalize at will the peculiar certainty of belief (Hua VIII, 368). In the case of belief in life's meaningfulness, the practical as-if, the energetic pursuit of the absolute ought in the face of *Tod und Teufel*, etc. the springs of faith cannot be from a logical necessity, but they must come from some other source. But is the necessity of being true to oneself a sufficient source?

It would seem that Husserl moves in the direction of holding that when confronted with the theme of surds and irrationalities then ultimately belief in the meaning of the world, in the sense of the pursuit of a meaningful life of theory and action, is a matter of grace.[31] The sense of grace here is quite unspecific and encompassing. It includes, as we have seen and as his diary from 1906-1908 indicates (Hua XXIV, 427), help from people with "big souls," the encounter with the Gospel presentation of Jesus, the acquaintance with people who are good and decent, "noble artists" who enflame our hearts, etc. In short whatever awakens us to, and holds open, the Idea, that can be regarded as "the grace of God."

> The world "strives" toward absolute goals, values, and it prepares the way for them in the hearts of humans. Humans can realize a divine world (*Gotteswelt*) in their freedom [which itself], of course, [is] motivated and disposed by godly grace, and thereby are they able to

---

[31] This parallels the basic theme of Husserl's philosophical theology, the *Wunder* of reason. See, e.g., Hua VII, 394.

strive for this godly world in highest awareness and will-power (Hua VIII, 258).

> I do not believe out of caprice but I believe from the necessity of being I and a member of humanity and being in regard to my actual surrounding a benificent agent. I can do no other than believe and in this disclosure of myself and the world to believe universally. Belief is the power of God. As long as I live in faith and in the direction of my calling there lives in me God's power (A V 21, 98b).

In short, the thematization of the contingency of rationality in life sets off motivation for a kind of eutopian, theological poetics and pragmatics.

What form this poetics will take is various. As we have already seen, Husserl believes the fictional presentation, in e.g., "idealistic art," of alternative eutopian human possibilities has the capacity for a transformative revelation. But there can well be a more explicit theological function. Aside from the grounds transcendental phenomenology has for the necessity of some sense of God in accounting for the foundations of consciousness, the reality of some sense of God seems eminently desireable from the meditation on the tensions between the heart and the mind and *logos* and *anangke*. The actuality of some sense of the divine provides the basis for faith which is "the Infinite Yes which overpowers the infinite No" (A V 21, 98a).

Thus faith and eutopian poetics provides a negative answer to the question of "whether the things we care for most are at the mercy of the things we care for least" (W.P. Montague). Faith in God, in the face of *Tod und Teufel*, need be neither spinelessness nor mushy sentimentality. If God be envisaged as what ultimately supports that which sustains the faith in our goals and striving, and therefore our striving, and as what preserves the achievements of this striving, it is a contradiction to desire that God not be.

Chapter 12

Biography as a Cultural Discipline

Mano Daniel
Florida Atlantic University

Abstract: *The biographical subject is the cultural object par excellence. By examining how biographers go about the task of studying and assembling their biographical texts, we will acquire a better appreciation of the nature of cultural objects and the appropriate philosophical and methodological strategies that facilitate the description and explanation of such objects. I begin by reflecting upon the practice of biographical writing in order to illuminate and explicate some of the salient features and theoretical concerns of biographical writing as experienced by practicing biographers. I then canvass and discuss ways in which biographies play relevant methodological and cultural roles.*

There is little systematic philosophical reflection on the practice of biographical writing, or life-writing, even though, as Jeffery Meyers reports, it is one of the "major literary genres of the twentieth century." This is surprising since the practice's popularity and pervasiveness has had a profound intellectual and theoretical impact. Biographies make up an significant proportion of books published in the humanities, social sciences and even in the history of science where they function not simply as works of reference, but as an important consequence of, and resource for, cultural research.

This paucity of theoretical reflection may be engendered by the polymorphous nature of the practice itself, which is more a site or region—"bounded on the north by history, on the south by fiction, on the east by obituary, and on the west by tedium"[1]—in which a cluster

---

[1] Michael Holroyd, "Literary and Historical Biography," in *New Directions in Biography*, edited by Anthony M. Friedson (Hawaii: University of Hawaii Press, 1979), 23.

M. Daniel and L. Embree (eds.), Phenomenology of the Cultural Disciplines, 297–317.

of assumptions are employed and techniques deployed to account for its object rather than a specific mode of inquiry with a specific methodology. In an age of academic specialization, the practice of biography exceeds traditional academic categories or classifications. Perceived as an anomaly, it is shunted to the periphery of theoretical attention. Biography is "generally considered a chameleon form: history, literary criticism, or what you read at night, and there seemed little need to go any further, theoretically then that."[2]

This unreflective stance is emblematic of the extent to which biography has become such an accepted phenomenon that its structure, rather than revealing the fundamental problems that it is trying to overcome, and the aims it is trying to accomplish, is seen as a transparent medium that provides straightforward answers to unambiguous questions. The ease with which practitioners and readers of biographies treat the practice as a innocuous medium is indicative of a certain myopia to the procedures and implicit assumptions of the discursive practice which are ignored or glossed. For instance, historians and literary theorists appear more concerned with claiming or reclaiming the practice as being within their respective domains than with exploring the contours of the practice in its own right. They thus often fail to appreciate the practice's tendency to interlace techniques from both. This defect is compounded by the practice's inherently fungible nature and receptivity to methods and research strategies developed in other theoretical endeavors in its task of inscribing a body of experiences connected to the life of a given individual. Hence, the healthy self-reflexivity which scrutinizes critically the practice's implicit and imported assumptions and procedures is done sporadically, and often in isolation. Faith in the veracity and accuracy of biography is thus often not warranted and the practice is shrouded by a shadow of suspicion.

This inattentiveness has also hindered the recognition of the practice as an eminent cultural discipline. After all, the terrain occupied by the

---

[2] Katherine Frank, "Writing Lives: Theory and Practice in Literary Biography," *Genre* 13 (Winter 1980), 499. Similarly, Leon Edel, in the introduction to the last volume of his *Henry James* writes: "I believe biography to be the most taken for granted—and the least discussed—of all the branches of literature. Biographies are widely read, but they are treated as if they came ready-made. . . . biographies are accepted as they come and relished for their revelations. . . . [But] questions of form, composition, structure are seldom raised." Leon Edel, *Henry James: The Master: 1901-1916* (New York: Lippincott, 1972), 19-20.

practice of biography is at the same time that of the socio-historical-cultural lived world. The biographer proceeds on the assumption that the biographee possesses a "personal life-history," which is a cultural rather than a 'natural' or biological concept. Only animate beings endowed with a historical consciousness and a sense of temporal continuity can have a life history that they, and others, can recognize as a unique and irreducible destiny. While we sometimes speak of a concept, a work of art, or even an institution, such as the judiciary system, as having a life-history, such usage is analogical, metaphorical or derivative. In fact, the biographical subject is the cultural object par excellence; that is, a product of a historical, cultural situation who nevertheless has an important say in how that life is to be lived. Accordingly, by reflecting on the procedures adopted and adapted for the study of persons, we are better able to appreciate and delineate the compass of the cultural world and the adequacy and limitations of inquiry directed at it.

The situation, however, is not totally bleak. The literary biographer Leon Edel has heralded the challenge of making biography "declare itself and its principles"[3]; that is, to confront, reform and transform the practice so that it may be made theoretically rigorous and methodologically fecund. Although Edel was focussed primarily on the practice of literary biography, he believed that the consequences of his analyses extended to other forms of the practice.[4] Edel's call for reform has been heeded. There is even an interdisciplinary quarterly, *Biography*, devoted to the subject that is published out of Hawaii. Even so, few philosophers have entered the fray.

My aim in this essay is to step into the breach and I see my task as two-fold. First, by reflecting on the practice, as practiced, I hope to delineate some of its distinctive features as well as to isolate and discuss some of the key assumptions that inform its aims and strategies. Accordingly, in the first section of this essay, I rehearse some of the salient methodological and practical difficulties that are encountered in the practice as well a number of the strategies offered to overcome or

---

[3] Leon Edel, "Biography and the Sciences of Man," in *New Directions in Biography*, edited by Anthony M. Friedson, (Hawaii: University of Hawaii Press, 1981), 5.

[4] Leon Edel, *Literary Biography* (Bloomington: Indiana University Press, 1973), 2 and Leon Edel, "The Biographer and Psycho-Analysis," *New World Writing*, edited by Stewart Richardson and Corlies M. Smith (Philadelphia: J. B. Lippincott Company), 52.

circumvent these difficulties. Biographies, as the product of the biographical process, also function as cultural objects. In the second section, I canvass a number of different accounts where biographies as cultural objects have been harnessed for methodological, heuristic, practical and political purposes. The link between the two parts is the claim that the practice of biography serves as a nexus of techniques aimed at the explication of the biographee as object, and, by understanding the context and circumstances that have led to the adoption of these techniques, we will be in a better position not only to provide a robust theoretical foundation for the practice, but also to appreciate the role of biography in the exploration and study of aspects of culture in general.

This essay is propaedeutic and hence necessarily suggestive, explorative and tentative. My aim is less to forward a theory of biography than to focus attention on the central yet polymorphous deployment of biography and to suggest that it should be considered a discipline in its own right by advocating the advent of a philosophy of biography devoted to such an inquiry. (Of course, to write biography does not require a philosophy of biography, although philosophy is indispensable for producing a philosophy of biography.)

I. Reflecting on the Practice of Biography

A biography, typically, is a presentation in words of a specific life—the "life of one individual who actually existed at a historically delineated moment in time."[5] As James Clifford puts it "biography contracts to deliver a self."[6] There are, of course, many kinds of biographies; for example, literary biographies (i.e., biographies of writers), cliobiographies (i.e., biographies of historians), political biographies, religious biographies, etc. Traditionally, biographies were written about exceptional individuals drawn from the upper strata or mainstream of society. This was especially true of Victorian biographers that for the most part picked their subjects because of the latter's public benefaction or excellence. This motive

---

[5] Michael Scriven, "Sartre on Flaubert: Problems of Biography." *Degré Second-* 2 (1978), 217. In order to distinguish biography from autobiography, it is further necessary to specify that the biography be written by an individual who is not the main subject of the biography.

[6] James Clifford, "Hang Up Looking Glasses at Odd Corners: Ethnobiographical Prospects," in *Studies in Biography*, edited by Daniel Aaron (Cambridge, M.A.: Harvard University Press, 1978), 44.

persists today. Phyllis Rose points out, however, that it can often embody an ideological bias which functions as "a tool by which the dominant society reinforces its values" and, hence, as a genre "it is much more elitist then the novel."[7] Spurred by egalitarian and pluralistic motivations, this traditional catchment for biographies has, in recent years, been extended to included "marginalized" individuals such as women and those of ethnic backgrounds who have led interesting or unusual lives or who are viewed as representative of a particular group's experiences. There have even been attempts to extend the genre by the writing of group biographies and, to use biographical techniques to describe the profile of a type. (For example, one can view Sartre's description of anti-Semitism as an attempt to provide a composite profile of the anti-semitic personality.[8])

The wide spectrum of biography is a consequence of its long and variegated history. Its origins lie in the hieroglyphics of Egypt and the fragments of early Greek literature but it only in Rome during the first century A. D. that it began to exhibit its distinctive form, and develop as a professional endeavour. Initially, biography was viewed either as a didactic endeavour that provided exemplars or models of moral virtue (Plutarch) or as an attempt to capture the complex and comprehensive portrait of a subject's life (Suetonius). During its Medieval incarnation the biographical task assumed the form of ecclesiastical exhortation, i.e., hagiography, and to a lesser extent, political encomium. The concern for individuals and the renewed interest in history, biography and autobiography that begun during the Renaissance flourished during the eighteenth and nineteenth centuries as represented by the writings of Sir Thomas More, William Roper and George Cavendish. It led Samuel Taylor Coleridge, in 1810, to coin the phrase "the age of personality"[9] to signal the historical and cultural context of individualism (a nineteenth century term) in which modern biography, in contradistinction to

---

[7] Phyllis Rose, "Fact and Fiction in Biography," in *Writing of Women: Essays in a Renaissance* (Middleton, Connecticut: Wesleyan University Press, 1985), 68.

[8] cf. Jean-Paul Sartre, *Anti-Semite and Jew*, translated by George J. Becker (New York: Grove Press), 1960.

[9] Samuel Taylor Coleridge, "A Prefatory Observation on Modern Biography," *The Friend*, January 25, 1810, 338-39.

hagiography, was beginning to assert itself.[10] The Victorian version was supplanted eventually in the early part of this century by the appearance, in large part influenced by the biographies of Lytton Strachey and Froude, of what has come to be known as "New" or "Modern" biography.

Modern literary biography's emergence in the 1920's can be attributed to the interest generated by Harold Nicolson, the vision and work of Virginia Woolf and advances in scientific, philosophical, psychological and, especially, psychoanalytic techniques. Its distinctiveness was marked by a shift in focus from an exaltation of the individual to an intimate act of critical evaluation performed through examining, describing and explaining the "inner," or "interiority" of the individual and its relation to her manifested actions and deeds. This new form of humanistic expression was heavily influenced by historiographical advances in the nineteenth century and by the assumptions and devices utilized with the advent and prominence of novels. As Edel points out, the "modern biography is as modern as the novel."[11]

Since a comprehensive treatment of biography is beyond the scope of this essay, I will, following Edel, focus on the practice of "modern" biography. As a product of advances procured by the age of Enlightenment and in the social disciplines inspired by its wake, it provides a good glimpse of the discursive practice's ability to "present a unified life, . . . reveal this unity with specific anecdotal evidence, and . . . demonstrate change, development, and/or growth with the passage of time."[12] It is thus an exemplary embodiment of many of the theoretical and cultural assumptions that undergird the study of the practical/cultural world.

As an act of reconstitution or resuscitation, a biography needs to be historically accurate, factually credible and internally coherent. The biography may aim at a sense of coherence either by a factual pattern arranged in terms of a chronological axis or by an interpretive pattern based upon a sense of the inner life of the subject. It should reflect the unrelenting veracity of the biographer to mold the "granite-like solidity" of truth and "the rainbow-like intangibility" of personality "into one

---

[10] Virginia Woolf, "The Art of Biography," in *Biography Past and Present: Selections and Critical Essays*, edited by William H. Davenport and Ben Siegal (New York: Charles Scribner's, 1965), 165.

[11] Edel, *Literary Biography*, 5.

[12] Peter Nagourney, "Literary Biography," *Biography* 1.2 (1978), 88.

seamless whole"—to present the "fertile fact; the fact that suggests and engenders."[13] Put differently, the task is to describe "who" the subject is rather than simply "what" she was.

If biography is conceived as a search for a historically verifiable truth, then it is a historiographical problem that typically attempts to recreate or reconstruct a person in terms of acts and events. The historian qua historian views the biographical subject as the sum total of her actions and deeds in the context of her historical situation. If biography is also seen as an attempt at understanding the biographee's personality, or "interiority," it becomes an interpretive problem. It will involve adopting assumptions about how overt behavior is to be understand and explained in terms of talk about motives, causes, intentions and projects. It becomes an artistic problem as well to the extent that it is seen as an attempt to portray creatively the personality as an integral unity. Moreover, it will involve the use of the tropes of metaphor and metonymy and such notions as proportion and unity. While each task can be seen as independent of each other, most biographies are a product of all three tasks in various degrees of attentiveness.

Typically, since there is no direct access to the biographical subject, the biographer proceeds indirectly. This process can be viewed as a four-fold.[14] First, it would require a fact-finding enterprise that catalogues and chronicles the major events of the subject through the preparation of a morphology of that life by delving into the documentary evidence, literary artifacts, and other anecdotal material. Biographical material falls into two main categories: primary and secondary. Primary materials relate to the documents that originate during the lifetime of the biographee and are written by, to, and about him or her. Secondary documents consist of background materials that help flesh out the context and period of which the life took place.

Second, it would require an investigation of the psychological make up of the subject and the articulation of the manner by which the subject understood and acted in her world. This will often involve having to navigate elusive labyrinths of personal disguise, by masks and personae,

---

[13] Virginia Woolf, "The Art of Biography," 171.

[14] Cf. Gail Porter Mandell, *Life into Art: Conversations with Seven Contemporary Biographers* (Fayetteville: The University of Arkansas Press, 1991) and *The Craft of Literary Biography*, edited by Jeffrey Meyers (New York: Schocken Books, 1985).

or chosen obscurity. Third, it would require a *Zeitgeist* exploration that tries to incorporate the historical, cultural, economic, etc., contexts as they pertain to the subject. In order to appreciate the historical context the biographer must often develop a professional competency in the era in which his subject lived and died and thus requires exhaustive, painstaking research in the pertinent archival and printed sources. Finally, it would require an articulation of existential statements that helped shape and explain the subject's individual, idiosyncratic characteristics.

One of the clearest and most perspicuous attempts at forwarding practical dictums that can be of use to the biographer in the exercise of his craft is offered by Edel who, influenced by his preference for psychoanalytic techniques, offers four principles. One, "the biographer must learn to understand man's ways of dreaming, thinking and using his fancy." Two, "the biographer must struggle constantly not to be taken over by their subjects, or to fall in love with them. The secret to this rule is to learn to be a participant-observer." Three, it is incumbent upon the biographer to "discover certain keys to the deeper truths of his subject—keys . . . to the private mythology of the individual." Four, since "[e]very life takes its own form," a "biographer must find the ideal and unique literary form that will express it."[15]

Only after the historical documentation has been collected, collated and analyzed can the biographer undertake the task of reconstruction. Only subsequently can the biographer recognize patterns emerging from the life of the subject which can then be used to structure the text. This retelling is highly selective, and imaginative. As Edel puts it,

> the biographer is allowed to be as imaginative as he pleases, so long as he does not imagine his facts. Saturated with facts, he may allow himself all the adventures of literary artifice, all the gratifications of story-telling—save those of make-believe.[16]

Biographers are not simply fact-gatherers. They must also interpret, judge and present the material. The writing of a biography is necessarily not just exigetical, but often isogetical. A biographer, fettered by fact, still invents her form and, through discursive techniques, such as the use of the narrative, metaphor, metonymy, flash-back scenes etc., and other dra-

---

[15] Leon Edel, "Biography and the Science of Man," 8-10.
[16] Leon Edel, "Biography: A Manifesto," *Biography* 1.1 (1978), 1.

matic techniques, directs the reader's impressions, images and interpretation of the subject. Such transformations, caused by the exigencies of language and the act of composition, ideally alter the shape but not the legitimacy of fact. As Desmond MacCarthy's famous dictum puts its, "A Biographer is an artist under oath." Biography, aesthetically speaking, is guided by the twin impulses of realism and idealization, although it is necessarily more conservative than fiction since its task is primarily one of description and explanation.

Hence, it is important to recognize the interpretive nature of the enterprise. A biography is a structured narrative that utilizes figurative language to organize, chronicle and "emplot" the story that constitutes a life history. This occurs through uniting discrete facts of the life with certain modes of plot structure so that the parts form a new whole identified as 'story'. Since biography uses figurative language and tropes as a means for representing or expressing a life, questions can be directed at the viability, reliability and veracity of these tropes. As such, biography is not an "innocent" form but a concerted strategy to achieve a particular end. As these strategies are a product of methodological and philosophical assumptions, the claim that a life can be captured and represented in a biography must be scrutinized and circumscribed in light of the techniques that contribute to its form.

Practicing biographers have been very sensitive to the methodological problems implicit in it's practice and have been actively engaged in attempts to formulate a Linnaean classification of biography. Harold Nicolson, whose *The Development of English Biography* is probably the most rigorous attempt to arrive at a definition of "the elastic category" of biography, classifies biography as either "pure" or "impure." The main causes that contaminate a biography are "the desire to celebrate the dead," "the desire to compose the life of an individual as an illustration of some extraneous theory or conception," and the "undue subjectivity in the writer."[17] Paul Murray Kendall, who notes that "modern biographies display an infinity of gradations," proceeds to differentiate among "eight perceptible types": from "the radical left appears the novel-as-biography, almost wholly imaginary . . . to works of such high specific gravity that they are little more than compilations of source-materials. Hovering above the center of the scale appears the radiant-plumaged "super-biography,"

---

[17] Harold Nicolson, *The Development of English Biography* (London: The Hogarth Press, 1968), 9-10.

which seeks to be both ultimately literary and ultimately scientific.[18] James
L. Clifford, in his admirable *From Puzzles to Portraits*, suggests "a series
of five categories for biography"—"objective," "scholarly-historical," "artistic-scholarly," "narrative," and finally "subjective"—although he makes
clear that his preference is for "artistic-scholarly,"[19] which he takes to
represent the happy medium between historical science and literary art.

The classification systems advanced above emphasize a methodological
median that biographies ought to adopt. Biographers acknowledge that
there is an irreducible subjective or interpretive element both in the
subject matter and in the production of biographies. This recognition of
irreducible subjectivity has manifested itself in the internecine battle to
determine whether biography is an "art," or "craft" or a "science." Woolf,
for example, advocates a melioristic position between these antipodean
options by arguing that the biographer is "a craftsman, not an artist; and
his work is not a work of art, but something betwixt and between."[20]

This interpretive element is also evident in the production of
biographies since biographers have agendas that influence their choice
and decision of subject, the angle that they adopt and the conventional
form in which they choose to depict their stories. Such productions will
also involve the adoption of philosophical assumptions about the nature
and viability of their enterprise. One particularly tantalizing and seductive
temptation is to adopt what Young-Bruehl calls the "essentialist assumption"—"the temptation to try to capture the subject as the subject really
was, to catch the subject in moment of truth, to reveal what was hidden
even from those close to the event, even from the subject himself or
herself."[21] This assumption can manifest itself in a number of different
forms. In the majority of academic biographies, the attempt is to discover
the key theoretical concern or the object of the subject's intellectual or
spiritual quest. Once located and an account advanced to explain the
adoption of this pivotal insight or central principle, the biographer then

---

[18] Paul Murray Kendall, *The Art of Biography* (New York and London:
Garland Publishing, 1985), 126-127.

[19] James L. Clifford, *From Puzzles to Portraits* (Chapel Hill: The University of
North Carolina Press, 1970), 83-89.

[20] Virginia Woolf, "The Art of Biography," 170.

[21] Elisabeth Young-Bruehl, "The Writing of Biography," in *Mind and Body
Politic* (New York: Routledge, 1989), 125. The following classification of the forms
of biographies and parts of its description are borrowed from this essay.

uses it to explain the subject's various theoretical inquiries and intellectual development. As such, it is an attempt to unite the "Life and Thought" of the subject into a coherent integral whole. The second form is also guided by the need for synthesis or coherence. It, however, stresses the strong traits of character, or dispositions, and seeks to demonstrate how they harmonize into a whole, or why they clash. It frequently involves a novelistic sketch of the subject doing what the subject essentially did. This is biography as personality portraiture.

The other two forms attempt to exhibit the spectrum of a life through the prism of a part; that is, such biographies are synecdochical rather than synthetic. The third form, favored by psychobiographers, attempt to represent the subject as essentially some deed, event, or crisis, that provides the key to understanding the life; i.e., the establishment of what Edel calls the "life-myth," or pivotal episode that structures the bio-graphee's often implicit sense of identity. Sartre, for example, in his biography of Jean Genet, locates this mythical key in a boyhood episode of Genet in which the child was labelled a thief. For Sartre, thiefhood became a sort of triumph that Genet incorporated as his identity, which he proceeded to champion and cherish. By construing Genet as transfixed by this childhood memory, in which a child is extinguished and a thug rises from its ashes, Sartre portrays Genet as a dead man suitable for biographical framing. (Note, although I have used Sartre's biography as an instance of this form of writing biography, it can be argued, per-suasively I believe, that Sartre's other philosophical commitments prevent him from succumbing to this essentialist assumption.)

The fourth form, which appears to be in decline, attempts to charac-terize the subject as a representative of an age, or some ideal. The subject is presented as a symbol, or an embodiment of something larger, as vestige of the past, a symptom of the present or a harbinger of the future or, personified as Liberty, Wisdom or Justice. The attempt here is to present the subject of this sort of biographical portraiture within a panoramic painting.

The demand to deliver a comprehensive account of the biographee that collaborates and confirms the thesis of the text that the biographer is offering will invariably emphasize closure and progress towards individuality, rather than openness, discontinuity and ambiguity. Hence, the temptation to claim that the biographee has been discovered in her essence and hence authorize the interpretation of the life. Resisting the essentialist temptation, however, does not vitiate the biographical enterprise although it does reveal that the task is more complex then envisioned.

A result of the process, the biography is an achievement that is not measured solely by its isomorphic relation to the often elusive historical subject, but by its ability to provide a satisfying, coherent and unifying account of the evidence that pertains to the biographical subject. Its literary or interpretative character should not, however, suggest that it is bereft of methodological rigor. The biographical subject is best construed as a locus which insists and, as a methodological minimum, is constrained by two general hermeneutical axioms for arriving at an accurate histori-cal/interpretive description: 1) "As in literary interpretation, historical truth is based on constructing an account one believes to be a plausible and meaningful context in which to place various data"[22]; 2) "No agent can be eventually said to have meant or done something which he could never be brought to accept as a correct description of what he had meant or done."[23]

Viewed philosophically, the biography is, in effect, an attempt to provide a formal solution to certain methodological and philosophical problems implicit in the attempt to reconstruct responsibly the life-history of an individual and to provide meaning for an individual's life by transmitting personality and character through prose. These include epis-temological, ontological and historiographical concerns. The practice of biography is ill-served by the adoption of the masquerade that it is an exact science. Rather, the task must be seen in the somber realization that completeness is an idealization since a biography "is a record in words of something that is as mercurial and as flowing, as compact of temperament and spirit, as the entire human being."[24] Hence, from a practical point of view, one is wise to heed the warning Carlos Baker offers in his biography of Hemingway, that "No biography can portray a man as he actually was. The best that can be hoped for is an approximation, from which all that is false has been expunged and in which most of what is true has been set forth . . . . If Ernest Heming-way is to be made to live again, it must be by virtue of a thousand

---

[22] Stuart L. Charme, *Meaning and Myth in the Study of Lives*, (Philadelphia: University of Pennsylvania Press, 1984), 150.

[23] Quentin Skinner, "Meaning and Understanding in the History of Ideas," *History and Theory* 8 (1969), 28.

[24] Edel, "Biography and the Science of Man," 2.

pictures, both still and moving, a thousand scenes."[25] Because biographies always exceed the boundaries of the purely factual, by making inferences and positing connections between diverse events in the individual's life, they are always approximations and provisional and hence subject to revision. The implicit monistic impulse to capture the life as it really was must thus be tempered. Objectivity is not necessarily compromised, but must be pursued by a clear recognition that interpretive techniques which have the capacity to enlighten and broaden our understanding of an individual life can affect, compromise or even distort the veracity of the portrait.

II. The Relevance of Biography as a Cultural Phenomenon

Opinion regarding the practical value of biography has been mixed. Its detractors argue that the practice is simply an occasion for pernicious vilification, which may explain the strong antipathetic reaction towards biographies as gossipy, intrusive, prying and predatory. Biography has been called "a disease of English literature" (George Eliot); practicing biographers are disparaged as "psycho-plagiarists" (Nabokov) and as "always superfluous" and "usually in bad taste" (Auden). Nevertheless, in the main, biographies are lauded for their salubrious, even therapeutic, qualities. Samuel Johnson was voicing a common view when he effused:

> no species of writing seems more worthy of culmination than biography, since none can be more delightful, none can be more certainly enchain the heart by irresistible interest, or more widely diffuse instruction to every diversity of condition.[26]

Johnson's laudatory endorsement of the genre is charming but beguiling since it portrays biography as a pleasant diversion and, assumes, rather than spells out its theoretical motivations and practical aims. Hence it is worth rehearsing some of its theoretical and cultural consequences. The psychologist Gordon Allport offers the following extensive, but by no means comprehensive, list of purposes for the practice:

---

[25] Carlos Baker, *Ernest Hemingway: A Life Story* (New York, 1969), vii.

[26] *Samuel Johnson: Selected Poetry and Prose*, edited by Frank Brady and W. K. Wimsett, (Berkeley: University of California Press, 1977), 182.

a. for exculpation or hagiography
b. as a literary-aesthetic creation
c. as a means of studying culture
d. as a means of studying one aspect of development (topical)
e. as an adjunct to therapy:
        changing person—e.g., psychiatry
        changing environment—e.g., vocational guidance
f. as illustration of a specific problem or mode of therapy
g. as illustration of a theory, e.g., to exemplify a conceptual framework
h. as a means of gaining maximum understanding of the individual[27]

A few comments on the above list are in order. Some of these purposes have been challenged, damned, declared otiose and rejected by reader and practitioner (for example (a)) while others ((e) and (f)) have been pursued vigorously by psychologists and psychoanalysis through psychobiography. Moreover, the list is a reflection of Allport's psychological bent. Allport could have offered additional purposes for the practice: in particular, the use of biography as a cultural resource for illuminating the historical context from and within which the individual enacts her socio-political role. Nevertheless, the list is useful since it draws attention to the use of biography and the understanding of individuals for the study of culture. Allport argued that the idiographic study of an individual through personal documents would not only afford an understanding of individual personality, but provide the foundation that would lead to discoveries of nomothetic generalizations about human personality. Adherents of this nomothetic quest are also to be found in sociology,[28] and anthropology.[29] What they share is a recognition of the important role of the individual in her cultural situation as the basis upon which the inquiry for nomothetic generalizations proceeds.

The foundational role of biography for historiography is advanced by Wilhelm Dilthey, who argued that the process of writing a biography is the plinth upon which the human sciences should be modelled. He writes:

---

[27] John A. Garraty, "Gordon Allport's Rules for the Preparation of Life Histories and Case Studies," *Biography* 4.4 (1982), 285.

[28] Cf. *Biography and Society: The Life History Approach in the Social Sciences*, edited by Daniel Bertaux (Beverly Hills, Sage Publications, 1981).

[29] Cf. Langness, L. L. and Gelya Frank *Lives: An Anthropological Approach to Biography* (Novato, C.A.: Chandler & Sharp, 1981).

The biography represents the most fundamental historical fact clearly, fully, and in its reality. Only the historian who, so to speak, builds history from these life-units, who seeks, through the concepts of type and representation, to interpret social classes, associations, and historical periods, and who links together individual lives through the concept of generations, only he will be able to apprehend the reality of a historical whole in contrast to the lifeless abstractions which are usually drawn from the archives.[30]

While the use of biography for nomothetic purposes is widespread, it is by no mean ubiquitous. The political philosopher Hannah Arendt argues for the relevance of biography in its own right as a means of embodying and preserving individual uniqueness. As she points out:

The chief characteristic of this specifically human life whose appearance and disappearance constitute worldly events, is that it is itself always full of events which ultimately can be told as a story . . . establish a biography . . . . For speech and action . . . are indeed the two activities whose end result will always be a story with enough coherence to be told, no matter how accidental or haphazard the single events and their causation may appear to be.[31]

Arendt was a consistent champion of human plurality and resolute in her affirmation of the proposition that "men, not man, inhabit the world." As such, to garner an understanding of the public/political world it was a methodological error to aim solely towards an understanding of an abstract, universal individual or citizen. Wary of attempts to ignore or dissolve this irreducible plurality, she argued that one ought to confront and account for the diversity and isonomy that characterizes the public cultural realm.

For Arendt, the modern attempt, in mimicking the natural scientific impulse to generate generalizable laws about human life, hobbled its ability to appreciate, confront and account for the freedom of the individual. For example, arguing against the sociological view of the early Marx, she claimed that Marx had cavalierly dismissed this biographical

---

[30] Wilhelm Dilthey, *Selected Works, Vol. 1*. Edited by Rudolf A. Makkreel and Frithjof Rodi (Princeton, New Jersey: Princeton University Press, 1989), 85.

[31] Hannah Arendt, *The Human Condition*, (Chicago: University of Chicago Press, 1958), 97.

aspect by reducing Man to the biological capacity for labor by charac-
terizing him as a "species-being." As laborer, human existence is reduced
to its capacity to live and toil as an undifferentiated member of a
species. Individuality is marginalized as the allegedly "real" function of life
is the preservation and perpetuation of the human species. The
identification of the biological with the social was to obscure, even
occlude, the political dimension of human life. As Arendt puts it,
hyperbolically, "[i]f this inside were to appear, we would all look alike."[32]

Human beings wrench themselves, as least partially, out of this life
cycle by virtue of their individual rectilinear existence; that is, in terms
of a beginning (natality, or birth), middle, and end (death). It is by the
capacity to work—to fabricate, or make a human artifice—and to
act—which affirms and expresses individuality—that the biographical
individual is constituted. In political action man "communicate[s] himself
and not merely something—thirst or hunger, affection or hostility or
fear."[33] Consequently, she argued that there was nothing politically
interesting nor distinctive about the biological self per se. Human uniqueness is a public and biographical characteristic. Hence her insistence that
human beings have a biological or sociological and a biographical
existence.

Arendt argues further that life-stories not only provide a more
accurate account of the public/cultural world, but that they can also serve
as exemplars for moral and political behavior. Accordingly, she published
Men in Dark Times (1968) which comprises a series of encomiums or
testimonials of persons, such as Karl Jaspers, Rosa Luxemburg, Isak
Dinesen and Randall Jarrell, that she admired. Arendt argues that such
people provide "illumination" in difficult times. This not primarily because
they provide normative principles for action, but rather, because they are
people who continued to think, reflect and live actively. Their stories
provide exemplars of individuals striving to lead integral, unified lives and
are part of the practical task of "contemporary memorialization."[34]

While James Clifford argues that this presentation of a unique unified
life may be a conceit of biography, he nevertheless recognizes it politi-
cal/practical value. Even as we live in an age that champions the ideal

---

[32] Hannah Arendt, *Life of the Mind, Vol. I* (New York: Harcourt Brace
Javanovich, 1978), 29.

[33] Hannah Arendt, *The Human Condition*, 176.

[34] Elisabeth Young-Bruehl, "The Writing of Biography," 135.

of the integral, autonomous person, we possess few cultural archetypes that actualize this possibility. "We strain for an unlivable identity. The desired unity can at least be known vicariously, through the reading of biographies."[35]

These considerations point to a sociological/political lacunae that can be addressed by posing questions such as: how do biographies function in the public world, what is their role in shaping how we think, in the building of a tradition, how they help to bind a community, etc? As Elisabeth Young-Bruehl argues, biography is aptly suited for this task since:

> Biography in the twentieth century has taken over for people from all kinds of backgrounds—religious backgrounds, ethical backgrounds—the task of telling exemplary lives. It's a cultural task. But biographies don't serve any particular ethnicities or religions . . . . In a sense, biography is cosmopolitan. It's concerned with a life in the world, not some particular world—although the life may be lived in particular world. But it should go beyond that particular world, like a kind of cultural ambassadorship.[36]

One last view needs to be discussed in this section as it is an instance of the attempt to combine the theoretical and practical uses the practice. It concerns Sartre's post-war excursus into biography. Sartre's interest was motivated primarily by methodological considerations since he viewed the practice as providing an important local for the utilization, deployment and assessment of historical, philosophical and literary techniques. Although Sartre was concerned primarily in the freedom inherent in and characterized by the existential subject, his reflections on the relationship between the subject and the situation prompted him to reconceive the individual as both "totalizer and totalized." He argued that "individual" and "society" are "modern" as well as interdependent, interactive notions. Hence, to understand the historical situation that characterizes the cultural world and the notion of agency within it, we need to pay attention to both.

Sartre, of course, was not alone in this endeavour. Sociologists, especially those drawn to the interpretive side of the discipline and

---

[35] James Clifford, "Hang Up Looking Glasses at Odd Corners," 44-45.
[36] Gale Porter Mandell, *Life in Art*, 190.

associated in the main with the Chicago sociologists during the twenties and thirties, had made attempts to utilize biography as a tool in the a form of investigation that is termed "symbolic interactionism." In this respect, Sartre was echoing a key theoretical precept in symbolic interaction for the connection between social dynamics and historical change; that is, an interest in the relationship between individual and collective *praxis* and sociohistorical *change*.

What is distinctive in Sartre, however, is his attempt to utilize biography as the litmus test of a philosophical theory (or, an existential philosophical anthropology); that is, on its ability to explain a human life. Sartre held that while a person cannot be the object of conceptual knowledge, it does not follow that a person is therefore incomprehensible by arguing "that each moment in a series is *comprehensible* on the basis of the initial moment, though *irreducible* to it."[37] Hence, his excursus into existential psychoanalysis—the retroactive search for an original choice, the cardinal factor, by which each human being fabricates or fashions herself as a person, tells herself what and who she is, and adopts a characteristic stance toward the world—in *Being and Nothingness*. There, his brief discussion of Flaubert serves as a prelude to his biographical tome on the French writer. Indeed, the attempt to show that a man can be made comprehensible is the avowed aim of Sartre's last biography. Sartre was constantly trying to find and refine tools for understand individuals in their freedom. As Sartre put it in an interview:

> The most important project in the *Flaubert* is to show that fundamentally everything can be communicated, that . . . simply as a man like any other, one can manage to understand another man perfectly, if one has access to all the necessary elements. . . . my goal, to prove that every man is perfectly knowable as long as one uses the appropriate method and as long as the necessary documents are available.[38]

Consequently, in *Search for a Method*, Sartre proposes the strategy he called the progressive-regressive method: it is progressive, or synthetic, as it attempts "to recover the totalizing moment of enrichment which engen-

[37] Jean-Paul Sartre, *Critique of Dialectical Reason*, translated by Alan Sheridan-Smith (London: Verso, 1982), 15.

[38] Jean-Paul Sartre, "On The Idiot of the Family,"in *Life/Situations*, translated by Paul Auster and Lydia Davis (New York: Pantheon Books, 1977), 123.

ders each moment."[39] And, it is regressive, or analytic, as it must "at the start proceed as far as is possible for us in the historical particularity of the object."[40] As Sartre envisions it, he is "inventing a personage" when he uses hypotheses to construct a narrative that encompasses this dialectical movement and interrelationship between the individual and his cultural context. The result is the individual who is both totalizer and totalized. The question to be asked is who someone must have been in order to have within his field of possibilities the possibility of doing what he did.[41]

Taken together the the progressive/regressive aspects of the strategy yield man as the universal singular who is the product of the interrelationship between the individual and his culture. As Sartre writes:

> a man is never an individual; it would be more fitting to call him a *universal singular*. Summed up and for this reason universalized by his epoch, he in turn resumes it by reproducing himself in it as singularity. Universal by the singular universality of human history, singular by the universalizing singularity of his projects.[42]

Sartre was well aware that his procedures were unorthodox and flamboyantly pronounced his last biography a "true novel," thus declaring it a curious hybrid of philosophical, literary, historical and psychoanalytic techniques. Concerned that the reception of his project would be greeted with hostility and incredulity, he nevertheless believed that he would be vindicated by the methodologically rigorous techniques employed in the study:

> This is a fabrication, I confess. I have no proof that it was so. And worse still, the absence of such proofs—which would necessarily be singular facts—leads us, even when we fabricate, to schematism, to generality; my story is appropriate to infants, not to Gustave in particular. Never mind. I wanted to follow it out for this reason alone: the real explanation, I can imagine without the least vexation, may be precisely the contrary of what I invent, but in any case it will have to

[39] Jean-Paul Sartre, *Search for a Method*, translated by Hazel E. Barnes (New York: Vintage Books, 1968), 147.

[40] *Ibid.*, 140.

[41] cf. *Ibid.*, 141.

[42] Jean-Paul Sartre, *The Family Idiot* Vol. 1, ix.

follow the paths I have indicated and refute my explanation on the ground I have determined.[43]

Now, it would be easy to dismiss this false sense of bravado as any, even cursory, reading of the text will reveal that, at best, it is a flawed biography. It would, however, be a serious mistake to view Sartre "confession" as a renunciation of the biographical enterprise; as a sort of "reductio ad absurdum." Two considerations ought to dispel this view: First, even if Sartre's interpretation of Flaubert were inadequate, any alternative accounts would nevertheless have to produce an equivalent treatment of the life and work, dealing with or refuting the same issues, and hence traverse many of the same paths that he has explored. As such, the passage is, in effect, a disguised challenge to alternate explanations. Second, if one draws a distinction between historical and psychical reality, then the criteria for judging the reconstruction of psychical reality cannot proceed in a manner equivalent to that of historical veracity since there is a paucity of observable historical events. Psychoanalytic constructions are not reducible to historically observable events although they are constrained or fettered by them. As such, the question of biographical veracity cannot be answered simply by the adequacy or discovery of external facts but must also involve questions concerning the attempt to symbolize the quality of Flaubert's subjectivity.

Douglas Collins, who defends Sartre's turn to biography, argues that:

> the test of a system of ideas lies in its ability to perform in the real world, and this ability is best revealed in its capacity to reconstruct the life of an historical individual. A philosophical system is thus subordinate in interest to the biography it generates, because in the biography the system's success or failure is ultimately evaluated. Rather than being the bastardization of philosophy, biography is its legitimation.[44]

Similarly, the attempt to produce a biography of a unique individual is also to test the adequacy of the assumptions, resources and techniques employed and deployed in the inquiry directed at the cultural world.

---

[43] "On the Idiot of the Family," 132.
[44] Douglas Collins, *Sartre as Biographer* (Cambridge, MA.: Harvard University Press, 1980), 5.

Since the practice is situated within a larger cultural horizon that tries to dialectically understand the past through the particular lives of individuals, a study that locates itself within the terrain of the cultural world "that does not come back to the problem of biography, of history and of their interactions within a society has completed its intellectual journey."[45]

III. Conclusion

This attempt to explore and canvass aspects of the biographical enterprise was to subject the practice to a philosophical examination and explication of its terrain and contours. Since the achievements of the biographer's subject is located in the same cultural space as the biography commemorating that achievement, reflection on the enterprise was directed at two levels: the biographical practice as a cultural object, and the biography as a product which is also an object that performs an assortment of functions in the cultural world.

The analysis offered in this essay was necessarily sketchy but, if persuasive, provides a set of pathways, distinctions and classifications that can provide a methodological and substantive basis for a more detailed examination of the subject matter at hand. As Edel puts it, "[a] biographer is like a grinder of lenses. His aim is to make us see."[46] The problem faced by the philosopher is to determine the adequacy of the lenses, to recognize its capacity to distort, magnify, and clarify, and ultimately to determine the value of the lenses. This task is fraught with theoretical pitfalls but the promise of theoretical and practical rewards is great. Hence, the theorist of biography, argues Woolf, is well advised to "go ahead of the rest of us, like the miner's canary, testing the atmosphere, detecting falsity, unreality, and the presence of obsolete conventions."[47]

---

[45] C. Wright Mills, *The Sociological Imagination*, (New York: Oxford University Press, 1959), 6.

[46] Leon Edel, "Biography: A Manifesto," 3.

[47] Woolf, "The Art of Biography," 169.

Chapter 13

Alfred Schutz and the Project of Phenomenological Social Theory

David Carr
Emory University

Abstract: *A discussion of Schutz' phenomenological approach to social theory leads me to some fundamental doubts about his project. Is phenomenology's central concept, intentionality, conducive to the task of understanding relations among persons? My doubts are expressed through a historical account: I claim the concept of intentionality was devised as a response to questions about the relation between human experience and nature. Applying it to social relations, I argue, may be a case of employing it outside its proper sphere.*

The work of Alfred Schutz is the most impressive and important effort to date to establish a philosophical social theory on phenomenological foundations. His most likely rivals in this effort are probably Sartre and Merleau-Ponty, but their sources of inspiration are not only phenomenological, and it can be argued that social theory is not the true focus of their thought. Schutz' inspirations are not exclusively phenomenological either, of course—think of the importance of Bergson and Weber in his work—but there is no doubt that phenomenology, especially Husserl's, is dominant. And it is clear that social reality is the sole focus of Schutz' efforts. In several places he explicitly brackets the concerns and investigations Husserl called "transcendental," and which we might also call epistemological or metaphysical, in favor of the more modest and well defined goal of understanding the social world.[1]

---

[1] Alfred Schutz, *The Phenomenology of the Social World*, translated by G. Walsh and F. Lehnert (Evanston: Northwestern University Press, 1967), 97. (Hereafter PSR) *Collected Papers II: Studies in Social Reality*, edited by A. Broderson (The Hague: M. Nijhoff, 1964), 25. (Hereafter CP II).

M. Daniel and L. Embree (eds.), Phenomenology of the Cultural Disciplines, 319–332.

But the circumscribed focus and the decided methodological commitment of Alfred Schutz did not prevent him from producing a body of work which is extraordinary for its breadth, its richness, and above all for its undoctrinaire flexibility and openness. When we think of the relatively short career of this man, made even more difficult by his being uprooted and relocated in the new world, we are amazed by what he accomplished. Perhaps most important for those of us who consider ourselves part of the phenomenological tradition, Schutz has done the most to make good on phenomenology's promise as a method extending beyond philosophy into other disciplines. His influence on sociology, through his students and through his writings, has been immense.[2] Part of this is due to phenomenology itself, part to Schutz' ability to present it clearly and undogmatically to his English-speaking students and readers, part no doubt to his own flexibility in adapting himself to and addressing the concerns of colleagues in his adopted country and culture.

The features of Schutz' thought and writing I have just mentioned—its openness, its flexibility, its undoctrinaire and undogmatic character—lead me to think he would have responded well to sympathetic and constructive criticism. In my own work I have drawn much from Schutz, but I have also been critical of many features of his thought. It is inevitable that such criticism take place, and as is the case with the really important thinkers, the kinds of criticisms I have developed would not even have been possible if I had not already absorbed the basic outlines of Schutz' thought. I often think of Schutz what I think of Husserl, that he was not always true to his own best intentions and insights. But we owe those intentions and insights to him, and our careful and critical attention to them is the most important homage we can offer to their author.

All this is by way of justifying the fact that my intention here is not to provide exposition or commentary on Schutz' work but to engage it in critical debate. But I fear that my concern runs deeper now than the sort of internal criticism I have attempted in some of my previous work, for it is aimed not at Schutz' consistency with his own fundamental insights but the acceptability of those insights themselves. I am beginning to wonder if Schutz' work, while exploiting phenomenology's promise to the full, is not also revealing to us its limits. I wonder if the phenomen-

---

[2] See *Worldly Phenomenology: The Continuing Influence of Alfred Schutz on North American Human Science*, edited by Lester Embree (Washington: Center for Advanced Research in Phenomenology and University Press of America, 1988).

ological approach itself, rather than Schutz' faithfulness to it, is responsible for some of the shortcomings of his work. So I want to address myself to some of the basic tenets of phenomenology in what follows, which means that I will speak as much about Husserl, Heidegger, Sartre, Merleau-Ponty, and others as about Schutz. But Schutz' work is crucial here, since my worry is that phenomenology might for reasons of principle be incapable of an adequate social philosophy. The understanding of the social thus becomes for me the crucial test of the capacity of phenomenology to do what it wants to do.

To some extent my worries are influenced by the widespread questioning of the "philosophy of the subject" that has dominated recent continental philosophy. The motives and sources of this questioning are as diverse as the thinkers involved in it and are only partly prompted by concern for a philosophy of the social. In some cases, such as late Heidegger and Derrida, concern for the social seems decidedly lacking; in other cases, such as Habermas and, in a related but distinct way, Levinas, it is central. What these thinkers question in different ways is a reflexive philosophical method and what it presupposes about the nature of the reflecting subject. But many of them also share the view that this approach inevitably construes the relation between "man and world" in terms of instrumental or technological reason and thus of power. If this is so then the relation between persons will be conceived in this way as well. Those who find such a view of social relations objectionable may think that phenomenology, if it does not actually encourage such a view, at least offers us no conceptual resources for constructing an alternative.

Could this be true? Everyone knows that the first attempts by phenomenologists to comprehend the social were not great successes. The Fifth Cartesian Meditation is a tangle, and almost everyone agrees that it fails, though there is no agreement on why. The two great *opera magna* of existential phenomenology, Heidegger's *Being and Time* and Sartre's *Being and Nothingness*, fare no better. The former gives notoriously short shrift to *Mitsein*, and some read Heidegger as equating social existence with inauthentic existence. Sartre's famous account of "the look," brilliant as it is, seems to bear out perfectly the contention we are examining: relations between persons are inevitably relations of power and domination.

These shortcomings were recognized by critics from the start, of course, and the work of two other great phenomenologists, Merleau-

Ponty and Schutz, can be seen largely as attempts to remedy the situation, to show that phenomenology and its offshoots did not have to shortchange the social. Indeed, both thinkers seem convinced that phenomenology has something valuable to offer precisely in this domain.

But are they right? It might be argued that the best way to answer this question is to examine their work to see whether in fact they overcome the defects of their predecessors and to see how much they actually accomplish on their own. But what if their positive accomplishments were achieved, not because of but in spite of their commitment to phenomenology? What if they were hindered throughout by that very commitment, prevented by it from recognizing some very important things about the nature of the social? This is the possibility I would like to explore in the following.

Where should we look for the fundamental tenet, the heart and soul, as it were, of phenomenology? To me the answer to this question has always been clear: intentionality. In the *Logical Investigations* Husserl borrowed the concept from Brentano and improved on it, in ways that were partly critical of Brentano. But most importantly he made a decision which was fateful for the whole course of phenomenology: he accepted intentionality as a fundamental principle. By that I mean that instead of regarding intentionality as a fact that had to be explained or derived from something else, Husserl chose simply to *describe* the intentional relation, explore its many features and peculiarities—in short, to get straight on what intentionality is and entails. From that, it seems to me, everything else follows. In particular, the phenomenological reduction is nothing but the acknowledgment of the fact that the *consciousness-of* relation (or quasi-relation) carries no commitment one way or the other to the existence of the objects of such a consciousness. The reduction solemnizes, as it were, the object-neutrality of the intentional relation and frees us from caring about the ontological status of the objects of our experience so we can focus our philosophical attention on their meaning.

Now you may object to my saying that intentionality and the reduction are at the heart of phenomenology, since early Heidegger and early Merleau-Ponty, if not Sartre, are severely critical of both principles. But in my view these thinkers reinterpret but do not reject these notions. If we take intentionality to be characteristic, not narrowly of "consciousness" but of "human experience," if we see its essential feature that of "meaning-bestowing," and if we conceive of the task of philosophy as the description of the world as a tissue of meanings rather than a collection of facts to be explained, we have accepted what really counts about intentionality and the reduction—the rest is quibbling about the details. In this sense all the great patriarchs of the phenomenological movement —Husserl, Heidegger, Sartre, Merleau-Ponty—are committed to intentionality and the reduction.

But even more important than what all the phenomenologists *accept* is what they all reject. We can get at this by asking what is to my mind the most important question to ask about intentionality: to what question is it a response, what problem does it purport to solve? More precisely, what issues are resolved when intentionality, as we have broadly defined it here, is accepted as a principle rather than taken as something that has to be explained. Answering these questions correctly will enable us to understand intentionality better but will also raise questions about its usefulness for an understanding of the social world.

Brentano acknowledged the medieval origins of the concept of intentionality, and this early sense was preserved in Descartes' use of the distinction between the formal and the objective reality of an idea. Ideas by nature refer to things beyond themselves, and for Descartes ideas are mental entities, *pensées*. Thus, Husserl thought that Descartes already had hold of something like the modern sense of intentionality, and even that his principle of hyperbolic doubt had captured the essence of the epoché or reduction that should follow from it.[3] Indeed, all the philosophers of the early modern period, especially the British empiricists, since they used the term *idea* in roughly the Cartesian sense, had all they needed to recognize the centrality of the intentional relation.

But they all made what Husserl considered the wrong move. They lived, after all, in the age in which the new and exciting and acceptable way of understanding things was to derive them from their antecedents according to causal laws. They shifted their attention from the intentionality of ideas to the causes of ideas—or, more often, they confused the two questions, equating the intentional object of a thought with its causal origin. As is well known, this led philosophy in two incompatible but strangely complementary directions: from the viewpoint of Descartes' problem of proving the existence and nature of the external world, it led directly to Humean skepticism. If the causal relation is a contingent or external relation, any inference from our ideas to a world which causes

---

[3] Edmund Husserl, *The Crisis of European Sciences and Transcendental Phenomenology*, translated by D. Carr (Evanston: Northwestern University Press, 1970), 75-78.

them is still open to skeptical doubts. But Hume is able both to express
such doubts and to devaluate them at the same time, urging us to plunge
ahead with our causal investigations of the world whether or not we are
sure that it exists and that its causal order is universal and necessary.

Once we assume both points, the scientific investigation of nature can
proceed, unhindered by epistemological qualms—which of course it was
doing anyway, as we know. But given the assumption of the universality
of the causal order, all being must belong to that order, including mental
being. The contents of the mind must be considered entities or events
which are related to the rest of nature according to causal principles. It
would be even more convenient, of course, if we could eliminate their
apparently non-spatial character by reducing them to or deriving them
from the physical states of the brain and nervous system. For those
actually engaged in the natural causal investigation of human nature, such
a reduction is nothing less than a solemn obligation, a promise that must
be kept.

But this is to jump ahead. We should pause to appreciate the acute
malaise in which philosophy found itself at the apogee of the enlightenment, thanks to Hume. The bold attempt to validate our knowledge of
the external world on first-person, reflective principles, inaugurated by
Descartes, had collapsed; the mind, conversant only with its own ideas,
could find no sure way beyond them. Berkeley tried to make a virtue
of this sad situation with his subjective idealism, but convinced few.
Meanwhile, in spite of this, the exploration of nature proceeded by leaps
and bounds. Yet, consistent with the collapse of the Cartesian project,
it did this under the guidance of an assumption—the universality and
necessity of the causal order—which remained ever and always a mere
assumption, a conjecture it could never justify or validate. Furthermore,
this assumption carried with it a consequence which some, at least,
especially those still committed to the original Cartesian project, found
intolerable: the naturalization of the human subject, "man," as he was
called, his reduction to the status of an isolated and peculiar effect of the
vast causal order. This was intolerable because it seemed to rob "man"
of his freedom and dignity.

I have spent some time on these familiar developments in early
modern philosophy because I believe that they constitute the situation,
the problem to which Husserl was responding with his treatment of the
concept of intentionality. You may object that this situation, this critical
moment of malaise and ambiguity, was already faced by Kant and that

he was responding to it as well. This is true, and this could lead us in one of the two directions: either to argue, as some do, that intentionality is already present in Kant; or alternatively, that what counts in Husserl is not so much the concept of intentionality as his "transcendental" treatment of this concept, a treatment he inherited from Kant. One could profitably pursue either of these lines; the affinity between Kant and Husserl has been stressed by many, notably by Husserl himself. For the moment I would like to put aside this interesting question in order to focus on Husserl and to explain in what way his treatment of intentionality responds to the situation I have been speaking about.

If the historical-intellectual situation really was as I have described it, intentionality can be seen as a brilliant stroke, a way of killing at least two birds with one stone and saving the day for philosophy and above all for "man."

On the one hand it responds to the skeptical solipsistic problem. If "all consciousness is consciousness of something," then human experience essentially refers beyond itself, indeed is purely outside itself, in the world, with and at its objects. Sartre was right in insisting that if intentionality is taken seriously, there can be nothing left in consciousness, no "ideas," in the sense of modern philosophy, with which it is conversant, only the world outside, without which consciousness could not be what it is. Thus there can be no problem of establishing indirect contact with the external world, much less of inferring its existence from some intermediate entities, since our contact with it is direct. To this extent our knowledge of it is assured; we can be sure that what we know is really about *it* and not somehow merely about the contents of our minds.

But even more remarkably, while it puts us in direct touch with the natural order, the concept of intentionality at the same time liberates us from that very order. For it ascribes to us certain properties or features which have no place in the natural world: consciousness-of, reference, In-der-Welt-Sein, être-au-monde. As the phenomenologists are at pains to point out, our way of being in the world is totally incommensurable with the way in which natural objects coexist: they are in the world but they don't *have* a world; things don't mean anything to them. Only for a conscious being, a being in the world in Heidegger's or Merleau-Ponty's sense, do things have meaning. For such a being, for us, that is, nature is assured as our habitat and milieu, our world, our place, the phenomenon which is there over against us, but which we can know, whose mysteries and secrets we can penetrate, which we can make our own.

Yet at the same time we are not subject to nature. We do not allow ourselves to succumb to her or be controlled by her; intentionality makes our relationship with her the very sign and exercise of our freedom.

In short, intentionality arises in philosophy in response to a problem we have with the natural world, a situation in which our connection with it and our freedom from it seemed seriously jeopardized. And intentionality provides us with a vision of ourselves in which we reestablish contact with nature, gain control and mastery over her, and liberate ourselves from her all in one stroke.

Looked at in this way, intentionality, together with the whole philosophical development which culminates in it, is simply another facet of the struggle between modern "man" and nature for domination, control, freedom and dignity which goes by the name of technology. Modern philosophy has, since Descartes, been occasioned by the growth of modern natural science, and has responded to problems created by it. It has been the handmaiden of modern science, allowing science to do its work while preserving its author (man) from suffering indignities in the process. In short, it is all part of the story of becoming the masters and possessors of nature rather than allowing her to master and possess us. And intentionality can be seen as just the latest chapter in that story.

I should repeat that intentionality can only function in this way if it is treated as a principle rather than as something to be explained or derived. Philosophers like Daniel Dennett, who recognize the power and pervasiveness of intentionality in our concept of mind, still try to reduce it to a causal relation or the result of a causal relation, even though they realize how hard a task that is. Husserl and the phenomenologists simply refuse to acknowledge such a task, let alone undertake it.

But how can they fail to acknowledge its necessity? Is not the mind/brain part of nature? And is not everything in nature, including the mind, guided by the same causal principles? Of course it is, *if* we assume the universality and necessity of the causal order. But that is an assumption and nothing more. Once made, as it is always made by the scientist, it imposes the obligation to take in everything. But once we recognize as philosophers that it is only an assumption, we need not allow ourselves to be guided by it. It imposes no obligation on us.

Dennett, Sellars, Quine, and other scientific realists are quite clear on this by now. Their commitment to the universality of the causal order, and to the causal explanation of everything that falls under it, is just that, a commitment one takes on like a religious vow. It is stated "up front," as we say, like the dedicatory preface to the princes of the church and to the greater glory of God at the beginning of a seventeenth-century philosophical treatise. It is a decision, a confession of faith, and it is one that phenomenologists refuse to make. Instead, they choose intentionality and go on from there. Their decision, too, is a confession of faith, and again like a confession of faith it imposes obligations. The obligation in this case is to explore to its limits the intentionality of human experience, the meaning-bestowing character of consciousness or Dasein, and the human world as a complex of meaningful, i.e., intentional objects and entities.

And this is, of course, what the phenomenologists have done. The results, it seems to me, bear out my contention that the intentional approach is essentially designed to deal with our relation to nature. After his initial preoccupation with logical thinking, it was perception that really captured Husserl's attention and served as the guiding thread and paradigm for all his investigations. What is most striking about Heidegger's *Being and Time*, what is the thing that most captured the attention of its first readers? The distinction between *Zuhandenheit* and *Vorhandenheit*, of course, a fundamental revision of our way of thinking about how we relate to the *material* world around us. Indeed, the very concept of world, as Heidegger describes it, is the totality of reference (*Bewandtnisganzheit*) among such entities. The later Husserl's concept of *Lebenswelt* retains its links to perception and arises out of the context of questions about the epistemic status of natural science. For Merleau-Ponty, of course, all phenomenology is centered in the phenomenology of perception.

But can the concept of intentionality, which so clearly arises out of problems of the relation between man and nature, and is so useful in that context, prove equally useful when it comes to describing inter-subjectivity, relations among persons, social reality? Or does it instead prove to be a hindrance? Let us examine some features of the phenomenological approach to these topics with a view to answering this question.

The description of an intentional relation is typically a reflexive procedure in which each of us examines his or her own experience. In the phenomenology of intersubjectivity, thus, I reflect on my experience and my world and try to describe how the other turns up in that world. How does my intentionality bestow the meaning *alter ego* or "other person" on certain entities within my world, and what are the essential features of the entities that bear that meaning? In other words, the other is treated as a "phenomenon" in the sense of phenomenology, a *cogitatum* within the overall scheme Husserl describes as *ego-cogito-cotitatum-qua-cogitatum*. For Husserl the problem is how the other, understood as a *cogitatum cogitans*, can be grasped out there in the midst of the natural world of perceived objects, which includes human bodies. Many of the problems of the Fifth Meditation derive from the fact that, here as elsewhere, for Husserl perception is primary. The difficulty then is to understand how, in this vast surrounding world of *things*, I can locate *mon semblable*, a fellow ego peeping out of one of those objects and staring back at me.

Husserl's procedure here, whatever else its problems (and there are many), indicates clearly the manner in which the intentional approach is geared to our relation to the natural world, and that in two ways: first because it *starts* with perception and the world of things, including bodies; second, because the other, when he/she emerges, is defined almost exclusively in negative terms, in opposition or contrast to things. And the obvious question to be raised here is whether this approach can ever do justice to the pervasiveness and priority of our social being and the *a priori* character of our relation to others.

But there is the further question of whether, in spite of being contrasted with natural objects as perceived, the other person is not still being conceived in their image when he/she is treated as phenomenon or *cogitatum*. Another way of asking this question is to ask whether the problem of *knowledge* is not the central problem in a text like the Fifth Meditation, and whether the other is being considered chiefly as an object of knowledge. I have argued elsewhere[4] that it is a mistake to regard Husserl as addressing the classical problem of *solipsism* in the Fifth Meditation even though he explicitly uses that term. He is not trying to prove *that* others in fact exist. But he might be understood as tracing the origins of the "concept" *alter ego*, as if one could somehow begin without it and then, on the basis of perceptual experience, acquire such a concept.

The latter, I believe, is what Schutz understands Husserl to be doing when he speaks of Husserl's "attempt to account for the constitution of transcendental intersubjectivity in terms of the operations of the consciousness of the transcendental ego." As is well known, Schutz takes

---

[4] David Carr, *Interpreting Husserl* (Dordrecht: M. Nijhoff Publishers, 1987), 45ff.

Husserl to have failed at this because it cannot be done. "Intersubjectivity is not a problem of constitution which can be solved within the transcendental sphere, but is rather a datum (*Gegebenheit*) of the life-world . . . the fundamental ontological category of human existence in the world. . . ."[5] This ringing declaration is taken by many to be a great advance over Husserl and to express the move by which Schutz, Merleau-Ponty, and others liberate themselves from the transcendental version of phenomenology in favor of its "existential" version. This might be taken to mean that these philosophers are no longer concerned with the philosophical problem of our *knowledge* of other minds.

The fact remains that Schutz, throughout his career, presents himself as attempting to work out the foundation of the social *sciences*. And even Merleau-Ponty rarely discussed the problem of "relations with others" without linking it with the problem of the "sciences de l'homme." In such sciences, obviously, the philosophical problems of skepticism, solipsism, and concept derivation have no place, but the point *is* to take this "datum of the life-world" and subsume it under the sort of rigorous and wide-ranging concepts that make for scientific treatment.

Could it be that these thinkers, at least implicitly, construe the other principally as an object of knowledge, or potential knowledge? And could this be a further effect of the concept of intentionality, however flexibly it may be treated by these post-Husserlian phenomenologists?

To be sure, it is not only the concept of intentionality but also that of science that they treat flexibly. Schutz and Merleau-Ponty belong to a long line of philosophers (beginning with Dilthey and the neo-Kantians) who adamantly refuse to treat the *Geisteswissenschaften* or *sciences humaines* as continuous with the natural sciences. Nothing is farther from their intentions than to reduce the human world to a collection of objects or events to be treated according to causal relations or laws.

But there are features of the concept of science, even after we eliminate the reductionistic connotation of the English term and are guided by the broader scope of the term *Wissenschaft*, which are carried along by Schutz and which may stand in the way of our understanding of the social. For Husserl, of course, the other is still a *Gegenstand* in the broadest sense, which belongs to a domain or region of being, dominated by its own material *a priori* laws. And over against that object

---

[5] Alfred Schutz, *Collected Papers III: Studies in Phenomenological Philosophy*, edited by I. Schutz (The Hague: M. Nijhoff, 1967), 82.

and its domain is, of course, the scientist, whose job is to know his object in both its *a priori* and its empirical features by constructing a coherent body of knowledge or theory. As we know, Schutz' bold stroke was to introduce the Weberian ideal type between these two poles as a tool for knowledge of the social world. By this means he sought to give concreteness to an otherwise only sketchy phenomenological notion of what a social science might be, and at the same time to provide an epistemological underpinning for what Weber had actually accomplished.

But the question is whether this goal of founding social science is compatible with understanding our social being and our relations with others. No doubt Schutz wants to begin in the prescientific world of the everyday, just as Husserl wanted to find the origin of the natural sciences in the life-world. But Husserl is often reproached for having described life-world experience too much in terms of its orientation toward science, and this in spite of his own recognition of this danger. Is Schutz open to the same criticism? I think he is. I think he describes concrete relations between persons almost exclusively as the sorts of relations that can grow, given certain interests, into scientific knowledge. Intersubjectivity is for him too much a matter of typifying or classifying others with a view to predicting, in a very broad sense, their behavior. Even the term understanding (*Verstehen*), the slogan of a whole sociological tradition given philosophical depth by Heidegger and Gadamer, may be indicative of a serious problem here. Is our relation with others really best characterized as that of *understanding* them? I'm not sure. And I suspect that in spite of all disclaimers, the centrality of this term may derive from a conception that social being is essentially founded on an *epistemic* relation.

I would now like, in conclusion, to make more concrete my misgivings and my doubts about a phenomenological approach to the social by turning directly to Schutz' work. I mentioned earlier certain shortcomings that I have long perceived in Schutz' work. One of those is that he always seems to portray social relations as somewhat bland and cooperative, yet at the same time distanced and polite. That is, he seems to overlook two possibilities of social relation that lie on opposite ends of this harmonious middle ground, namely, conflict on the one hand and real community on the other. Far from being remote or rare contingencies, these are, it seems to me, the driving forces behind social reality and social dynamics. Nor are they simply opposite extremes: they are intimately involved, together and dialectically, in some of the most important social relations, like politics and love.

Now I used to think that Schutz had simply overlooked this, whereas Hegel, for one—with his masterful description of the trajectory from the struggle to the death through the relation of domination to the establishment of community--had fully appreciated it. But now I am asking whether this oversight might be due to the concept of intentionality at the heart of Schutz' project.

This can be seen in Schutz' description of "the world of directly experienced social reality"[6] in which he speaks of the face-to-face situation and the we-relationship. The face-to-face situation is founded on the "thou-orientation" which is that of being "aware of a fellow human being *as a person.*"[7] But Schutz' description, in spite of its important references to "a community of space and a community of time," i.e., to a common surrounding world,[8] is to me too observational. The other is a certain *type* of entity (a person) as distinct from the other entities within my visual field. The thou-orientation can be one-sided if the other does not notice my presence. When she does, the thou-orientation is reciprocal and the we-relation is established. In other words, the we-relationship of the face-to-face situation is that of my intending you as a person, taking you to be a person rather than something else, and your doing the same to me.

There are several problems with this. One is that Schutz seems to overlook the possibility that I could be thou-oriented toward the other without knowing that the other is thou-oriented toward me, yet according to Schutz the we-relation would still obtain. But more important is that Schutz' description of the thou-orientation is essentially that of subsuming a particular under a concept, which in this case puts too much emphasis on the others being like me, belonging to the same species, being my *semblable*, my fellow, my counterpart (*Gegenüber*) or mirror-image, rather than my antagonist or opponent. In other words, the emphasis is on the sameness rather than the otherness of the other.

Yet there are good reasons for thinking—and this was Hegel's insight, I believe—that only with an other who is recognized as an antagonist can I form a genuine community by overcoming the antagonism in a

---

[6] PSR, 163.

[7] PSR, 164.

[8] PSR, 163.

common project. And this means surpassing the face-to-face relationship toward an action or experience whose proper subject is the *we*. Here my relation with the other is not that of subject to object but one of participation or membership in the same community. And I'm not sure this relation can be described as an intentional relation at all. Schutz has described essentially a subject-object relationship where the object happens to be another subject, and then, under the title of the we-relationship, a reciprocal subject-object relation. But in my view this restricts the notion of the social relation by conceiving it on the model of that obtaining between a scientific observer and his scientific object.

Let me conclude by looking critically at Schutz' descriptions of two phenomena he rightly takes to be examples of the we-relationship—having a conversation[9] and making music together.[10] In each case Schutz' description focuses too much, it seems to me, on how one person, by means of external bodily signs, can reconstruct and predict the thoughts and expressions of another person. But I would suggest that in a genuine conversation the attention of the participants is directed not to each other but toward that which is being said, toward the conversation itself. And in making music together the point is the music; the point is to let it sound and express what it has to express. The common project of saying something or letting something be said, in a conversation, or of making music together, establishes a relation among persons which cannot be comprehended by the concept of intentionality.

[9] PSR, 104.
[10] CP II, 159ff.

Notes on Contributors

David Carr (Ph.D., Yale, 1966) is Professor of Philosophy at Emory University. He is the author of *Phenomenology and the Problem of History* (1974), *Time, Narrative and History* (1986), and *Interpreting Husserl* (1987) and translator of Husserl's *Crisis of European Sciences*.

Mano Daniel (Ph.D., University of Waterloo, 1992) is William F. Dietrich Fellow at Florida Atlantic University. His current interests are in the philosophy of biography, political theory and Hannah Arendt.

Lester Embree (Ph.D., New School for Social Research, 1972) is William F. Dietrich Emiment Scholar and Professor of Philosophy at Florida Atlantic University and President of the Center for Advanced Research in Phenomenology, Inc. He has authored, translated, and edited a number of books and articles chiefly in Husserlian phenomenology. His current interests are in the history and philosophy of science (cultural sciences specifically, archaeology in particular), technology, and enviromentalism.

James Hart (Ph.D., University of Chicago, 1972) is Associate Professor in the Department of Religious Studies at Indiana University. He is the author of *The Person and the Common Life* (1992) and, with Steven W. Laycock, co-edited *Essays in Phenomenological Theology* (1986). His current interests are in political theory, peace studies, philosophy of religion and German Idealism.

Don Ihde (Ph.D., Boston University, 1964) is Leading Professor of Philosophy at the State University of New York, Stony Brook. He is author of *Technics and Praxis* (1979), *Existential Technics* (1983), *Technology and the Lifeworld* (1990), *Instrumental Realism* (1991), *Postphenomenology* (1993), and *Philosophy of Technology* (1993). His current interests are in philosophy of science and technology in its social and cultural contexts.

333

Stanford M. Lyman is the Robert J. Morrow Eminent Scholar in Social Sciences at Florida Atlantic University. He was formerly Professor of Sociology at the New School for Social Research, and has lectured extensively throughout the United States, Europe, Africa, and Asia. He is the author of many books, including *The Black American in Sociological Thought* and, most recently, *Militarism, Imperialism, and Racial Accommodation: An Analysis and Interpretation of the Early Writings of Robert E. Park.*

Don E. Marietta, Jr. (Ph.D., Vanderbilt, 1958), is Adelaide R. Snyder Professor of Ethics and Professor of Philosophy at Florida Atlantic University. He has authored many articles in Environmental Ethics. His current interests are in environmental ethics and the phenomenological approach to meta-ethics and value theory.

Ullrich Melle (Ph.D., Heidleberg, 1980) is Professor of Philosophy at Catholic University in Leuven. He is a collaborator in the Husserl-Archives and editor of two volumes in the Husserliana series. He has published on Husserl, animal rights, and environmental philosophy. His current interest are in the philosophy and politics of radical environmentalism.

Algis Mickunas (Ph.D., Emory University, 1969) is Professor of Philosophy at Ohio University. He has co-authored, co-edited, and translated various works, has published articles in diverse disciplines, from philosophy to sociology, pedagogy, communications, comparative civilizations, semiotics, and film.

J. N. Mohanty studied philosophy in Calcutta and Göttingen. He is Professor of Philosophy at Temple University, Philadelphia. His most recent publication is *Reason and Tradition in Indian Thought* (1992).

Tom Nenon (Ph.D., Universität Freiburg, 1983) is Associate Professor of Philosophy and Director of the Humanities Center at Memphis State University. He co-edited volumes XXV and XXVI of the *Husserliana*; is the author of *Objektivität und Erkennmis*, as well as numerous articles on Dilthey, Gadamer, Heidegger, Husserl and Kant; and has translated Werner Marx's *Is There a Measure on Earth?* and *The Philosophy of F. W. J. Schelling* into English. His current research interests include questions of personhood and subjectivity, especially in Husserlian phenomenology; issues in the philosophy of the social sciences; and Heidegger's concepts of untruth.

Maxine Sheets-Johnstone (Ph.D., University of Wisconsin, 1963) is an independent scholar who was most recently Visiting Associate Professor at the University of Oregon. She is the author of *The Roots of Thinking, The Phenomenology of Dance* and *The Roots of Power: Animate Form and Gendered Bodies* (forthcoming). She is the editor of *Giving the Body its Due* and *Illuminating Dance: Philosophical Investigations*.

Osborne Wiggins is Associate Professor of Philosophy and of Family and Community Medicine at the University of Louisville. With John Z. Sadler and Michael A. Schwartz, he has edited *Philosophical Perspectives on Psychiatric Diagnostic Classification* (1993). He has also published articles on phenomenology, psychiatry, and ethics.

Richard M. Zaner (Ph.D., Graduate Faculty, The New School for Social Research, 1961) is Ann Geddes Stahlman Professor of Medical Ethics, Professor of Philosophy, and Professor of Religion (Ethics), at Vanderbilt University. He is the author of *The Problem of Embodiment* (1964), *The Way of Phenomenology* (1971), *The Context of Self* (1981), *Ethics and the Clinical Encounter* (1988), and *Troubled Voices* (1993). He has translated or co-authored books with Alfred Schutz, Aron Gurwitsch, and Wilhelm Dilthey, and has edited or co-edited a number of books. His current interests include the basic and practical features of clinical ethics, and the ethical and philosophical problems presented by the Human Genome Project.

Index of Names

Index of Topics

abortion, 40,
accommodation, 230,
accountability, 47,
Acheulian handaxes, 93f,
action, 23, 33, 70, 76,
African-American, 222,
alienation, 180,
anthropocentrism, 186,
Anti-semitism, 227, 301,
animals, 268,
anological thinking, 93, 95,
anthropologism, 81,
apocalypse, 171,
archaeology, 30,
archaisation, 161ff,
architecture, 17, 23,
art, 164,
artificial intelligence, 118,
Asians, 225,
assililation, 212f, 219f,
Aswan Dam, 196,
asymmetry, 42-44, 48, 61,
autonomy, 224,
awareness, 31, 148f, 239-241,
axiotic, 18, 25ff., 32,
being-in-the-world, 74,
belief, 129,
believing, 14,
biography, 297ff,
bodily movements, 24, 73, 76,
body, 86ff, 291,
bricolage, 258-260,
Buddhism, 152, 222, 270,
capitalism, 178,
Catholicism, 224, 237, 252,
chaos theory, 120,
charity, 145,
chimpanzees, 9f,
civil rights, 212, 217,
civilization, 222, 227, 266,
class, 222, 270,

clinical event, 42, 60ff,
clinician, 47, 49, 59,
cognition, 16f, 26,
cognitive science, 67ff, 116, 130, 132,
community, 177, 268,
competition, 181,
computers, 116f,
Confucianism, 152,
connectionism, 115ff,
consciousness, 107, 221, 225, 322, 325, 327,
constitution, 68, 80f, 140, 205, 207, 226,
context, 40, 60ff,
contingent, 82, 288f,
cooperation, 204,
corporeal semantics, 88, 95, 98, 101, 114,
corporeal invarients, 88, 90, 95,
craft, 3f.,
crisis, 150, 153, 185, 186,
criticism, 26,
Cultural Disciplines, 1ff., 10, 15, 16ff., 28, 32, 297ff,
culture, 6ff., 9, 11f, 14, 17, 25, 28, 89, 135ff, 174, 198, 222, 238, 251, 253f, 257, 259-262, 265f, 269, 271,
Dasein, 327,
death, 288f,
decision-making, 44,
deconstruction, 228, 234, 260,
democracy, 166,
description, 237,
dialogue, 58ff,
dispositions, 126, 127, 129, 130,
domination, 145, 331,
doom, 293f,
ecology, 172ff, 193ff,
economics, 195 199,

341

Contributions to Phenomenology

IN COOPERATION WITH

THE CENTER FOR ADVANCED RESEARCH IN PHENOMENOLOGY

1. F. Kersten: *Phenomenological Method. Theory and Practice.* 1989
   ISBN 0-7923-0094-7

2. E. G. Ballard: *Philosophy and the Liberal Arts.* 1989   ISBN 0-7923-0241-9

3. H. A. Durfee and D.F.T. Rodier (eds.): *Phenomenology and Beyond.* The Self and Its Language. 1989   ISBN 0-7923-0511-6

4. J. J. Drummond: *Husserlian Intentionality and Non-Foundational Realism.* Noema and Object. 1990   ISBN 0-7923-0651-1

5. A. Gurwitsch: *Kants Theorie des Verstandes.* Herausgegeben von T.M. Seebohm. 1990   ISBN 0-7923-0696-1

6. D. Jervolino: *The Cogito and Hermeneutics.* The Question of the Subject in Ricœur. 1990   ISBN 0-7923-0824-7

7. B.P. Dauenhauer: *Elements of Responsible Politics.* 1991
   ISBN 0-7923-1329-1

8. T.M. Seebohm, D. Føllesdal and J.N. Mohanty (eds.): *Phenomenology and the Formal Sciences.* 1991   ISBN 0-7923-1499-9

9. L. Hardy and L. Embree (eds.): *Phenomenology of Natural Science.* 1992
   ISBN 0-7923-1541-3

10. J.J. Drummond and L. Embree (eds.): *The Phenomenology of the Noema.* 1992
    ISBN 0-7923-1980-X

11. B. C. Hopkins: *Intentionality in Husserl and Heidegger.* The Problem of the Original Method and Phenomenon of Phenomenology. 1993
    ISBN 0-7923-2074-3

12. P. Blosser, E. Shimomissé, L. Embree and H. Kojima (eds.): *Japanese and Western Phenomenology.* 1993   ISBN 0-7923-2075-1

13. F. M. Kirkland and P. D. Chattopadhyaya (eds.): *Phenomenology: East and West.* Essays in Honor of J. N. Mohanty. 1993   ISBN 0-7923-2087-5

14. E. Marbach: *Mental Representation and Consciousness.* Towards a Phenomenological Theory of Representation and Reference. 1993
    ISBN 0-7923-2101-4

Contributions to Phenomenology

IN COOPERATION WITH
THE CENTER FOR ADVANCED RESEARCH IN PHENOMENOLOGY

Further information about our publications on *Phenomenology* is available on request.

Kluwer Academic Publishers – Dordrecht / Boston / London